Extra-Sensory Perception of Quarks

Extra-Sensory Perception of Quarks

By

Stephen M. Phillips, Ph.D.

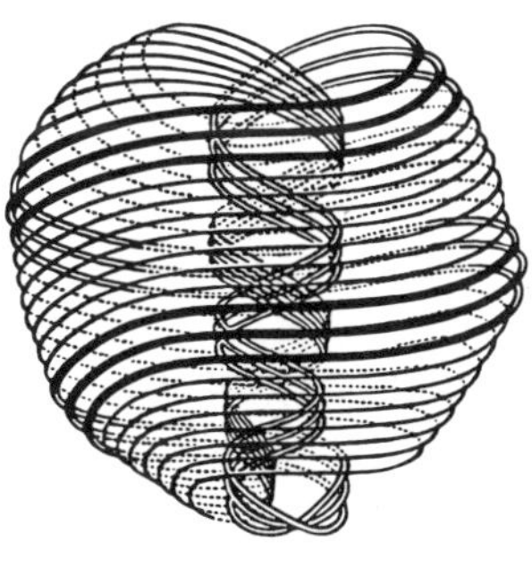

The Theosophical Publishing House

Madras, India/London, England

Wheaton, Ill. U.S.A.

For additional information write to: The Theosophical Publishing House, 306 West Geneva Road, Wheaton, Illinois 60187.

Published by the Theosophical Publishing House, a department of the Theosophical Society in America.

Library of Congress Cataloging in Publication Data

Phillips, Stephen M. 1946–
Extra-sensory perception of quarks.

Includes index.
1. Extrasensory perception. 2. Quarks—Miscellaneous.
I. Title.
BP573.E9P47 001.9 80-22048
ISBN 0-8356-0227-3

Printed in the United States of America
at the University of Chicago Printing Department.

HERBERT A. KERN, whose Foundation funded the publication of this book, was a research chemist familiar with Besant and Leadbeater's *Occult Chemistry* but was never able in his lifetime to see a link between its conceptualizations and those of modern science. The importance of Phillips' work may well lie in the implication that new efforts, scientifically addressed, might offer intuitive insights into other esoteric subjects only partially understood today by science. Herbert Kern was always fascinated by this possibility, so it is a pleasure to be able to dedicate the publication of this work to his memory.

Dedicated to my family

Contents

Preface

While studying several years ago in the United States of America as a physics graduate student, I came across one day a copy of the book *The Physics of the Secret Doctrine,* written by William Kingsland (London: Theosophical Publishing House, 1910). One page in particular captured my attention, for it displayed diagrams of the "atoms" of hydrogen, nitrogen, and oxygen, supposedly highly magnified through the use of a form of extra-sensory perception. The diagram of the "hydrogen atom" was especially curious and interesting, because I immediately recognized in it the physicist's model of a proton as a triangular cluster of three particles that he calls "quarks." On returning to England a few years later, I made inquiries about the source of these drawings and soon discovered many more curious things.

In this book, I have endeavoured to give a scientific explanation of a large collection of observations made by Annie Besant and C. W. Leadbeater during their investigations (1895–1933) of the atomic structure of matter. These two leading members of the Theosophical Society claimed that they could "see" atoms and molecules with the aid of a faculty of extra-sensory perception acquired by their training in yoga. Over the years, many Theosophists have never doubted that their work might finally prove to be of significance and of value to science (for a valuable introduction see E. Lester Smith, V. Wallace, and G. Reilly, *The Field of Occult Chemistry* [London: Theosophical Publishing House, 1934]). But its puzzling features and apparent contradiction of well-established facts of chemistry and physics have remained unexplained until recently. I have analysed the work of Besant and her colleague from the point of view of a physicist who is interested in the implications of parapsychology for physical science, and vice versa, and as one who is neither sceptical of the reality of extra-sensory perception nor firmly committed to a belief in it. I have tried to evaluate their work in an objective and unbiased way, being always aware during this task of the questionable status of their research but remaining sensitive to its possible merits wherever these can be shown to exist. I have ignored the issues that are raised by the substantiation in this book of the long-standing claims made by Besant and Leadbeater, since others are more able than I to comment on these matters. My intention is not to promote any doctrine or theory but, instead, to present to the reader much of the evidence that is available and to interpret it in terms of elementary particle physics, so that he may better judge for himself the validity of these claims.

I acknowledge very helpful discussions with Dr. E. Lester Smith, F.R.S. I am deeply grateful for his useful comments, kind advice, and constant encouragement. I wish to thank Edward Lussow for his editorial and design assistance and Christina Borges and Kathleen Miners for their excellent contribution in the reproduction of diagrams. Finally, I thank the Kern Foundation for a research grant.

STEPHEN M. PHILLIPS

Bournemouth, England
October 1979

Introduction

E. LESTER SMITH, D.Sc., F.R.S.

Some two hundred subatomic particles are now recognized by particle physicists. Many of them have been observed experimentally by means of the sophisticated technology of high-energy physics; the increasingly detailed understanding of subatomic events has permitted postulation of other particles such as the quarks. Scientists will find it hard to credit the claims of this book, to the effect that some observations in this field were made eighty years ago using quite different techniques. A form of extra-sensory perception that can be described as magnifying clairvoyance or micro-psi has been known to oriental yogis for thousands of years. The technique does not consist in actually magnifying the small object but, conversely, in "making oneself infinitesimally small at will," as it has been picturesquely described. It is mentioned, for example, in the yoga Sūtras of Patañjali (book 3, Sūtra 26): "Knowledge of the small, the hidden or the distant by directing the light of superphysical faculty" (Taimni's translation).

Some early members of the Theosophical Society undertook intensive yoga training under expert guidance and in due course acquired this faculty. In 1895 it was suggested to C. W. Leadbeater that he might direct it upon the atoms of the chemical elements. Annie Besant soon joined in what was to become a long series of investigations, lasting on and off for thirty-eight years. Starting with the light elements hydrogen, oxygen, and nitrogen, the researches were gradually extended to cover all elements known at the time, plus a few then undiscovered elements and isotopes. Also, a number of typical inorganic and organic compounds were investigated. The objects as "seen" in this manner were described to an assistant, and sketches were made and notes taken. The atoms appeared as highly structured bodies, giving the impression of definite external shapes, with the interior subdivided into compartments of spherical, ovoid, conical, and other geometrical shapes. These bodies in turn had smaller components within them, which in their turn were observed to contain very much smaller particles; these seemed to be the ultimate units of physical matter and were called "ultimate physical atoms." All the elements consisted in the last analysis of these particles, which were of two types only, one the mirror image of the other.

Because the atoms were in vigorous motion in all conceivable modes, it was necessary to slow them down by a special effort of will-power (psychokinesis) before accurate observation and counting of components were possible. The external and internal shapes were delineated by the volume of space swept out, so to speak, by the energetic movements of the subatomic particles, which appeared to create tenuous "walls" of a nature that could not be determined.

The observers found that they could facilitate the examination by applying psychokinesis to dismember the atoms, in stepwise fashion, into smaller groupings of ultimate particles; at each step a considerably greater power of "magnification" was needed. A great deal of this work was brought together in the first and second editions of *Occult Chemistry* (1908 and 1919), and the remainder was included in a rewritten third edition in 1951. The fourth edition will be reprinted in 1980.

To appreciate the difficulties in relating these researches to the contemporary state of science, it is sufficient to consider hydrogen. Its atom as "seen" by E.S.P. contained eighteen of the ultimate physical atoms, grouped into six spheres of three apiece, these spheres appearing to be arranged at the corners of interlacing triangles. No subatomic particles were known then—or indeed are known now—eighteen of which could make up the greater part of a hydrogen atom, possibly its nucleus. As understanding of the nucleus progressed, the possibility of reconciliation with *Occult Chemistry* seemed to recede rather than improve. Those few scientists who came across the book felt justified in rejecting it as fantasy after cursory inspection. The still fewer who had some sympathy with parapsychology were simply baffled; some of them knew the investigators and respected their integrity.

The first pointer to a possible reconciliation came when quarks were postulated, requiring subdivision of the proton into three bound quarks. But between three and eighteen there still remained a factor of six to be bridged. This feat has been achieved by Dr. Stephen Phillips, the author of this book. He is an English physicist educated at Cambridge University, where he received his B.A. and M.A. in theoretical physics; at Cape Town University, where he received his M.Sc.; and at the University of California, where he researched for his Ph.D. in particle physics. In 1979 his paper entitled "Composite Quarks and Hadron-Lepton Unification" appeared in *Physics Letters* (and has been incorporated in this book). This suggests an improvement in quark theory involving further subdivision of each quark into three subquarks, for which the name *omegon* is proposed. This theory thus provides for nine omegons per proton; the still persisting factor of two was bridged by imperative reinterpretation of the E.S.P. data. Besant and Leadbeater claimed to "see" the atom exactly as it was;

they could not have known that the very act of focusing their attention upon it and checking its "wild gyrations" psychokinetically must inevitably cause perturbation. Phillips has carefully analysed the nature of this perturbation and concludes that it would induce the fusion of two atoms of an element into a plasma of free omegons and quarks, which then interact to form stable, quasi-nuclear systems of bound particles. The new patterns derived by application of the rules of theoretical physics tally perfectly with the diagrams of *Occult Chemistry*. With hindsight it can be seen that there are many pointers to this doubling-up phenomenon in the text of *Occult Chemistry*.

It stretches credulity to concede that omegons, conceived by Phillips in the mid-1970s, and the physicist's quarks had been "seen" extra-sensorily some eighty years previously, and possibly much earlier by Indian yogis. But the effort may prove rewarding because Phillips has unified the two schemes not merely in outline but in the most precise and convincing detail. Science will continue on its own course, but there is a tendency now to find parallels, and perhaps inspiration, in the ancient wisdom of the East, as for example, in Fritjof Capra's *Tao of Physics* and J. P. M. Whiteman's *Philosophy of Space and Time*. The present book provides a big impetus to this trend.

CHAPTER 1

Historical Background

Nothing is too wonderful to be true, if it be consistent with the laws of Nature, and in such things as these, experiment is the best test of such consistency.

Michael Faraday

INTRODUCTION

What does matter ultimately consist of? During the twentieth century, the answer given by physical science to this question has undergone several major changes as rapid technological advances have enabled man to acquire knowledge of the microcosmos at levels farther and farther beyond the range of his senses. For example, high-resolution electron microscopy can record images of atoms that are about one-hundred-millionth of an inch in diameter, and high-energy particle accelerators can boost charged particles to speeds almost equal to that of light (about 186,000 miles per second) in order to probe deep inside nuclear particles extending over distances less than one-hundred-thousand-billionth of an inch. But, to his surprise, man has discovered that he may study the paradoxical features of the microcosmos revealed by his detecting and measuring apparatus only if he discards the common language in which he expresses his sensory experience of the world at large. Nuclear and atomic particles cannot be visualized and described in such terms. Their future behaviour cannot be predicted with certainty. Prior to quantum theory, physical science was dominated by Cartesian dualism and by the mechanistic philosophy of Newtonian physics, which depicted the universe as a void interspersed with atoms whose predictable motions could be measured, in principle, with arbitrary precision without ever mentioning the human observer or the procedure that he used to make the measurement. But quantum theory has invalidated the notion that the microcosmos is just a miniature version of the macrocosmos, filled with visualizable particles that obey the same deterministic laws. Its description must take into account the intervention of the observer. Every physical measurement constitutes an interaction with the micro-object and affects its properties, which, therefore, cannot be defined or discussed separately, without reference to the act of measurement. Processes occurring in atoms and nuclei have to be analysed in probabilistic terms that properly take account of the unpredictable disturbance inherent in the measurement process.

The new picture of the microcosmos is one of sets of chance events that are mutually interconnected to give the illusion of order and equilibrium to the world of human experience. But the discovery of this picture has not altered the basic methodology of science, even though it has imposed limits on the extent of possible knowledge of the microcosmos. The scientist mentally reconstructs what may be responsible for his data and tests his theories and their implications by further experimentation. If the data happen to be too imprecise to be able to distinguish between rival theories, scientific understanding can become ambiguous and uncertain for a while. If a crucial experiment or measurement is unavailable or unfeasible in terms of available technology, scientific understanding may also become polarized, with different schools of thought emerging in the scientific community. This is an endemic tendency of high-energy physics, where new ideas and models frequently outstrip technological capacities to verify their predictions. In his search for truth, man's fertile imagination sometimes makes nature appear ambiguous. The possibility that man can obtain knowledge of the microcosmos directly, without instrumental aids, is one that most scientists and nonscientists alike would treat with extreme scepticism. But various people have claimed in the past to possess a faculty of perception that made microscopic objects become visible to them. This book will present some observations made with this faculty and will evaluate their claims critically within the framework of contemporary elementary particle physics. Evidence and arguments will be given for the general validity of these claims. They do not make the faculty less remarkable (rather the contrary), neither do they make it easier for the impartial reader to accept such an incredible possibility. But their merit at least warrants serious consideration of the evidence, and this is all that the author requests from the reader in the present work.

HISTORICAL BACKGROUND

Frequent reference is made in the extensive literature of Indian yoga to extraordinary mental abilities ("siddhis") that may be acquired through the practice of yoga. A few of these, such as telepathy and precognition, are well known in Western culture and have been scientifically investigated under controlled laboratory conditions by parapsychologists, although their claims to have succeeded in demonstrating irrefutably the occurrence of these forms of extra-sensory perception still remain controversial. In the first authoritative and systematic exposition of yogic practice—the yoga Sūtras of Patañjali, written about 400 B.C.—the physiological and psychological results of meditation, concentration, and contemplation are prescribed in detail. In Aphorism 3.26 of the Sūtras,[1] it is stated that the yogi can acquire "knowledge of the small, the hidden or the distant by directing the light of superphysical faculty."

२६. प्रवृत्त्यालोकन्यासात् सूक्ष्मव्यवहितविप्रकृष्टज्ञानम् ।

Pravṛtty-āloka-nyāsāt sūkṣma-vyavahita-viprakṛṣṭa-jñānam.

The Sanskrit name for this siddhi is *animā*.[2] The yogi can develop an inner organ of perception that displays "knowledge of the small" in a visual form. He exercises the siddhi while he is in an altered state of consciousness in which he experiences visual images of objects too small for human sight to discern, his perception being from a point of view in space that gives him the illusion that he has shrunk to a size commensurate with the objects that he is viewing.

The experience of a person in this state is not that of a passive spectator peering down a microscope. Instead, it is characterized by a vivid, subjective sense of actually being in the microcosmos, of being suspended in space amid particles in great dynamical activity. While functioning in this state, he retains full control of his intellect and can converse normally with people around him, being able to describe to them what he "sees." But he has to apply certain techniques of yogic meditation as he concentrates on the interior of some specimen substance that is placed in front of him. He experiences images whether his eyes are open or closed, but in practice his concentration is aided by keeping his eyes closed so as to eliminate distracting images due to his normal sight. The images experienced are three-dimensional, may be coloured, and usually exhibit rapid, complex motion. By a deliberate exertion of his will, the observer can slow down the movement of any chosen image and keep an apparently stationary image in his field of vision for an indefinite period of time. The length of time that such observation can be sustained seems to be limited only by the degree of mental tiredness that results from his functioning in this altered state of consciousness. The sizes of images can be varied at will, and there appears to be no upper limit to the level of magnification attainable, although a practical limit is set by the ability of the person and by the strain felt by him when he views highly magnified images. Unlike some other forms of extra-sensory perception, this state can be induced or terminated whenever the observer wishes.

In chapter 2 it will be argued that the images correspond to real, physical objects present in the examined substance at the time they are experienced, so that the experience is one of extra-sensory perception of the objective world at the atomic and nuclear level. It differs from other altered states of consciousness in which a person encounters hallucinatory images either pertaining to his own self at the subconscious level or arising from deeper, mythological sources whose language is symbolic and universal. This type of quasi-visual experience will be referred to as "micro-psi vision" (or "micro-psi," a term compounded from *micro*, denoting smallness, and "*psi* faculty" or "*psi* ability," terms used in parapsychology to denote extra-sensory perception). A person functioning in this state will be called a "micro-psi observer." It must be emphasized here that it is only for the sake of convenience that micro-psi is assumed to be a form of extra-sensory perception. In making this assumption, there is no intention to imply that micro-psi violates any laws of physics. No explanation of its modus operandi will be attempted in this book.

Two leading members of the Theosophical Society—Annie Besant and Charles W. Leadbeater—claimed to possess this siddhi and to be able to "see" atoms in their normal state. How they supposedly acquired such an ability through the practice of yoga is a matter that will not be pursued here, because issues relating to the early history of the Theosophical Society are not the subject of discussion of the present work. From 1895 until 1933, Besant and Leadbeater examined all the elements from hydrogen to uranium, as well as an assortment of organic and inorganic compounds. In this vast task, they were assisted by their colleague C. Jinarajadasa, who acted as recorder during their experimental sessions. The well-known chemist Sir William Crookes, a friend of both the investigators, provided specimens of some elements. Various minerals were studied in a museum in Dresden, Germany.

The first substance chosen for examination with micro-psi was gold, but it was soon found to be too complex for superficial description. Next Besant and Leadbeater studied air, and they published in the Theosophical journal *Lucifer* (November 1895) diagrams and accounts of what they assumed to be atoms of hydrogen, nitrogen, and oxygen, highly magnified by micro-psi vision (the question of how elements were identified will be discussed in chap. 4). In 1907, fifty-nine more elements were examined, variations being noted in the supposed atoms of neon, argon, krypton, xenon, and platinum. Slight differences among the atoms of an inert gas were not, at first, interpreted in terms of isotopes. This was because the existence of isotopes was not suspected by science at the time. Soddy first gave the name "isotopes" to atoms of an element differing in mass in 1913, five years after the two investigators had published in the Theosophical journal *The Theosophist* (vol. 30) their discovery of a variation of neon. In volume 29 of this journal they published details about three elements which, they claimed, were unknown to science, and they gave them the names "Occultum," "Kalon," and "Platinum B." They also described a group of three transition elements ("X," "Y," and "Z"), which they believed science had not discovered. Radium was examined in 1908, and a diagram supposedly depicting its atom appeared in volume 30 of *The Theosophist.* A summary of the initial research was published in the book *Occult Chemistry.*[3] In 1909, twenty more elements were studied, notably so-called Illinium, which was later recognized to be promethium (discovered by science in 1945), and the element "C," which was later found to be actinium (discovered by science in 1898). A second edition of *Occult Chemistry* was published in 1919,[4] but it contained no material in addition to that in the first edition, and it did not refer to the work carried out after 1907.

Benzene, methane, and other organic compounds were examined in 1922, and supposed descriptions of their molecules were published in 1924. Included here were diagrams depicting the molecule of water and the unit cell of the atomic lattice of sodium chloride. In the following year, a model of the crystal structure of diamond, constructed from extensive micro-psi observations, was published in volume 46 of *The Theosophist.* The plane-sheet structure of graphite, with its hexagonal arrays of carbon atoms, was described in 1926. More material was published in *The Theosophist* in 1932. This included descriptions of the atoms of element "87" (called "francium" by science in 1939), element "85" (later, in 1940, to be given the name "astatine" by science), and element "91" (isolated in 1921 and called "protoactinium"). Element "43" had been recorded in 1909 as "Masurium" and given its correct position in the periodic table. It was re-examined in 1932, five years before technetium was detected by science. An element of atomic weight 2 was reported by the investigators in 1932 and given the name "Adyarium," its discovery being made at Adyar in Madras, India. It was regarded as a new element and was not identified with deuterium, which had been discovered by Urey, Brickwedde, and Murphy in the previous year. This happened because they had interpreted as the molecule of deuterium another object, observed during their earlier examination of the gases released by the electrolysis of water. In the final year of their research, Besant and Leadbeater published a study of all the inert gases and certain of their isotopes. They also reported the existence of two forms of what they regarded as the atom of hydrogen, three isotopic varieties of oxygen, and two species of ozone. All material accumulated during their work over thirty-eight years was compiled and published in an enlarged third edition* of *Occult Chemistry*.[5] This will be analysed in the present work. Significant aspects of more recent investigations using micro-psi will be discussed in chapter 5. The Appendix contains a chronology of all published research, with references.

GENERAL MICRO-PSI DESCRIPTION OF MATTER

When the micro-psi observer concentrates on the interior of a crystal or a metal, a specimen of which is placed before him, he often claims to notice a grey mist or fine haze of light enveloping it, sometimes followed by discharges of rays or streams of "points of light" shooting out from the material in all directions. With greater magnification, the mist becomes particulate, being found to consist of myriads of similar points of light in chaotic motion. The observer may catch sight of globes of scintillating points, tossing and waving about as if blown by an invisible wind. On closer examination, these are found to consist of very closely packed groups of points of light held in rigid geometrical configurations such as tetrahedra, octahedra, and hexagons. Some may exhibit periodic motion as if forced to move in orbits; others may fly across the field of vision in cascades, like showers of meteors in the night sky. Small groups of points of light appear joined by thin, luminescent streams of energy, or "lines of force." Provided that not too high a level of magnification is used, a global organization and pattern emerge for the points of light. Collectively, they create the impression of a rigid, geometrical figure. Its shape is characteristic of a particular element in the substance being examined with micro-psi vision, for this form (and only this form) is observed for the element, irrespective of whether the specimen is in a solid, liquid, or gaseous state and of whether the element is in a pure state or in a chemical compound. This object appears to the micro-psi observer to be the basic structural unit of which the substance is composed. It will be called a "micro-psi atom" (or M.P.A.). Besant and Leadbeater classified all M.P.A.'s into seven types:

1. *Spike Group*. The unit consists of a number of identical, spike-shaped projections containing groups of points of light. These spikes vary in number from one to sixteen (depending on the element) and radiate symmetrically from a central globe, which also contains groups of lighted points (fig. 1.1, *a*).
2. *Dumb-bell Group*. The M.P.A. has a central rodlike section containing a row of groups of points of light. At each end there is a globe with other clusters of lighted points in it. The two globes are identical, and twelve funnel-shaped forms radiate from each of them. The funnels are sym-

* All subsequent references apply to this edition unless otherwise specified.

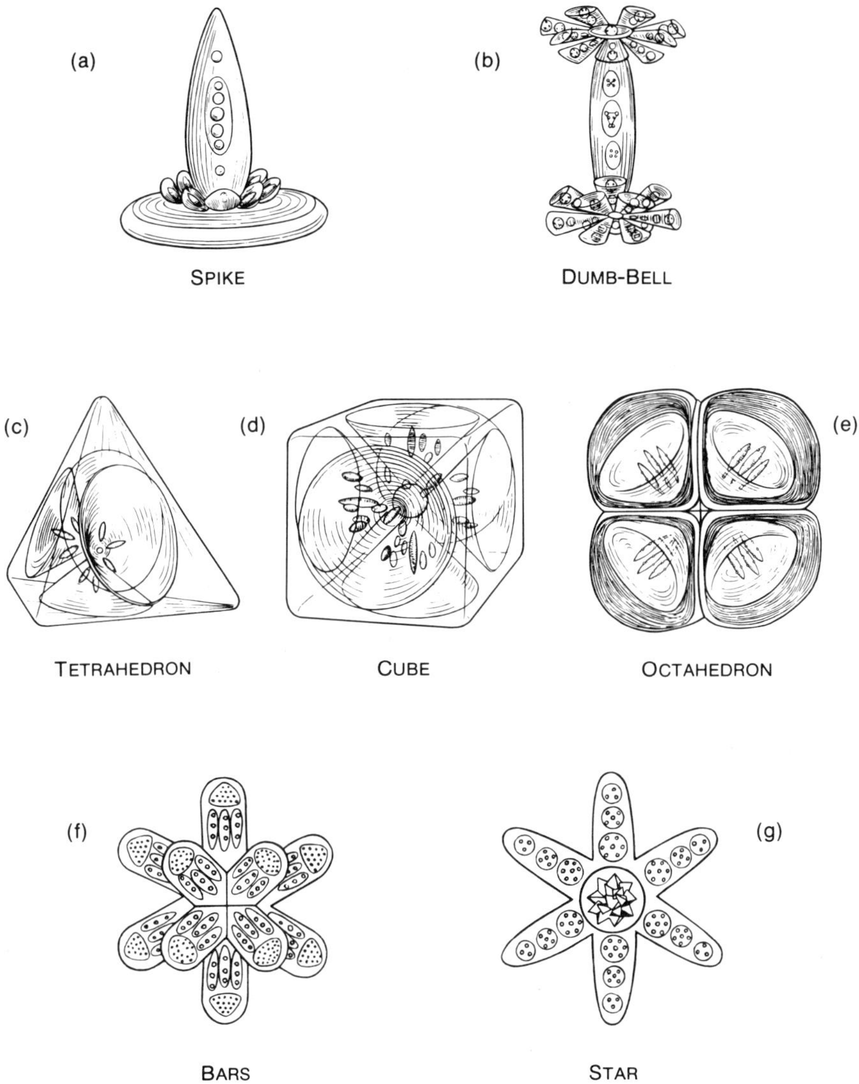
(a)
Spike
(b)
Dumb-Bell
(c)
Tetrahedron
(d)
Cube
(e)
Octahedron
(f)
Bars
(g)
Star

Fig. 1.1

metrically arranged about the centre of a globe like flower petals, each pointing slightly upwards or downwards, alternatively. The whole unit has the appearance of a dumb-bell (fig. 1.1, *b*).

3. *Tetrahedron Group*. The M.P.A. has four funnels that contain ovoid bodies and open out on the faces of a regular tetrahedron. They may radiate from a central globe enclosing groups of points of light. In three elements belonging to this type of M.P.A., there are also four spikelike forms directed towards the corners of the tetrahedron (fig. 1.1, *c*).
4. *Cube Group*. The M.P.A. consists of six funnels, containing ovoid bodies and opening on the faces of a cube. It may also have a central globe. In the case of two elements, the M.P.A. has eight spikes as well. These point towards the corners of the cube (fig. 1.1, *d*).
5. *Octahedron Group*. The M.P.A. is an array of eight funnels that open on the faces of a regular octahedron. It is rounded at the octahedral corners and a little depressed between the faces in consequence of this rounding. It has the appearance of a "corded bale." All elements except one have a central globe, and two elements display in addition six spikes that point to the corners of the octahedron (fig. 1.1, *e*).
6. *Bars Group*. The M.P.A. consists of fourteen bar-shaped projections. They radiate from a common point towards the corners of a cube and towards the centres of its faces. All bars have the same composition of groups of lighted points. There is no central globe (fig. 1.1, *f*).
7. *Star Group*. The M.P.A. has the form of a flat, six-pointed star, like a starfish. Each of the six "arms" of the star has the same composition of lighted points (for a given element). The arms radiate from a central sphere containing five intersecting, tetrahedral arrays of identical groups of points of light (fig. 1.1, *g*).

It should be emphasized that, in all the geometric forms mentioned above, nothing resembling actual tetrahedra, cubes, etc., is visible to the micro-psi observer. These figures represent merely the geometrical scaffolding supporting the funnels, spikes, and other features. The micro-psi observer may change the perspective in which he seems to view an M.P.A., but its intrinsic shape does not change during the period of observation. Diagrams such as figure 1.1 are not drawn to scale, so only qualitative features have significance. This is because the observer has difficulty in judging relative sizes of bodies and in comparing their distances apart. Also, structural features of widely varying size require different degrees of magnification, so the diagrams reproduced in this book are not necessarily realistic representations of micro-psi images in terms of scale.

Every M.P.A. is described as being enclosed in a "hole" or "cell," whose "walls" are usually spherical, although sometimes ovoid in shape. The latter appear to the observer to be transparent, like glass or stretched plastic film. Inside the M.P.A. various aggregates of lighted points also are enclosed in cells. They generally cluster together in space in a highly symmetric distribution. As examples, figures 1.2, 1.3, and 1.4 show the arrangement of lighted points in the central globes of the M.P.A. of barium, strontium, and radium, respectively. It should be understood that the reported arrays are in three-dimensional space, so that sectors of circles in

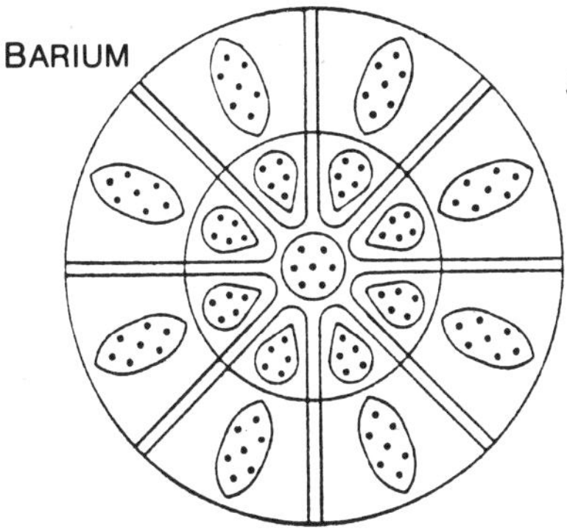

Fig. 1.2

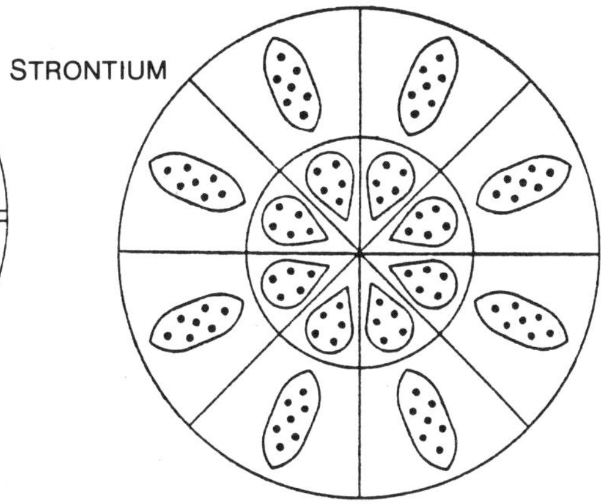

Fig. 1.3

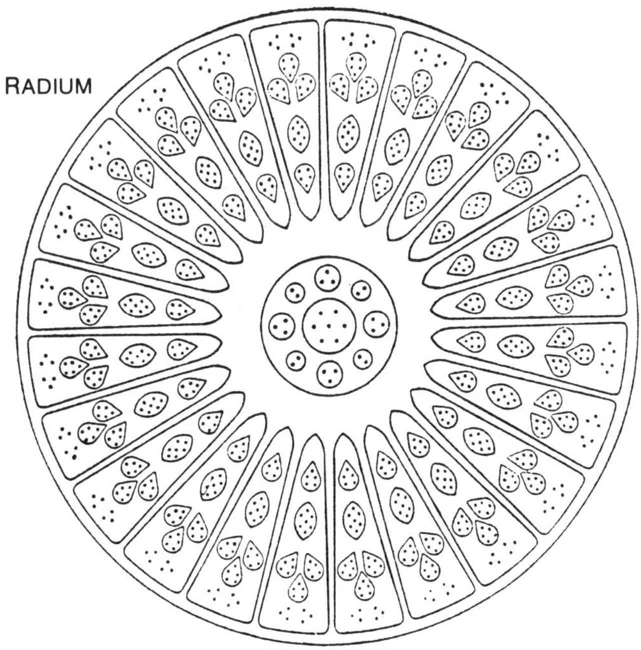

Fig. 1.4

these (and similar) diagrams actually depict segments of spheres. The surfaces of funnels, bars, and so on, are reported to have a similar transparent quality: the observer can "look" through them and notice other M.P.A.'s nearby. They are not so much surfaces as membrane-like boundaries separating the interior of the M.P.A. from the ambient space. The same is true for the cells enclosing groups of lighted points. Their shapes are related generally to the spatial arrangement of these points and are static, although they can be deformed by neighbouring cells. Examples of groups of lighted points and their cells are shown in figure 1.5.

A number of elements have M.P.A.'s that do not belong to any of the seven types listed above. These are hydrogen, helium, nitrogen, and oxygen. The M.P.A. of hydrogen (fig. 1.6) was, according to the investigators, "seen to consist of six small bodies, contained in an egg-like form. . . . It rotated with great rapidity on its own axis, vibrating at the same time, the internal bodies performing similar gyrations. The whole atom spins and

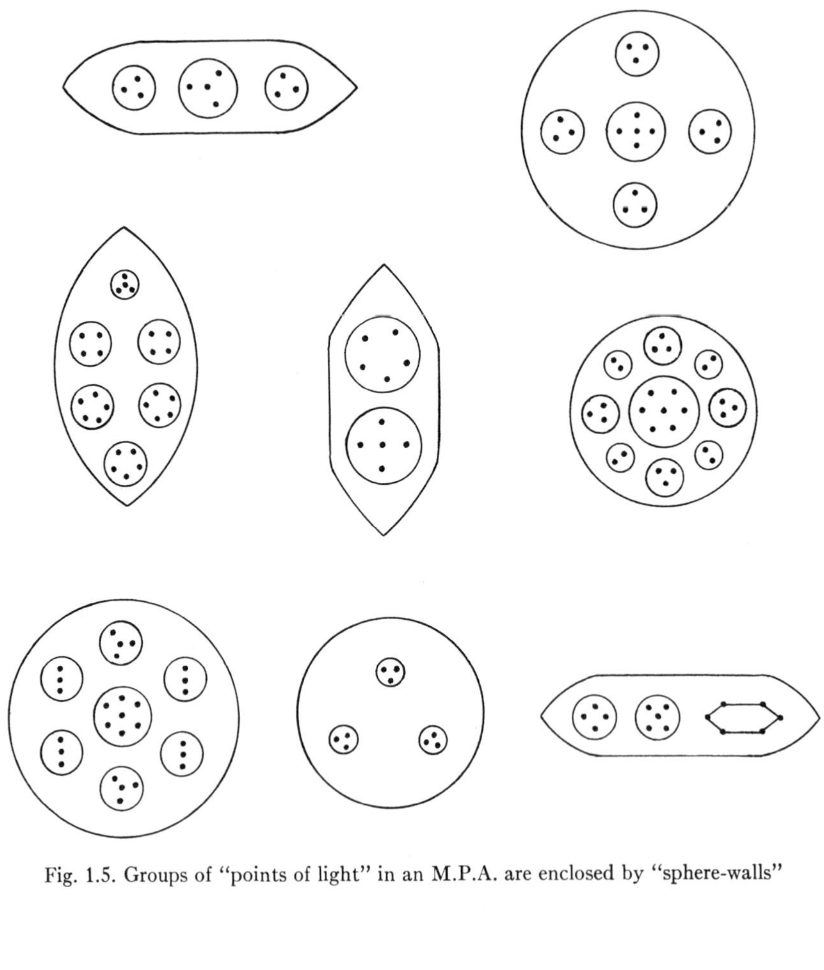

Fig. 1.5. Groups of "points of light" in an M.P.A. are enclosed by "sphere-walls"

quivers and has to be steadied before exact observation is possible. The six little bodies are arranged in two sets of three, forming two triangles that are not interchangeable."[6] The six bodies are not identical. Each contains three points of light, but in two bodies they are arranged in a line, while in the other four they are arranged in a triangle. The two triangular arrays of bodies are quite distinct. They interpenetrate and never separate while they are observed. The M.P.A. of helium is ovoid (fig. 1.7). Two tetrahedral arrays of bodies revolve round an egg-shaped central body, which consists of two groups of points of light in the shape of triangles. These spin on their axes while revolving around each other. In the case of nitrogen (fig. 1.8) the M.P.A. is a sphere, in the middle of which floats a balloon-shaped body. This contains six smaller spheres in two horizontal rows and a long ovoid body between them. Below it is another ovoid body, and around it are four spheres. The oxygen M.P.A. is ovoid, and within it two bodies, arranged like the coils of a stiff spring, revolve at high speed around each other in opposite directions (fig. 1.9). Five brilliantly lit points appear on the crests and hollows of the spirals. Between them are small groups of two lighted points that are evenly interspaced along the spirals like beads on a necklace. Rotating very rapidly, the coils create for the micro-psi observer an illusory impression of a rigid, cylindrical surface.

Repeated reference has been made above to "points of light" inside M.P.A.'s. Higher magnification reveals that these are particles, in the sense that they have a highly localized inner structure. Because these objects appeared to Besant and Leadbeater to be the most elementary forms of physical matter, they called them "ultimate physical atoms"* (U.P.A.'s), claiming that they were the fundamental, indivisible constituents of the atom known to physics and chemistry. They described U.P.A.'s in terms that were very different from the materialistic notions prevalent in science at that time (1895). Instead of being the "solid, massy, impenetrable, movable particles" conjectured by Sir Isaac Newton, the individual U.P.A.'s consist of ten currents of energy flowing in separate, closed, continuous curves (fig. 1.10). The paths taken by these currents spiral down over the surface of a sphere, making two and a half revolutions, and return to the top of the sphere in an inner spiral about the central axis. Three of the whorls appear to be "thicker" and "brighter" than the other seven. The top of the U.P.A. is open and slightly depressed inwards, giving it a heart-shaped outline. Besant and Leadbeater detected two types of U.P.A.: in the "positive" type, the currents flow from the flattened end to the more pointed end in a clockwise sense (as seen from the flattened end); in the "negative" variety, the spiral flow is anticlockwise. In structure the two

* They also referred to them as "Anu," from the Sanskrit word for atom.

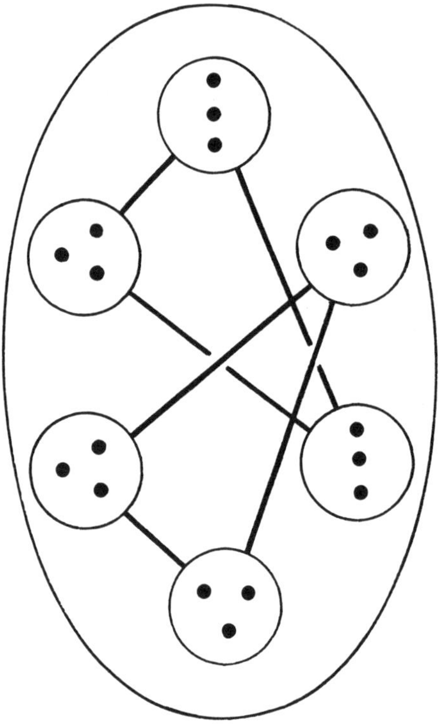

Fig. 1.6. M.P.A. of hydrogen

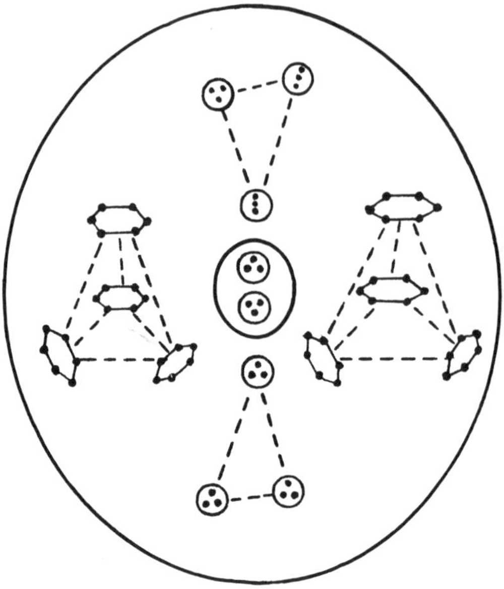

Fig. 1.7. M.P.A. of helium

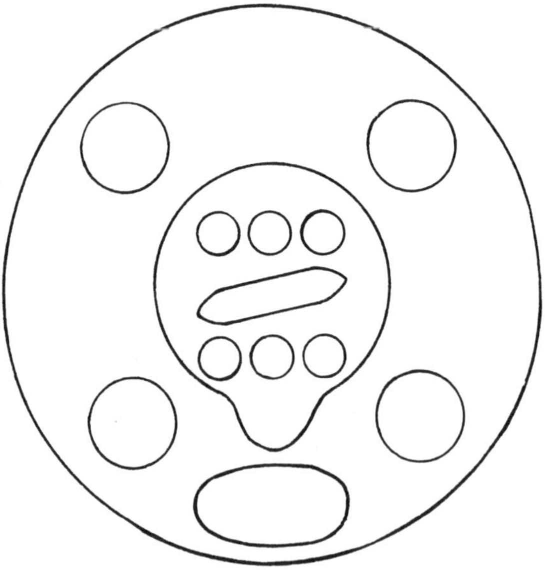

Fig. 1.8. M.P.A. of nitrogen

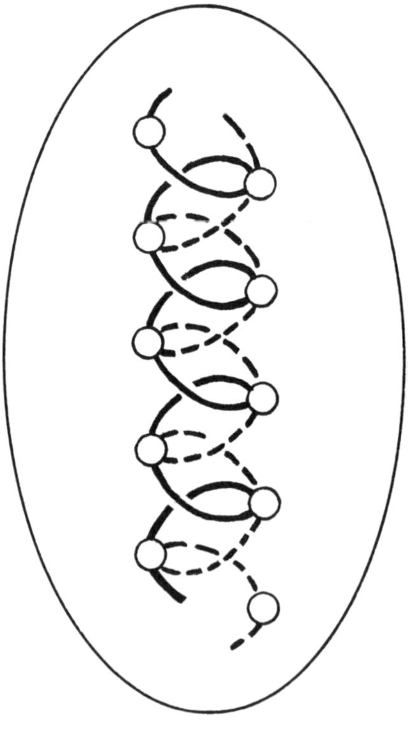

Fig. 1.9. M.P.A. of oxygen

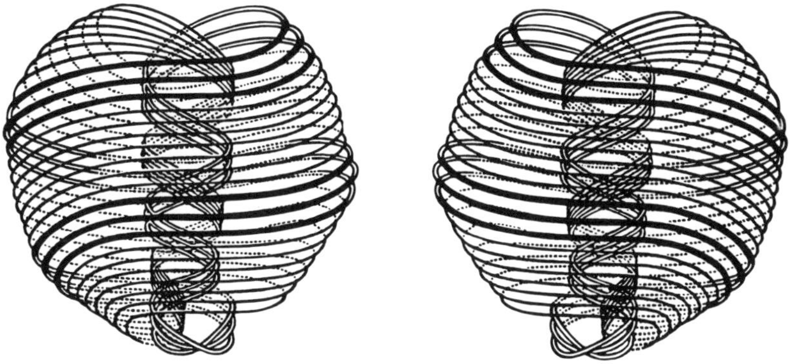

Fig. 1.10. The two chiral forms of the "ultimate physical atom"

types are enantiomorphs, that is, chiral or non-superposable mirror images of each other. "Lines of force" flow in at the depressed end of the U.P.A. and flow out at the more pointed end. They connect it with other U.P.A.'s and are responsible for the cohesion of groups of these particles. Three proper motions are detected in the U.P.A.: spinning of the whole spiral form about its central axis; pulsation, that is, periodic expansion and contraction of the outer spiral; and precessional rotation of the axis of spin in a circle, that is, motion like that of a wobbling, spinning top. Firsthand accounts of other properties of U.P.A.'s are discussed in chapter 6.

REFERENCES

1. Yoga Aphorisms of Patañjali, translated by S. Prabhavananda and C. Isherwood, in: How to Know God (London: Allen & Unwin, 1953); also: The Science of Yoga, by I. K. Taimni (Adyar, Madras, India: Theosophical Publishing House, 1965).

2. Yoga, by Ernest Wood (Pelican Books, 1965), p. 78.

3. Occult Chemistry, by Annie Besant and C. W. Leadbeater (1st ed.; Adyar, Madras, India: Theosophical Publishing House, 1908).

4. Occult Chemistry, by Annie Besant and C. W. Leadbeater (2d ed.; London: Theosophical Publishing House, 1919).

5. Occult Chemistry, by Annie Besant and C. W. Leadbeater (3d ed.; Adyar, Madras, India: Theosophical Publishing House, 1951). All references are to this edition unless specified otherwise.

6. Ibid., p. 10.

CHAPTER 2

Epistemological Aspects of Micro-Psi Vision

What we observe is not nature itself, but nature exposed to our method of questioning.

Natural science does not simply describe and explain nature; it is part of the interplay between nature and ourselves.

Werner Heisenberg

This chapter discusses certain philosophical objections that could be raised against the validity of micro-psi observations in general and of those of Annie Besant and C. W. Leadbeater in particular.

The most critical point of view concerning the work of Besant and Leadbeater is that their observations are not genuine but, instead, are the product of elaborate fraud. The following questions must be directed to a critic with this objection:

1. How did Besant and Leadbeater conceive of fabricating minor variations in the M.P.A.'s of some elements before the concept of isotopes was available to them? This concept was introduced to science by Soddy some six years after the two Theosophists published a report of an observed variation in the M.P.A. of neon.[1]
2. Besant and Leadbeater claimed to find approximate agreement between the chemical atomic weight of an element and the "number weight," defined as

$$\text{Number weight} = \frac{\text{Number of U.P.A.'s in the M.P.A.}}{18}$$

(the M.P.A. of hydrogen contains 18 U.P.A.'s, so its number weight is 1). They considered (wrongly, as will be shown later) that this agreement had significance. But the error can amount to several units in the case of some elements, and it could have been reduced if U.P.A. populations were, in fact, chosen so as to provide good agreement. Why, therefore, did they not fabricate these numbers in order to obtain exact or almost exact agreement between their calculation of number weights and known chemical atomic weights? The correspondence of these numbers could have been made much more impressive than it is, if their claim to be able to describe atoms was indeed fraudulent.

3. If the Theosophists derived their published work from scientific research already known to them, why did they fabricate descriptions of many molecules that differ significantly from the known structures? For example, why did Besant and Leadbeater describe the molecule of benzene as octahedral in shape[2] and not as hexagonal—a fact well known to chemists at the time as well as to themselves?
4. How could two individuals possessing only a layman's knowledge of physics invent descriptions of particle behaviour that vividly portray the Larmor precession of spinning, charged particles in magnetic fields—descriptions that needed a knowledge of the intrinsic spin of particles but were published prior to the experimental discovery of the spin of the electron and the proton? For example, precessional motion of "Hydrogen Triangles" (identified later in this work as protons) was described[3] during observation of the OH group in 1924, before Uhlenbeck and Goudsmit proposed that the electron had an intrinsic angular momentum and before the spin of the proton was experimentally detected. Also, how could the investigators fabricate an observation of the alignment, parallel to an external, static electric field, of the spin axes of various U.P.A.'s? It is highly unlikely that they could have known that spinning magnetic charges possess electric dipole moments (the concept of magnetic charges is not explicitly referred to in their published work), nor could they have decided to make magnetic monopoles out of U.P.A.'s, because their observation of spin alignment in electric fields appeared in the first edition of *Occult Chemistry* (published in 1908), whereas magnetic monopoles were first seriously discussed by Dirac in 1933.[4]
5. This book presents numerous examples of exact correspondence between facts and ideas of contemporary elementary particle physics and micro-psi observations published as long ago as 1895. If the latter are merely fabricated, all these remarkable similarities can be only coincidental—a conclusion that lacks credibility.

These arguments (particularly the last) indicate that fraud is not a possibility that need be seriously considered. Micro-psi may be interpreted in two ways. In the first, the "subjective" view, micro-psi images refer to the subjective, inner world of dreams, fantasies, and hallucinations that are studied by psychology and not to the objective world studied by physical science. They are merely events that occur in the brain of the observer, having no causal connection with any object in the world at large (or, in particular, with the material that the observer examines).

Suppose that the images are hallucinations and, therefore, of interest only to the psychologist and not to the physicist. Since they persisted for over thirty-eight years in the case of Besant and Leadbeater, they would have had to be pathological in origin. But images of the same type were recorded by both individuals, one always being consistent with the other. Moreover, this explanation fails to take account of the yogic tradition behind this siddhi. If micro-psi images are not quasi-visual representations of objects in the physical world, the following question arises: how can hallucinations, which are self-induced and causally unconnected with the material specimen that the subject examines, be manipulated by an experimenter who is working with him? There are on record various experimental investigations in which the images experienced by a subject were disturbed by another person passing an electric current through the specimen or by his changing the current from A.C. to D.C.[5] How can hallucinations of non-existent objects be altered by the experimenter—moreover, in just the manner that would be expected if they were, indeed, real objects that were responding to the altered experimental conditions that he had introduced? Suppose, for example, that micro-psi images were archetypal symbols of Jung's collective unconscious. How can such psychological entities uniquely belonging to this altered state of consciousness be affected by the switching on or off of an electric current? Micro-psi observers have recorded that U.P.A.'s and other particles behave in electric and magnetic fields in a manner consistent with their having properties like electric charge, spin, and electric dipole moments. If these supposed symbols do not refer to objective, physical particles, how can they be made by external fields to act as such? Scientific knowledge available to Besant and Leadbeater between 1895 and 1933 would not have enabled them to anticipate such dynamical behaviour.

The second way of interpreting micro-psi—the "realist" viewpoint—considers that micro-psi records objective phenomena viewed under great magnification and all images are causally linked to objects in the examined substance. According to the basic thesis of this book, the realist interpretation is the correct one. It may be argued that, since it remains unexplained how the magnification is achieved, it cannot be known whether the images are accurate or distorted representations of microscopic objects. By showing the obvious similarities between certain micro-psi pictures and models of particle physics, it is hoped to remove any doubt that the reader may have concerning the intrinsic accuracy of micro-psi vision.

The criticism that the work of Besant and Leadbeater is a product of fraud cannot be sensibly defended. The problems associated with the "subjective" point of view make it untenable. It cannot satisfactorily explain why hallucinations that are not causally connected with objective reality bear striking similarities to modern theories of particle physics.

The following interpretation of micro-psi is proposed: Micro-psi vision is extra-sensory perception of physical, microscopic objects that are present in space-time during the period of their active observation. It must not be assumed that these objects are in the same state before, during, and after the observation. The possibility must be allowed that micro-psi is inherently perturbative and that, as a way of obtaining information about the quantum world, it is subject to the restriction placed on physical knowledge by Heisenberg's Uncertainty Principle. Although the involvement of extra-sensory perception is explicitly assumed, this assumption is made for the sake of convenience and not out of necessity. It is possible that all forms of human experience commonly known as "psychic" have a physical basis in reality in terms of processes that science can study. By regarding the micro-psi experience as a form of extra-sensory perception, we wish neither to affirm nor to deny this point of view. Instead, we are stating what is a well-known fact: atoms are too small to be seen with unaided human sight.

If the above hypothesis is true and micro-psi can provide valid knowledge about objects that may be invisible even to the electron microscope, the faculty presents a difficult problem for scientific understanding. In this regard, of course, it is similar to other forms of extra-sensory perception, such as telepathy and precognition, which have received to date no convincing explanation. It may be argued, however, that the question of whether micro-psi images refer to real objects in space-time cannot be separated from the problem of how they arise in an altered state of consciousness. If a theory cannot explain the latter, then it cannot answer the former. But it would be unreasonable to reject solely for this reason a theory or conceptual map of these images that made no reference to their causation. It is unknown how the neural machinery of the brain reconstitutes into a whole visual picture the impulse discharges in the optic nerve fibres, for nowhere in the brain can be found neurons responding specifically to even a small zone of the retinal image or to the observed picture. Yet no one doubts the objective reality of what he sees because he is ignorant of the final stage of the brain process leading to visual experience. No theory of micro-psi will be offered in this book, the task of which is to map out, in terms of contemporary elementary particle physics, the strange features and unfamiliar terrain of the world experienced by the micro-psi observer.

The hypothesis that micro-psi provides accurate information about the structure of microscopic objects implies that observational error is a meaningful possibility. The sustained retention of complex images within the field of vision of the observer is a strenuous mental activity for him, according to his testimony. He may make mistakes when he simultaneously applies his intellect to quantitative description of these images. For example, he may miscount the number of particles he sees as "points of light." Unlike errors due to faulty measuring instruments (errors that are systematic or constant and are correctable), errors of human observation are likely to occur at random and to be unpredictable. Consequently, when one applies a theory to data that are derived from observations prone to error, one should not expect exact agreement in every instance. Indeed, a theory whose predictions agreed with every datum would be just as unsatisfactory as one that gave poor agreement. The crucial issue is whether the agreement is good enough as a whole for the data to be seen as supporting the theory. In other words, is the probability of this measure of support having arisen by chance so low that such a possibility may be reasonably discounted? This question is settled in chapter 4 by the application of a statistical significance test to the theory of M.P.A.'s presented in this book.

Finally, it should be pointed out that, in the task of identifying micro-psi phenomena, one needs to be sensitive to the following problem: the observer must describe what he sees in non-technical, everyday language that conveys very inadequately his impression of complex and bizarre images. One must distinguish between firsthand reporting by the observer and accounts that amount only to arbitrary interpretations that could be false or misleading. The reports of Besant and Leadbeater concerning their investigations of the chemical elements may be expected to be coloured by the popular notions of Victorian science with which they were familiar. Their testimony should be examined with both these considerations in mind.

REFERENCES

1. "Meta Neon," The Theosophist, vol. **30,** part 1, December 1908.
2. The Theosophist, vol. **45,** April 1924.
3. Ibid.
4. P. A. M. Dirac, Proc. Roy. Soc. (London), **A133,** 60 (1931).
5. Investigations (1956–59) by Geoffrey Hodson and Dr. D. D. Lyness (unpublished).

CHAPTER 3

Quark Theory—Old and New

It seems to me that God in the beginning formed matter in solid, massy, hard, impenetrable, movable particles, of such sizes and figures, and with such other properties, and in such proportions to space as most to conduce to the end for which He formed them; and that these primitive particles being solids, are incomparably harder than any porous body compounded of them; even so very hard as never to wear or break to pieces; no ordinary power being able to divide what God himself made in the first creation.

Sir Isaac Newton

QUARKS FOR EVERYONE*

Suppose that a person is given an assortment of strawberries, cherries, and other fruit and that he is asked to make as many different selections from it as possible, taking any two or three fruits at a time. The number of possible choices available to him depends on the number of varieties of fruit in the assortment and not on the total number of fruits in it. Next suppose that all these choices are made correctly and that the fruit in each selection is crushed and thoroughly mixed. A second person, deprived of his sense of smell and his sight, is then asked to determine how many varieties of fruit were in the original assortment by tasting each mixture and by identifying the fruit in it. He can perform this task successfully only if no variety is missing in the selections given to him. He cannot be certain of this, but he can deduce what fruits were in the assortment only from the flavours that he detects. Although, in principle, he does not need to taste every sample of two or three fruits, in practice it is essential that he does. This is because in some samples one flavour may be masked by a stronger one, while in others the mixing of flavours may render them tasteless.

The high-energy physicist, who examines and identifies the debris scattered in all directions by the collisions of very fast-moving particles, finds himself in an analogous situation. Over the last thirty years, more than two hundred different types of particles have been discovered, although they

* Readers who are proficient in quark theory may ignore this introduction and proceed to the next section.

possess only a few "flavours." The nature of the fundamental relationship among them has not been easy for the physicist to uncover, but he is aided by various experimental facts and laws that he has inferred from them. Some of these are now summarized. A particle flavour is assigned a number that can assume only certain discrete values. These distinguish different particles with the same flavour. Examples are isospin, strangeness, and charm. A particle with a certain flavour and its antiparticle have flavour quantum numbers of the same numerical value but of opposite sign. A particle has a flavour quantum number of value zero if it does not have the appropriate flavour. These quantum numbers are additive: the flavour quantum number of a group of particles with different flavours is the sum of the values taken by the number for each particle. There are two classes of particles: the hadron, which interacts with other hadrons through the "strong" force (responsible for the binding together of protons and neutrons in nuclei), and the lepton, which interacts with hadrons and other leptons through a much weaker force—the "weak" interaction (responsible for the decay of radioactive nuclei). Hadrons weakly interact with one another, but leptons do not strongly interact with one another or with hadrons (as far as is as yet known). The flavour quantum numbers of hadrons are unchanged by their strong interaction but are altered by their weak interaction. Electrically charged hadrons and leptons share the "electromagnetic interaction" (a force of intermediate strength that is responsible, for example, for the binding of electrons in atoms). Each class of particles is assigned another quantum number that is unchanged by all three types of force. This is the "baryon number" for hadrons and the "lepton number" for leptons. Hadrons have a lepton number of zero, and leptons have a baryon number of zero.

All hadrons detected up to 1974 could be grouped in certain families or "multiplets" according to their flavours and baryon numbers. The latter were found to be just those allowed if all hadrons were assembled from a set of three fundamental particles called "quarks" (u, d, and s) with different flavours, taken either three at a time or as a quark-antiquark pair. Despite intensive search, no particles have been found that are other combinations of quarks—for example, single quarks, pairs of quarks ("diquarks"), or three quarks and an antiquark. Since hadrons differ widely in mass, one quark (the s quark with the "strangeness" flavour) has to be heavier than the other two. The original triplet of quarks proposed by the physicists M. Gell-Mann and G. Zweig in 1964 was increased to four when, in 1974, a heavy, remarkably stable hadron (the "J/ψ") was found. It could not be accommodated in the long-established families of hadrons, and it signalled the existence of a new hadron flavour ("charm"), provided by a new quark (the c quark). More recently, the detection of another long-lived hadron ("upsilon") has indicated that a quark with a fifth flavour exists. Called "bottom" (b), it has led to the confident expectation that a sixth quark, "top" (t), with a new flavour should exist as well. This is because a pattern has emerged wherein quarks appear to be related in pairs: (u, d), (c, s), (t, b), and so on, in a way that is mirrored by corresponding pairs of leptons: (ν_e, e^-), (ν_μ, μ^-), (ν_τ, τ^-), and so on. This parallelism (which may extend beyond the three pairs of known leptons to as yet undiscovered leptons and quarks) suggests that some property is shared by leptons and hadrons. Since both exhibit electromagnetic and weak forces, this property must be related to these forces. An obvious suggestion is that leptons have flavour as well as quarks and that quarks and leptons weakly interact by changing their flavour. The pattern emerging from high-energy physics is that successive pairs of quarks increase in mass. Only the two lightest quarks (u and d), which constitute the protons and neutrons in nuclei, and the first lepton pair (ν_e, e^-), which is involved in the decay of radioactive nuclei, are used in the construction of the universe. The other pairs appear briefly only in high-energy processes. Protons consist of two u quarks and one d quark; neutrons are composed of two d quarks and one u quark. The u and d quarks have electric charges of $2e/3$ and $-e/3$, respectively (e is the magnitude of the charge of an electron).

With a few unconfirmed exceptions, the detection of free, single quarks has not been reported. Leptons can be created in the laboratory and their properties measured very accurately. But information about quarks has to be obtained indirectly by making measurements on the various three-quark groups or quark-antiquark pairs created by the collisions of particles, by using leptons to probe the interior of protons and neutrons, or by looking at the products of the annihilation of electrons and positrons (their antiparticles). The forces that hold quarks together appear to imprison them as well as to restrict their power of combination to just groups of three or to quark-antiquark pairs. Are separate mechanisms responsible for these features, or does one mechanism account for both? The theory outlined in the next section proposes one mechanism.

Suppose that the person referred to at the beginning of this chapter had been asked to select fruit at random from the assortment and then to paint the fruits either red, blue, or green, and that another person had to make as many different groups of three colours from them as possible, without regard for their variety. The number of distinct choices that he could make is ten. The number is, of course, independent of how many different fruits there were in the assortment. One of the choices consists of a red-, a blue-, and a green-coloured fruit, but these may be of any variety. According to a recently proposed theory of strong interactions, quarks have three "colours" (red, blue, and green), a property, of course, not to be understood in its literal sense. Just as the electromagnetic forces between electrically charged particles are due to their charges, so the strong force between coloured quarks is due to their colours. A quark changes its colour (but not its flavour) by emitting or by absorbing zero-mass particles called "gluons." These are passed to and fro between quarks, coupling them together to form systems of either three differently coloured quarks or a quark of a certain colour and an antiquark, whose colour is a mixture of the other two colours. Since equal mixtures of red, blue, and green make a white or colourless combination, free hadrons are called "colourless." The exchanged gluons interact strongly with one another as well as with quarks, and the resultant force acting on quarks does not weaken with their distance apart but is of constant strength, thereby imprisoning them permanently inside hadrons. Quarks remain hidden from direct experimental detection because the forces between them do not permit a coloured quark to break loose and exist as a free, independent particle.

The colour theory, or "quantum chromodynamics" (Q.C.D.), is currently emerging as a successful (though not yet established) theory of strong forces. But it was originally proposed in an ad hoc fashion in order to solve a technical problem of quark theory. Since Q.C.D. did not originate in a unified theory of hadrons and leptons, it did not explain why leptons do not have colour and so do not interact strongly: the empirical division of elementary particles into separate families of leptons and hadrons remained unexplained. The possession of colour by quarks and the lack of colour in leptons are shown in the next section to be consequences of treating fundamental hadrons and leptons as members of a single family of particles.

A UNIFIED HADRON-LEPTON THEORY

In the paper reproduced below, the author proposed a new theory of matter.[1] It is summarized here for the general reader. The remainder of this chapter is devoted to a more technical discussion of particular aspects of the theory that are relevant to the micro-psi observations described in subsequent chapters.

It is proposed that quarks are not discrete, fundamental objects but, instead, are composite, tightly knit clusters of three particles called "omegons."* Protons and neutrons, which are each made up of three quarks, therefore contain nine omegons. These basic hadrons have ten different flavours, and each determines its own type of quark. Experiments have already established the existence of half this number of flavoured quarks.

* Pronounced *ō-mee-guns*.

The two lightest omegons (labelled $\mathcal{P}$ and $\mathcal{N}$) make up the u and d quarks from which the proton and neutron are assembled. The u quark consists of two $\mathcal{P}$ omegons and one $\mathcal{N}$ omegon; the d consists of one $\mathcal{P}$ and two $\mathcal{N}$ omegons. A proton (with two u quarks and one d quark) contains five $\mathcal{P}$ and four $\mathcal{N}$ omegons; a neutron (with one u and two d quarks) contains four $\mathcal{P}$ and five $\mathcal{N}$ omegons. $\mathcal{P}$ and $\mathcal{N}$ omegons carry electric charges of $+5/9$ and $-4/9$, respectively (in units of the charge of an electron). Omegons have nine "colour-shades"—dark, medium, and light shades of the colours red, blue, and green. A quark of a given colour is composed of omegons with dark, medium, and light shades of that colour. Omegons are trapped inside quarks by the same mechanism that confines quarks inside observable hadrons. Because flavour and colour-shade properties are given equal dynamical status, particles with a tenth colour are predicted to exist. These are leptons, which also have ten flavours, each corresponding to its omegon partner. Five are electrically neutral (neutrinos), and five have the charge of an electron. Two massive leptons remain to be detected, according to the theory.

COMPOSITE QUARKS AND HADRON-LEPTON UNIFICATION*

S. M. PHILLIPS

205, Malmesbury Park Road, Bournemouth, Dorset BH8 8PX, England

Received 5 February 1979

A unified hadron-lepton theory, based on the symmetry group $SU(10)_{flavour} \times SU(10)_{colour}$ is presented. It predicts that quarks are composite and there exist five generations of singly flavoured quarks mirrored by a finite heavy lepton sequence.

We postulate that the basic indivisible constituents of matter belong to the fundamental fermion multiplet $F = f^{(i,A)}$ (row index $i = 1, 2, \ldots, m$ denotes flavour; column index $A = 1, 2, \ldots, m$ denotes colour), the (m, m) representation of the unified hadron-lepton symmetry group $SU(m)_f \times SU(m)_c$. ($SU(m)_f = SU(m)_{flavour}$, $SU(m)_c = SU(m)_{colour}$.) Generalising a suggestion by Pati and Salam [1], we propose that the mth colour state of each fermion is a lepton. Present experimental evidence for three quark generations, each mirrored by a lepton and its associated neutrino, suggests the following generalisation: fermion flavour states consist of N SU(2) doublets

* Permission by the North-Holland Publishing Company to reprint this paper is gratefully acknowledged.

with electric charges $(Q, Q-1)$ $(Q = 0$ or $2/3)$, gauged for weak and electromagnetic interactions according to the Weinberg-Salam $U(1) \times SU(2)$ symmetry group. Accordingly, $m = 2N$. The electric charge operator Q, being a sum of $SU(2N)_f \times SU(2N)_c$ generators, is:

$$Q = Q_f + Q_c \,. \tag{1}$$

Q_f and Q_c are traceless $2N \times 2N$ matrices whose elements $Q_f^{(i,A)}$ and $Q_c^{(i,A)}$ are flavour and colour contributions to the charge of $f^{(i,A)}$. In particular, $f^{(2N-1,2N-1)}$ is a hadron with flavour charge $Q_f^{(2N-1,2N-1)} = Q_H = 2/3$ or $-1/3$ and $f^{(2N,2N)}$ is a lepton with flavour charge $Q_f^{(2N,2N)} = Q_L$. Assuming that different colour states of a hadron with a given flavour have the same charge, then:

$$0 = \mathrm{Tr}\,(Q_f) = \sum_{n=1}^{2N} Q_f^{(n,n)} = (N-1)/3 + Q_H + Q_L \,. \tag{2}$$

Solutions are: $N = 2$ $(Q_L = 0$ or $-1)$, $N = 5$ $(Q_L = -1$ or $-2)$, $N = 8$ $(Q_L = -2$ or $-3)$, etc. $N = 2$ corresponds to the Pati-Salam quark-lepton unification theory $SU(4)_f \times SU(4)_c$ [2], which excludes t and b quarks; $N \geq 8$ leads to composite quark theories of greater complexity than that with $N = 5$, which we now choose. The (10, 10) representation of $SU(10)_f \times SU(10)_c$ consists of ten leptons $L^i = f^{(i,10)}$ and ten fundamental hadrons, each in nine colour states: $\omega^{i\alpha} = f^{(i,\alpha)}$ $(i = 1, 2, \ldots, 10;\ \alpha = 1, 2, \ldots, 9)$. We call the latter "omegons"‡[1] and shall show that bound systems of three such hadrons have the quantum numbers of quarks.

$SU(10)_c$ is allowed to be spontaneously broken by the Higgs vacuum (with the Higgs field in the self-adjoint representation of $SU(10)_c$), to leave $U(1) \times SU(9)_c$ as an exact local gauge symmetry. For $N = 5$, the values $Q_L = -1$ or -2 are the flavour charges of the fifth lepton doublet. Known leptons have electrical charge $Q = 0$ or -1, so we require: $Q_c^{(2N-1,2N)} = Q_c^{(2N,2N)} = 1$. Assuming $Q_c^{(i,A)}$ is independent of i, then $Q_c^{(i,2N)} = 1$ and Q, Q_f and Q_c have the form:

$$Q = (1/9)\begin{bmatrix} 5 & 5 & \ldots & 5 & 0 \\ -4 & -4 & \ldots & -4 & -9 \\ \cdot & \cdot & & \cdot & \cdot \\ \cdot & \cdot & & \cdot & \cdot \\ \cdot & \cdot & & \cdot & \cdot \\ 5 & 5 & \ldots & 5 & 0 \\ -4 & -4 & \ldots & -4 & -9 \end{bmatrix}, \tag{3}$$

‡[1] From "omega," the last letter of the Greek alphabet.

$$Q_f = (1/3)\begin{bmatrix} 2 & 2 & \cdots & 2 & -3 \\ -1 & -1 & \cdots & -1 & -6 \\ \cdot & \cdot & & \cdot & \cdot \\ \cdot & \cdot & & \cdot & \cdot \\ \cdot & \cdot & & \cdot & \cdot \\ 2 & 2 & \cdots & 2 & -3 \\ -1 & -1 & \cdots & -1 & -6 \end{bmatrix},$$

$$Q_c = (-1/9)\begin{bmatrix} 1 & 1 & \cdots & 1 & -9 \\ 1 & 1 & \cdots & 1 & -9 \\ \cdot & \cdot & & & \cdot \\ \cdot & \cdot & & & \cdot \\ \cdot & \cdot & & \cdot & \cdot \\ 1 & 1 & \cdots & 1 & -9 \\ 1 & 1 & \cdots & 1 & -9 \end{bmatrix}. \qquad (3\ cont.)$$

F consists of five leptons with their associated neutrinos:

$$\begin{pmatrix}\nu_e \\ e^-\end{pmatrix}, \quad \begin{pmatrix}\nu_\mu \\ \mu^-\end{pmatrix}, \quad \begin{pmatrix}\nu_\tau \\ \tau^-\end{pmatrix}, \quad \begin{pmatrix}\nu_\theta \\ \theta^-\end{pmatrix}, \quad \begin{pmatrix}\nu_\kappa \\ \kappa^-\end{pmatrix},$$

mirrored by ten omegons:

$$\begin{pmatrix}\mathcal{P} \\ \mathcal{N}\end{pmatrix}, \quad \begin{pmatrix}\mathcal{C} \\ \lambda\end{pmatrix}, \quad \begin{pmatrix}\mathcal{J} \\ \mathcal{B}\end{pmatrix}, \quad \begin{pmatrix}\mathcal{E} \\ \mathcal{J}\end{pmatrix}, \quad \begin{pmatrix}\mathcal{G} \\ \mathcal{H}\end{pmatrix},^{*}$$

with charges (5/9, −4/9), each of which exists in nine colour states that make up the fundamental representation of $SU(9)_c$. A colour nonet of a given flavour consists of three triplets: $(\omega^1, \omega^2, \omega^3)$, $(\omega^4, \omega^5, \omega^6)$ and $(\omega^7, \omega^8, \omega^9)$ (flavour indices are suppressed). The members of a triplet are distinguishable by having three degrees of freedom called "colour-shades" (CS): "light," "medium" and "dark." Different triplets are distinguishable by their colour. Colour-shade triplets are physically distinct 3-dimensional representations of $SU(3)_{\text{colour-shade}}$ ($SU(3)_{cs}$). We define three $SU(3)_{cs}$ scalar wavefunctions

$$q^a = (3!)^{-1/2}\epsilon_{\alpha\beta\gamma}\omega^\alpha(1)\omega^\beta(2)\omega^\gamma(3) \qquad (a = 1, 2, 3)\ , \tag{4}$$

for a bound system of three omegons (labelled 1, 2, 3), with $\alpha, \beta, \gamma = 1, 2, 3$ $(a = 1)$, $\alpha, \beta, \gamma = 4, 5, 6$ $(a = 2)$ and $\alpha, \beta, \gamma = 7, 8, 9$ $(a = 3)$. $\epsilon_{\alpha\beta\gamma}$ is the completely antisymmetric tensor ($\epsilon_{123} = \epsilon_{456} = \epsilon_{789} = 1$). The q^a constitute a triplet of states which we propose to identify as the colour states of quarks [3]. Coloured quarks are colour-shadeless composite systems of three

* Omegon symbols have been changed in the article reproduced here.

colour-shaded omegons. Note that CS triplets are not representations of $SU(3)_c \times SU(3)_{cs}$, where $SU(3)_c$ is the familiar colour gauge group of quarks. Omegons do not possess quark colour, but have three "shades" of them. To support this suggestion, we now show that $SU(3)_f$ quarks also emerge from our unified theory. Each of the nine omegon CS states obeys the following 10-dimensional generalisation of the Gell-Mann–Nishijima relation:

$$Q_f = I_3 + \tfrac{1}{2}(N + S + C + T + B + I + E + G + H)\,, \tag{5}$$

where $Q_f = \text{diag}\,(2/3, -1/3, -1/3, \ldots, 2/3, -1/3)$ is the flavour charge, $I_3 = \text{diag}\,(1/2, -1/2, 0, \ldots, 0)$ is the third component of isospin, $N = \text{diag}\,(1/3, 1/3, \ldots, 1/3)$ is the baryon number, $S = \text{diag}\,(0, 0, -1, 0, \ldots, 0)$ is strangeness, $C = \text{diag}\,(0, 0, 0, 1, 0, \ldots, 0)$ is charm, etc. T, B, I, E, G and H are additional quantum numbers conserved by strong interactions, having values 1, −1, 1, −1, 1 and −1 for the $\mathcal{T}$, $\mathcal{B}$, $\mathcal{E}$, $\mathcal{I}$, $\mathcal{G}$ and $\mathcal{H}$ omegons, respectively, otherwise 0. We note that $\mathcal{P}$, $\mathcal{N}$ and λ form an $SU(3)_f$ triplet, as well as having three sets of three colour-shades, i.e., they are a (3, 3) representation of $SU(3)_f \times SU(3)_{cs}$. The isodoublet $(\mathcal{P}\mathcal{P}\mathcal{N})$ and $(\mathcal{P}\mathcal{N}\mathcal{N})$ and the isoscalar $(\mathcal{P}\mathcal{N}\lambda)$, members of the colour-shadeless (8, 1) multiplet in $(3, 3) \times (3, 3) \times (3, 3)$, have the SU(3) quantum numbers of u, d and s quarks, respectively. Thus, colour SU(3) quarks emerge as ω-ω-ω composite particles. More generally, a CS triplet forms the (10, 3) representation of $SU(10)_f \times SU(3)_{cs}$. All quarks, as colour-shadeless bound states, belong to $SU(3)_{cs}$-singlet representations in $(10, 3) \times (10, 3) \times (10, 3)$. Among them are ten singly flavoured quarks with charges (2/3, −1/3): u = $\mathcal{P}\mathcal{P}\mathcal{N}$, d = $\mathcal{P}\mathcal{N}\mathcal{N}$, s = $\mathcal{P}\mathcal{N}\lambda$, c = $\mathcal{P}\mathcal{N}\mathcal{C}$, t = $\mathcal{P}\mathcal{N}\mathcal{T}$, b = $\mathcal{P}\mathcal{N}\mathcal{B}$, i = $\mathcal{P}\mathcal{N}\mathcal{I}$, e = $\mathcal{P}\mathcal{N}\mathcal{E}$, g = $\mathcal{P}\mathcal{N}\mathcal{G}$ and h = $\mathcal{P}\mathcal{N}\mathcal{H}$. Baryons with symmetric spin-unitary spin states are: $\mathrm{B} = (3!)^{-1/2} \times \epsilon_{abc}q_i^a q_j^b q_k^c$; mesons are: $M = (3)^{-1/2}q_i^a \bar{q}_j^b$ (a, b, c = 1, 2, 3; i, j, k = 1, 2, . . . , 10). Four more flavourless, long-lived, heavy mesons of upsilon type are predicted, $\mathrm{i\bar{i}}$, $\mathrm{e\bar{e}}$, $\mathrm{g\bar{g}}$ and $\mathrm{h\bar{h}}$. The gauge group for omegon strong interactions is $SU(9)_c$. The 80 gauge fields are made up of three colour-shade-changing $SU(3)_{cs}$ octets of "super gluons," exchange of which binds omegons inside quarks without changing the latter colour states, and 56 gluons, which are exchanged between omegons in different quarks, changing their colour. Unlike in the Pati-Salam scheme, there are no exotic gluons strongly coupling fundamental hadrons with leptons. The proton is predicted to be absolutely stable, therefore. Also, baryon number and lepton number $L = n_e + n_\mu + n_\tau + n_\theta + n_\kappa$ are separately conserved. Two new heavy leptons are predicted: θ^- and κ^-. They should appear in the electron-positron annihilation reactions: $e^+ + e^- = \theta^+ + \theta^-$, $e^+ + e^- = \kappa^+ + \kappa^-$.

Non-abelian generalisations of the Nielsen-Olesen model [4], providing permanent confinement of quarks and saturation of quark forces in zero-triality states, are very attractive in the context of a composite quark theory. The condition for topologically distinct Nielsen-Olesen vortices and Dirac monopoles to exist such that the former cannot be transformed into one another by continuous gauge transformations is that the global gauge group should be multiply connected [5]. If this group is SU(9)/Z_9, where the cyclic group

$$Z_9 = (I_9, \nu I_9, \nu^2 I_9, \ldots, \nu^8 I_9) \qquad (\nu = \exp(2\pi i/9)) , \tag{6}$$

is the centre of SU(9), then nine distinct vortices exist, since SU(9)/Z_9 is 9-fold connected. One corresponds to the ground state of the vacuum and consists of no vortex; the other eight correspond to non-equivalent magnetic monopoles of monopole moment $g_0, 2g_0, \ldots, 8g_0$ ($g_0 = 1/2e$). This is because, with the Higgs field belonging to an adjoint representation of SU(9), single-valuedness modulo $2\pi/9$ of its phase implies that vortex flux (and magnetic charge, therefore) is defined only modulo 9 (in Dirac units). Thus, nine and only nine SU(9) monopoles form a magnetically neutral system when embedded in a superconducting Higgs vacuum and bound by Nielsen-Olesen vortices. Similarly for three SU(3) monopoles in a vacuum with broken SU(3) symmetry. Omegons are bound in quarks by three super-gluon octets of SU$(3)_{cs}$. With omegons as SU(9) monopoles, a neutral system of nine monopoles can cluster in three groups of three, bound internally and externally by Y-shaped strings that are connected via vortex bifurcation. In this way, the Meissner effect would confine both omegons inside quarks and quarks in hadrons. Identification of quarks as clusters of three SU(9) monopoles guarantees zero triality for bound states of quarks. This empirical rule is a dynamical consequence of an SU(9) vacuum, which confines fermions with magnetic charge (omegons) but not fermions with zero magnetic charge (leptons). In this scheme, mesons are quark-antiquark pairs, joined by three strings, not by one. Quark compositeness improves the success of the SU(6) results: $\mu_p/\mu_n = -1.50$ and $\mu_\Lambda/\mu_p = -0.33$, for proton ($\mu_p$), neutron ($\mu_n$) and $\Lambda(\mu_\Lambda)$ magnetic moments. To show this, we assume omegon magnetic moments are proportional to their electric charge:

$$m_\omega = \delta(e_\omega/e) , \tag{7}$$

(e_ω and e are omegon and electron charges, δ is a scale parameter). Three omegons in a quark have an antisymmetric colour-shade wavefunction. Assuming their orbital motion is non-relativistic, then SU(6) symmetry is valid for Fermi-Dirac omegons in u, d and s quarks. Hence, these quarks have symmetric SU(6) wavefunctions. Calculated u, d and s magnetic moments are:

$$\mu_u = \langle u | m_u^z | u \rangle = (8/9)\delta , \quad (8)$$

$$\mu_d = \langle d | m_d^z | d \rangle = -(7/9)\delta , \quad (9)$$

$$\mu_s = \langle s | m_s^z | s \rangle = -(4/9)\delta . \quad (10)$$

We find:

$$\mu_p = (53/27)\delta , \quad \mu_n = -(4/3)\delta , \quad \mu_\Lambda = -(4/9)\delta ,$$

giving

$$\mu_p/\mu_n = -1.47 \quad (-1.46, \text{ measured value}) , \quad (11)$$

$$\mu_\Lambda/\mu_p = -0.23 \quad (-0.26) .^{\ddagger 2} \quad (12)$$

Agreement is better for composite quarks than for point-like quarks. We predict: $\mu_u = 1.26\ \mu_N$ ($0.67\ \mu_N$ for point-like quarks), $\mu_d = -1.10\ \mu_N$ ($-0.33\ \mu_N$) and $\mu_s = -0.63\ \mu_N$ ($-0.33\ \mu_N$). The omegon g-factor is: $g_\omega = (27/53)\ 2.79\ (M_\omega^*/M_p)$, where M_ω^* is the effective mass of a bound omegon in an external magnetic field. Evidence for composite quarks would appear in the effect of finite quark size on hadron form factors. This would become important at large values of momentum transfer. The relativistic magnetic form factor $F^M(t)$ of a bound system of three spin-$\frac{1}{2}$ particles is [7]:

$$F^M(t) = \alpha^{-1} F(-t/\alpha) \sum_{j=1}^{3} Q_j \epsilon_j F_j^M(t) , \quad (13)$$

where $\alpha = 1 - t/4m^2$, m is the composite particle mass, $F(t)$ is its form factor, Q_j is the charge, ϵ_j is the expectation value of the third component of spin and $F_j^M(t)$ is the magnetic form factor of the jth constituent. If these are also composite systems of three particles, then:

$$F_j^M(t) = \alpha_j^{-1} F_j(-t/\alpha_j) \sum_{\beta=1}^{3} Q_{j\beta} \epsilon_{j\beta} \mathfrak{F}_{j\beta}^M(t) , \quad (14)$$

where $\alpha_j = 1 - t/4m_j^2$, m_j is the mass of the jth composite particle, $F_j(t)$ is its form factor and $Q_{j\beta}$, $\epsilon_{j\beta}$ and $\mathfrak{F}_{j\beta}^M(t)$ are as above for the βth particle in the jth composite. Then:

$$F^M(t) \sim |t|^{-2} \mathfrak{F}^M(t) \quad (|t| \to \infty) , \quad (15)$$

($\mathfrak{F}^M(t)$ is the omegon magnetic form factor), in agreement with SLAC data [8] which, for the proton, indicate $F^M(t) \lesssim |t|^{-2}$, as $|t| \to \infty$. The observed

‡2 We take the value $\mu_\Lambda = (-0.73 \pm 1.6)\ \mu_N$ [6].

approximate $|t|^{-2}$ behaviour of the proton magnetic form factor could be due to Lorentz contraction of both finite-sized proton and quark wave-packets, indicative of composite quarks.

I thank Professor A. Salam for his encouragement.

References

[1] J. C. Pati and A. Salam, Phys. Rev. D8 (1973) 1240.
[2] J. C. Pati and A. Salam, Phys. Rev. D10 (1974) 275.
[3] M. Gell-Mann, Acta Phys. Austriaca Suppl. 9 (1972); H. Fritzsch, M. Gell-Mann and H. Leutwyler, Phys. Lett. 47B (1973) 365.
[4] H. B. Nielsen and P. Olesen, Nucl. Phys. B61 (1973) 45; P. Olesen, Phys. Lett. 31B (1974) 494.
[5] H. C. Tze and Z. F. Ezawa, Phys. Rev. D14 (1976) 1648.
[6] Particle Data Group, Phys. Lett. 33B (1970) 1.
[7] A. L. Licht and A. Pagnamenta, Phys. Rev. D2 (1970) 1156.
[8] D. H. Coward et al., Phys. Rev. Lett. 20 (1968) 292.

QUANTUM NUMBERS OF THE OMEGONS

	Q/e	N	I_3	Y	S	C	T	B	I	E	G	H
$\mathcal{P}$	5/9	1/9	1/2	1/9	0	0	0	0	0	0	0	0
$\mathcal{N}$	−4/9	1/9	−1/2	1/9	0	0	0	0	0	0	0	0
λ	−4/9	1/9	0	−8/9	−1	0	0	0	0	0	0	0
$\mathcal{C}$	5/9	1/9	0	1/9	0	1	0	0	0	0	0	0
$\mathcal{T}$	5/9	1/9	0	1/9	0	0	1	0	0	0	0	0
$\mathcal{B}$	−4/9	1/9	0	1/9	0	0	0	−1	0	0	0	0
$\mathcal{I}$	−4/9	1/9	0	1/9	0	0	0	0	−1	0	0	0
$\mathcal{E}$	5/9	1/9	0	1/9	0	0	0	0	0	1	0	0
$\mathcal{G}$	5/9	1/9	0	1/9	0	0	0	0	0	0	1	0
$\mathcal{H}$	−4/9	1/9	0	1/9	0	0	0	0	0	0	0	−1

QUANTUM NUMBERS OF THE LEPTONS

	Q/e	n_e	n_μ	n_τ	n_θ	n_κ	L
ν_e	0	1	0	0	0	0	1
e^-	−1	1	0	0	0	0	1
μ^-	−1	0	1	0	0	0	1
ν_μ	0	0	1	0	0	0	1
ν_τ	0	0	0	1	0	0	1
τ^-	−1	0	0	1	0	0	1
θ^-	−1	0	0	0	1	0	1
ν_θ	0	0	0	0	1	0	1
ν_κ	0	0	0	0	0	1	1
κ^-	−1	0	0	0	0	1	1

NOTE.—$L = n_e + n_\mu + n_\tau \pm n_\theta + n_\kappa$ = lepton no.; n_e = electron lepton no.; n_μ = muon lepton no.; n_τ = tau lepton no.; n_θ = theta lepton no.; n_κ = kappa lepton no.

The dynamical contents of the Omegon Model are summarized below.

UNIFIED FLAVOURDYNAMICS

The flavour symmetry of the gauge group $SU(10)_f$ is spontaneously broken by the Higgs mechanism into the compound symmetries of $U(1) \times SU(2)_w \times SU(5)_g$. This divides the ten flavour states into five left-handed $SU(2)_w$ doublets,

$$L^j = \tfrac{1}{2}(1 + \gamma_5) \begin{pmatrix} f_1^j \\ f_1^j \end{pmatrix} \qquad (j = 1, 2, 3, 4, 5)\ ,$$

that form the (2, 5) representation of $U(1) \times SU(2)_w \times SU(5)_g$. There are five generations of fermions:

$$(1) \begin{pmatrix} \nu_e & \mathcal{P} \\ e^- & \mathcal{N} \end{pmatrix} \quad (2) \begin{pmatrix} \nu_\mu & \mathcal{C} \\ \mu^- & \lambda \end{pmatrix} \quad (3) \begin{pmatrix} \nu_\tau & \mathcal{J} \\ \tau^- & \mathcal{B} \end{pmatrix} \quad (4) \begin{pmatrix} \nu_\theta & \mathcal{E} \\ \theta^- & \mathcal{I} \end{pmatrix} \quad (5) \begin{pmatrix} \nu_\kappa & \mathcal{G} \\ \kappa^- & \mathcal{H} \end{pmatrix},$$

similarly electrically charged members of which are gauged according to $SU(5)_g$—the "generation" or "horizontal" symmetry group. The corresponding quark doublets of the weak isospin,

$$(1) \begin{pmatrix} u \\ d \end{pmatrix} \quad (2) \begin{pmatrix} c \\ s \end{pmatrix} \quad (3) \begin{pmatrix} t \\ b \end{pmatrix} \quad (4) \begin{pmatrix} e \\ i \end{pmatrix} \quad (5) \begin{pmatrix} g \\ h \end{pmatrix},$$

have the omegon flavour constituents: $u = \mathcal{P}\mathcal{P}\mathcal{N}$, $d = \mathcal{P}\mathcal{N}\mathcal{N}$, $c - \mathcal{P}\mathcal{N}\mathcal{C}$, $s = \mathcal{P}\mathcal{N}\lambda$, $t = \mathcal{P}\mathcal{N}\mathcal{J}$, $b = \mathcal{P}\mathcal{N}\mathcal{B}$, $e = \mathcal{P}\mathcal{N}\mathcal{E}$, $i = \mathcal{P}\mathcal{N}\mathcal{I}$, $g = \mathcal{P}\mathcal{N}\mathcal{G}$, and $h = \mathcal{P}\mathcal{N}\mathcal{H}$. The ninety-nine vector gauge bosons of the flavour gauge group $SU(10)_f$ make up the $SU(2)_w$ and $SU(5)_g$ multiplets in the decomposition of the direct product:

$$(2, 5) \times (\bar{2}, \bar{5}) = (1, 1) + (3, 1) + (1, 24) + (3, 24)\ .$$

The $SU(5)_g$ singlet (3, 1) is a weak isospin triplet from which the photon, Z^0, and $W^\pm$ are constructed according to the Weinberg-Salam theory of electro-weak interactions; (1, 24) consists of twenty-four neutral vector bosons mediating flavour-conserving, weak transitions between like charged members of different generations of fermions; (3, 24) determines seventy-two charged and neutral vector bosons mediating charged and neutral current interactions between different generations of omegons and leptons.

UNIFIED CHROMODYNAMICS

The colour symmetry of the gauge group $SU(10)_c$ is spontaneously broken by the Higgs mechanism into the reduced symmetry of $U(1) \times SU(9)_c$. The nine colour states of omegons form an $SU(9)_c$ nonet. The tenth colour state, an $SU(9)_c$ singlet, is a lepton. The symmetry of $SU(9)_c$ is further spontaneously broken by the Higgs mechanism into the compound symmetry of $SU(3)_c \times SU(3)_s$ (c = colour, s = shade). The nine "colour-shade" states $\omega^{a\alpha}$ (a = 1 [red], 2 [blue], 3 [green]; α = 1 [light], 2 [dark], 3 [medium]) form the (3, 3) representation of $SU(3)_c \times SU(3)_s$. The three shade triplets are written

$$\omega^{1\alpha} = \psi^\alpha , \qquad \omega^{2\alpha} = \theta^\alpha , \qquad \omega^{3\alpha} = \phi^\alpha .$$

ψ^α, θ^α, and ϕ^α have, respectively, the colour isospin $T_{3c} = 1/2, -1/2$, and 0 and the colour hypercharge $Y_c = 1/3, 1/3$, and $-2/3$. ψ^1, ψ^2, and ψ^3 have the shade isospin $T_{3s} = 1/2, -1/2$, and 0 and the shade hypercharge $Y_s = 1/3, 1/3$, and $-2/3$, respectively (similarly for θ^1, θ^2, and θ^3 and for ϕ^1, ϕ^2, and ϕ^3). Of the antisymmetric $SU(3)_s$ singlet states in the reduction

$$(3, 3) \times (3, 3) \times (3, 3) = (1, 1) + 2(8, 1) + (10, 1) + 2(1, 8) + (1, 10) + 4(8, 8) + 2(10, 8) + 2(8, 10) + (10, 10) ,$$

the $SU(3)_c$ decuplet (10, 1) alone is symmetric in the colour gauge indices. The Exclusion Principle restricts such states with three omegons in non-relativistic S orbitals to being symmetric in combined spin, weak isospin, and generation gauge indices. Three of these shadeless states may be shown to form a 3-dimensional representation of SU(3) and are identified with the "colour" states of quarks. Colour SU(3) emerges as an effective, non-gaugeable symmetry of the strong interactions of composite quarks. The ninety-nine vector gauge fields associated with $SU(10)_c$ are made up of nineteen superheavy bosons, responsible for lepton-omegon interactions that do not conserve baryon and lepton number, and the eighty vector gauge fields of $SU(9)_c$, which comprise the following $SU(3)_c$ and $SU(3)_s$ octets:

$$(3, 3) \times (\bar{3}, \bar{3}) = (1, 1) + (8, 1) + (1, 8) + (8, 8) .$$

The octet (1, 8) is made up of "supergluons" that couple superstrongly to omegons, confining them in $SU(3)_s$ singlet states to form quarks; (8, 1) and (8, 8) are nine octets of "gluons" whose exchange between omegons is responsible for the strong interaction between quarks as bound states of three omegons.

MAGNETIC MONOPOLE/HIGGS VORTEX STRUCTURE OF HADRONS

In their topological classification of Nielsen-Olesen vortices and their end-point Dirac magnetic monopoles in non-Abelian gauge theories with any compact Lie group, Tze and Ezawa[2] showed that the necessary condition for the existence of vortices in the superconducting Higgs field is that the global invariance group acting on the vectors of the physically realized representations should be multiply connected. According to a theorem of group theory, all N-ply connected Lie groups having the same Lie algebra as a given simply connected group G are of the form $G \simeq \bar{G}/\Delta$, where Δ is some invariant discrete subgroup of $\bar{G}$. Δ is isomorphous to the fundamental group of G, $\pi(G)$. This is a finite, discrete Abelian group of order N that classifies all topologically distinct vortices allowed by G. To its identity element—the class of closed curves in the group manifold of G that are homotopic to the null path—corresponds the vacuum state of no vortex and no monopole. To the $N - 1$ non-trivial classes of homotopic curves, there correspond $N - 1$ types of vortices and monopoles. The fundamental group of the simply connected universal covering group G is of order 1, consisting only of the identity element, and so no vortex exists for such a group. If the global gauge group is the N-fold connected $G = \mathrm{SU}(N)/Z_N$, where the cyclic Abelian group

$$Z_N = (I_N, \Omega I_N, \Omega^2 I_N, \ldots, \Omega^{N-1} I_N) \tag{1}$$

is the centre of $\mathrm{SU}(N)$ ($\Omega = \exp(2\pi i/N)$ is the Nth primitive root of unity; I_N is the $N \times N$ unit matrix), then Z_N is the fundamental group of G and its elements Ω^m ($m = 0, 1, \ldots, N - 1$) define topologically distinct vortices of monopoles of charge m (in Dirac units of $1/2e$). Since the powers m are defined only modulo N, so are the magnetic charges and the magnetic flux in their vortices. Dirac's quantization condition

$$eg = m/2 \qquad (m = 0, 1, 2, \ldots, \infty) \tag{2}$$

for Abelian magnetic monopoles whose charges are the homotopic invariants of the fundamental group of the infinitely connected global group $\mathrm{U}(1)$, is generalized to

$$eg = m/2 \ (\mathrm{mod}\ N) \qquad (m = 0, 1, 2, \ldots, N - 1) \tag{3}$$

for non-Abelian monopoles with the global invariance group G. Vortex flux $4\pi g$ is quantized in units of $2\pi/e$. Vortices carrying flux quanta m and m (mod N) correspond to the same homotopic class and can be continuously gauged into each other. This means that they are physically indistinguishable. Vortices carrying N flux quanta do not exist, for they can be continuously gauged into the vacuum state of no vortex and no monopole ($m = 0$). The magnetic charge m is a homotopic invariant of the fundamental group $\pi(G) \simeq Z_N$. Its $N - 1$ non-trivial values mirror the $N - 1$ non-trivial homotopic classes of closed curves in the N-fold connected group manifold of $\mathrm{SU}(N)/Z_N$. The physical consequence of the definition by modulo N of magnetic charge is that two types of vortex configuration are topologically stable: mesonic single, finite vortices with a monopole and antimonopole at their ends (fig. 3.1) and baryonic N-legged finite vortices (fig. 3.2) that terminate in Dirac monopoles. SU(N) vortices and their end-point monopoles have a topological "plurality." Their bonding power is saturated only for systems of N monopoles.

Ezawa and Tze also showed[3] that this property is possessed by "colourless" monopoles, that is, monopoles whose flux is an SU(N) scalar and, therefore, gauge-invariant. Only N colourless monopoles, embedded in an SU(N) superconducting Higgs vacuum that implements spontaneous breakdown of the gauge symmetry and renders gauge fields short-range, can be confined by vortices in a topologically stable bound state. Equivalently, only systems of N identical SU(N) monopoles emit no flux and are magnetically neutral. This is seen by considering the n-legged vortex configuration in figure 3.3. The ith vortex carrying flux Φ_i (in units of $2\pi/e$) terminates on a monopole of charge $m_i = m$, where

$$\Phi_i = m \ (\text{mod}\ N) \tag{4}$$

Fig. 3.1

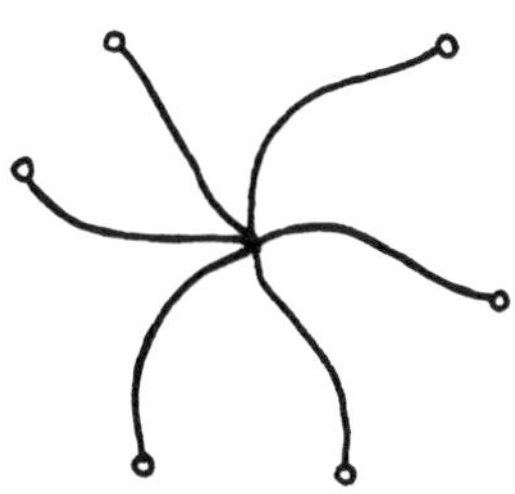

Fig. 3.2

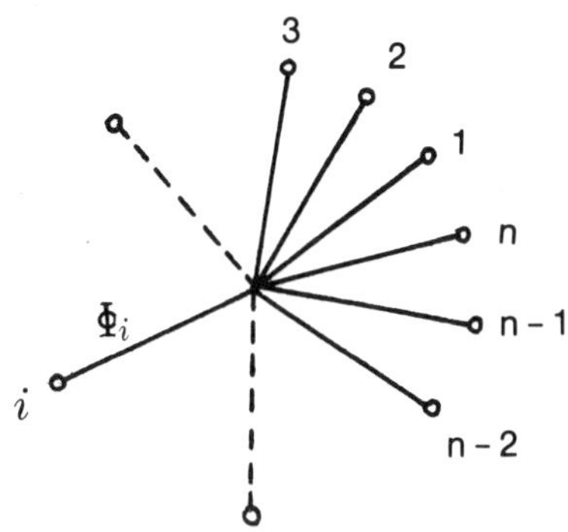

Fig. 3.3

The total flux emanating from the n monopoles is

$$\Phi = \sum_{i=1}^{n} \Phi_i = nm \ (\text{mod}\ N)\ . \tag{5}$$

No flux emanates from the system, provided that $n = N$. In general, the condition for magnetic neutrality is

$$\sum_{i=1}^{N} m_i = 0 \ (\text{mod}\ N)\ . \tag{6}$$

This is the condition for the bound system of monopoles to be an SU(N) singlet state, as now shown.

When a colour SU(N) vector $\Psi = \text{col}\ (\psi_1, \psi_2, \ldots, \psi_N)$ is transported once around an infinitesimal closed loop l encircling the Dirac string of a colourless SU(N) monopole of charge $g = m$ (mod N), beginning and ending at the same point $\boldsymbol{x}$ in the normal vacuum, its resulting linear transformation

$$\Psi' = U_L(x, x)\Psi \tag{7}$$

belongs to a connected Lie group—the holonomy group at $\boldsymbol{x}$. It may be shown[4] that $U_L(x, x) \in Z_N$, so that

$$\Psi' = \Omega^m \Psi \, . \tag{8}$$

The vacuum state $g = 0$ (mod N) corresponds to the identity element of Z_N. When the colour vectors $\Psi(i)$ ($i = 1, 2, \ldots, N$) of a system of N colourless SU(N) monopoles of charge $g_i = M_i$ (mod N) are transported around their Dirac strings, the antisymmetric colour SU(N) singlet wave-function of the system

$$\Psi_N = (N!)^{-1/2} \begin{vmatrix} \psi_1(1) & \psi_1(2) & \ldots & \psi_1(N) \\ \psi_2(1) & \psi_2(2) & \ldots & \psi_2(N) \\ \cdot & \cdot & & \cdot \\ \cdot & \cdot & & \cdot \\ \cdot & \cdot & & \cdot \\ \psi_N(1) & \psi_N(2) & \ldots & \psi_N(N) \end{vmatrix} \tag{9}$$

is transformed to

$$\Psi'_N = \Omega^{\sum_{i=1}^{N} M_i} \Psi_N \, . \tag{10}$$

Since $\Omega^N = 1$, $\Psi'_N = \Psi_N$ provided that

$$\sum_{i=1}^{N} M_i = 0 \text{ (mod } N) \, , \tag{11}$$

which is equation (6). Only colour SU(N) singlet bound states of N colourless SU(N) monopoles are magnetically neutral.

In the Omegon Model, SU$(10)_c$ is the colour gauge symmetry group whose spontaneous breakdown due to the Higgs mechanism leads to the division of fermions into hadrons and leptons. Its fundamental representation comprises nine omegon colour states (hadrons), each with colour electric charges of $-e/9$. This colour nonet is the fundamental representation of colour SU(9), the local invariance group for the strong interactions between omegons. It also includes a colour SU(9) singlet (i.e., a lepton) of colour electric charge $+e$. These charge assignments are not arbitrary, since it has been shown[5] that, if the larger group SU($N + 1$) is spontaneously broken to leave U(1) × SU(N) as the residual exact gauge symmetry, then

$$\exp(i4\pi g Q_c) = \operatorname{diag}(\Omega^M I_N, 1) \quad (M = 1, 2, 3, \ldots, N-1)\,, \tag{12}$$

where Q_c is the colour electric charge operator (a traceless generator of SU($N+1$)), g is a possible magnetic charge, and $\Omega = \exp(i2\pi/N)$. Therefore, for $g = \pm M(1/2e)$,

$$\exp(\pm i2\pi M Q_c/e) = \operatorname{diag}[\exp(i2\pi M/N)I_N, 1]\,, \tag{13}$$

so that

$$Q_c = (\pm e/N)\operatorname{diag}(1, 1, \ldots 1, -N)\,. \tag{14}$$

In the Omegon Model, $N = 9$. The colour SU(9) nonet is made up of three "shade" triplets: $\boldsymbol{\psi}$, $\boldsymbol{\phi}$, and $\boldsymbol{\theta}$, forming fundamental representations of SU(3)$_s$ that are distinguished by their colour SU(9) quantum numbers. Omegons are magnetic monopoles with the 9-fold connected global invariance group SU(9)/Z_9, where

$$Z_9 = (I_9, \Omega I_9, \Omega^2 I_9, \ldots, \Omega^8 I_9) \quad (\Omega = \exp(i2\pi/9)) \tag{15}$$

is the centre of SU(9). Embedded in a broken colour SU(9) superconducting vacuum, their Dirac strings become physical vortices, carrying flux quanta defined modulo 9 and forming a baryonic nine-legged configuration of monopoles M_i (fig. 3.4) that is a colour SU(9) singlet and magnetically neutral, provided that

$$\sum_{i=1}^{9} M_i = 0 \pmod 9\,. \tag{16}$$

Z_9, the cyclic group of order 9, has only one subgroup, namely, Z_3, the cyclic group of order 3:

$$Z_3 = (I_3, \omega I_3, \omega^2 I_3) \quad (\omega = \Omega^3 = \exp(i2\pi/3))\,. \tag{17}$$

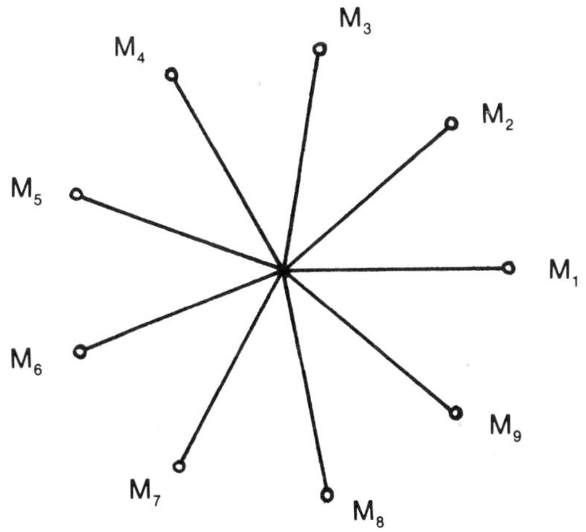

Fig. 3.4

This is the fundamental group of $SU(3)_s/Z_3$, which is triply connected and for which, therefore, two non-trivial types of vortices are defined—those whose end-points are monopoles of charge 1 (mod 3) and 2 (mod 3). In being colour SU(9) monopoles, omegons are $SU(3)_s$ monopoles as well. As such, they are the ends of Y-shaped vortices carrying flux quanta defined modulo 3 (fig. 3.5). Consequently, omegons are the joint end-points of nine-legged SU(9) vortices and Y-shaped $SU(3)_s$ vortices, that is, they cluster into three groups of three, bound externally by the former and internally by the latter. Figure 3.6 shows the proposed vortex configuration in a baryon.

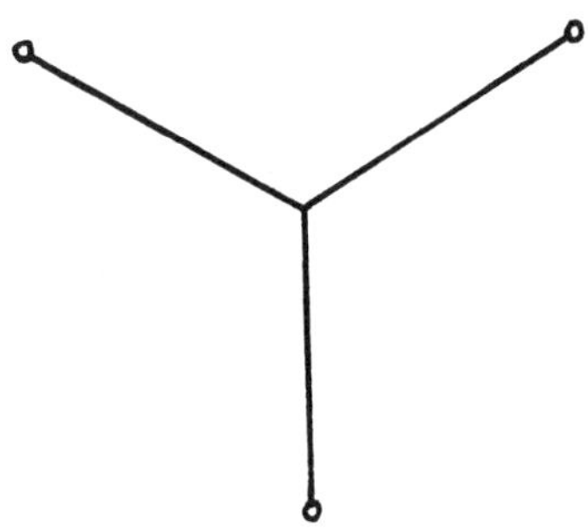

Fig. 3.5

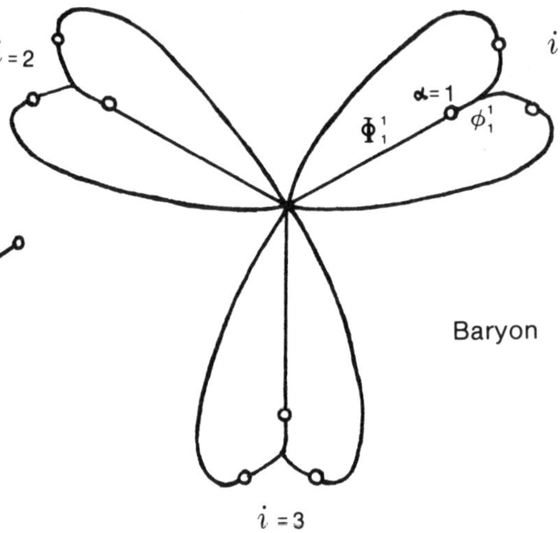

Fig. 3.6

The $SU(3)_s$ and colour SU(9) vortices terminating on the αth omegon in the ith group carry flux

$$\phi_i^\alpha = m_i^\alpha \,(\text{mod } 3) \,, \tag{18}$$

$$\Phi_i^\alpha = M_i^\alpha \,(\text{mod } 9) \,, \tag{19}$$

respectively. The ith group is an $SU(3)_s$ singlet, provided that

$$\sum_{\alpha=1}^{3} m_i^\alpha = 0 \,(\text{mod } 3) \,. \tag{20}$$

The total flux emanating from it is

$$\Phi_i = \sum_{\alpha=1}^{3} \Phi_i^\alpha = 3\bar{M}_i \,(\text{mod } 9) \,, \tag{21}$$

where

$$\bar{M}_i = \frac{1}{3}\sum_{\alpha=1}^{3} M_i^\alpha . \tag{22}$$

Therefore,

$$\sum_{i=1}^{3} \bar{M}_i = 0 \ (\text{mod } 3) , \tag{23}$$

using equation (15). Comparison with equation (11) with $N = 3$ indicates that this is the condition of magnetic neutrality for a colour SU(3) singlet bound state of three SU(3) monopoles. Each cluster of three SU(9) monopoles acts, effectively, as an SU(3) monopole whose charge is the arithmetic mean of their charges.

This result can also be derived as follows: Each omegon in a group is in a colour-shade state belonging to one of the SU(3)_s triplets: $\boldsymbol{\psi}$, $\boldsymbol{\phi}$, and $\boldsymbol{\theta}$. According to the Omegon Model, a baryon is a colour SU(9) singlet, with each of its nine omegon constituents in a different colour-shade. The colour SU(9) vector for the αth omegon in the ith group is

$$\Psi(i, \alpha) = \text{col}\,(\boldsymbol{\psi}(i, \alpha), \boldsymbol{\phi}(i, \alpha), \boldsymbol{\theta}(i, \alpha)) . \tag{24}$$

The ith bound state exists in three colour-shadeless (SU(3)_s singlet) states $q^A(i)$ ($A = 1$, 2, and 3 is a non-gaugeable index) that are distinguished by their SU(3)_c isospin (T_{3c}) and hypercharge (Y_c) quantum numbers:

$$q^1(i) = (3!)^{-1/2}\epsilon_{abc}\psi^a(i, 1)\psi^b(i, 2)\psi^c(i, 3) \qquad (T_{3c} = 3/2, \quad Y_c = 1) , \tag{25}$$

$$q^2(i) = (3!)^{-1/2}\epsilon_{abc}\phi^a(i, 1)\phi^b(i, 2)\phi^c(i, 3) \qquad (T_{3c} = -3/2, \quad Y_c = 1) , \tag{26}$$

$$q^3(i) = (3!)^{-1/2}\epsilon_{abc}\theta^a(i, 1)\theta^b(i, 2)\theta^c(i, 3) \qquad (T_{3c} = 0, \quad Y_c = -2) , \tag{27}$$

(ϵ_{abc} is the completely antisymmetric tensor, with $\epsilon_{123} = 1$). When transported once around an infinitesimal loop encircling the Dirac string of an SU(9) monopole of charge M_i^α ($\neq 0$ (mod 9)), the colour vector $\Psi(i, \alpha)$ is transformed to

$$\Psi'(i, \alpha) = \Omega^{M_i^\alpha}\Psi(i, \alpha) , \tag{28}$$

according to equation (8). Applying this transformation to each monopole in the baryon,

$$q^A(i) \rightarrow q^{A\prime}(i) = (\Omega)^{\Sigma_{\alpha=1}^{3} M_i^\alpha} q^A(i) = \omega^{\bar{M}_i} q^A(i) , \tag{29}$$

so that

$$\boldsymbol{q}'(i) = \omega^{\bar{M}_i}\boldsymbol{q}(i) . \tag{30}$$

According to equation (8), similar transport of a colour SU(3) vector $\boldsymbol{q}$ around the Dirac string of a monopole of charge m ($\neq 0$ (mod 3)) results in the phase change

$$\boldsymbol{q}' = \omega^m \boldsymbol{q} \qquad (m = 1, 2) . \tag{31}$$

Equations (25), (26), and (27) define the components q^A of $\boldsymbol{q}$, provided that

$$\bar{M}_i = N \pmod 3 \qquad (N = 1, 2)\,. \tag{32}$$

Therefore,

$$\sum_{i=1}^{3} \bar{M}_i = 0 \pmod 3\,, \tag{33}$$

which is equation (23). Colour-shadeless bound states of three omegons have thus been shown to transform under equation (28) according to the centre Z_3 of colour SU(3) and to simulate colour SU(3) monopoles with charges $\bar{M}_i$, the end-points of an equivalent Y-shaped vortex configuration shown in figure 3.7. The flavour quantum numbers of these states have been shown[6] to be those of u, d, c, s, t, and b quarks (and of four more predicted to exist). Coloured, flavoured, composite quarks emerge from the Omegon Model.

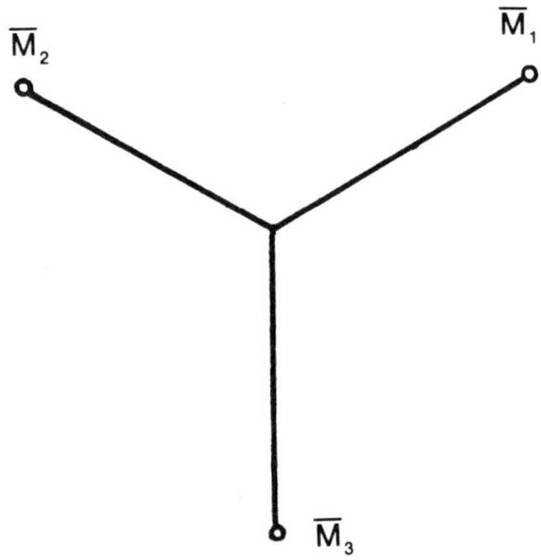

Fig. 3.7

Baryon colour wavefunctions

$$B = (3!)^{-1/2}\epsilon_{ABC}q^A(1)q^B(2)q^C(3) \tag{34}$$

transform under equation (30) as

$$B \rightarrow B' = (\omega)^{\sum_{i=1}^{3}\bar{M}_i}B = B\,, \tag{35}$$

where we have used equation (33). Therefore, baryons are colour SU(3) singlets. Meson colour wavefunctions

$$M = 3^{-1/2}q^A(1)\bar{q}^A(2) \tag{36}$$

transform as

$$M \rightarrow M' = (\omega)^{\bar{M}_1+\bar{M}_2}M = M\,, \tag{37}$$

provided that

$$\bar{M}_1 + \bar{M}_2 = 0 \pmod 3\,, \tag{38}$$

that is,

$$\sum_{\alpha=1}^{3} (M_1^\alpha + M_2^\alpha) = 0 \text{ (mod 9)} , \tag{39}$$

using equation (22). A meson consists of a quark made up of three SU(9) monopoles M_i^α and an antiquark composed of three antimonopoles of opposite polarity ($M_2^\alpha = -M_1^\alpha$). Figure 3.8 shows the proposed vortex configuration in a meson. The three SU(9) vortices are equivalent to a single finite SU(3) vortex with end-point monopoles of charge $\bar{M}_1$ and $-\bar{M}_1$ (fig. 3.9). The total flux emanating from a quark is, by equations (21) and (32),

$$\begin{aligned} \Phi_i &= 3N \text{ (mod 9)} \\ &= 3 \text{ or } 6 \text{ (mod 9)} , \end{aligned} \tag{40}$$

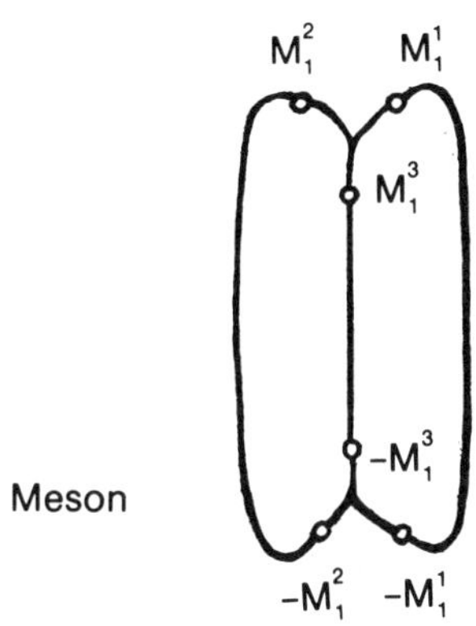

Fig. 3.8

Fig. 3.9

that is, a quark has an SU(9) magnetic charge of 3 or 6 (mod 9). Because of this charge, free quarks (and, of course, free omegons) would emanate infinitely long vortices with infinite energy and so cannot exist in the superconducting colour SU(9) vacuum, which supports only magnetically neutral bound states of omegons having finite vortices and, therefore, finite energy. Only colour SU(9) singlet states are free and observable, namely, three-quark baryons and quark-antiquark mesons. Free omegon-antiomegon bound states are forbidden because they cannot be finite, closed $SU(3)_s$ strings as well as finite, closed colour SU(9) strings.

According to the Omegon Model, "colour" is a global property of quarks arising from their compositeness in terms of three colour SU(9) magnetic monopoles. Colour states are colour-shadeless, differing in their colour SU(9) quantum numbers but being $SU(3)_s$ singlets. A quark is bound to the other quarks in a baryon or to the antiquark in a meson by a set of three colour SU(9) strings that simulate an equivalent colour SU(3) string. The theory predicts that colour SU(3) is *not* a fundamental gauge symmetry that should be incorporated in "grand unified theories" with U(1) × $SU(2)_w$ for electro-weak forces and with SL(2, *C*) for gravitation (although $SU(3)_s$ is the gauge symmetry of the superstrong interaction confining three omegons in a quark). Instead, colour $\mathring{SU}(3)$ is an *effective* gauge symmetry of the total interaction between composite quarks, one that is actually mediated by the gauge fields of the primary colour SU(9) coupling between omegons.

MAGNETIC CHARGE COMPOSITION OF BARYONS

The condition that quarks are $SU(3)_s$ singlet bound states of three $SU(3)_s$ monopoles is

$$m_i \equiv \sum_{\alpha=1}^{3} m_i^\alpha = 0 \pmod 3 . \tag{41}$$

Since $SU(3)_s$ monopole moments are defined by modulo 3, $m_i^\alpha = \pm 1, \pm 2$, and $m_i = 0, \pm 3, \pm 6$. $SU(3)_s$ monopoles couple to a gauge-invariant $SU(3)_s$ scalar field tensor in an Abelian-like manner, obeying Dirac's equation for an Abelian monopole.[7] Omegons have an electric dipole moment of Dirac form. The corresponding operator for a non-relativistic system q_i ($i = 1, 2, 3$) of three Dirac monopoles of charge g_i^1, g_i^2, and g_i^3 in a bound state with zero orbital angular momentum is

$$\boldsymbol{P}_i = -\sum_{\alpha=1}^{3} (g_i^\alpha \hbar/2M_\alpha c)\, \boldsymbol{\sigma}_\alpha . \tag{42}$$

$\mathcal{P}$ and $\mathcal{N}$ omegons in u and d quarks are bound in symmetric spin/unitary spin states, so that

$$\langle \mathrm{q} | P_i^z | \mathrm{q} \rangle = -(\hbar/2Mc)\bar{g}_i , \tag{43}$$

where

$$\bar{g}_i = \frac{1}{3}\sum_{\alpha=1}^{3} g_i^\alpha , \tag{44}$$

and q = u or d ($\mathcal{P}$ and $\mathcal{N}$ omegons are assumed to have the same mass M). The nucleon (N) electric dipole moment (E.D.M.) operator is

$$\boldsymbol{P} = -\frac{\hbar}{2Mc}\sum_{i=1}^{3} \bar{g}_i \boldsymbol{\sigma}_i . \tag{45}$$

Therefore,

$$\langle \mathrm{N} | P^z | \mathrm{N} \rangle = -(\hbar/2Mc)\bar{g} , \tag{46}$$

where

$$\bar{g} = \frac{1}{3}\sum_{i=1}^{3} \bar{g}_i = \frac{1}{9}\sum_{i,\alpha=1}^{3} g_i^\alpha \,. \tag{47}$$

The experimental upper limits[8] for the nucleon are

$$\text{E.D.M.} \lesssim (7 \pm 9) \times 10^{-21} e \text{ cm} \qquad \text{(proton)}$$

$$\lesssim 5 \times 10^{-23} e \text{ cm} \qquad \text{(neutron)}\,.$$

The omegon mass would have to exceed 10^{10} GeV in order for the theoretical nucleon E.D.M. to be less than these upper limits. This would imply that quarks are massive, contrary to current estimates of about 350 MeV for the u and d. It is inferred that the E.D.M. is exactly zero, that is,

$$\sum_{i,\alpha=1}^{3} g_i^\alpha = 0 \tag{48}$$

or (in Dirac units)

$$\sum_{i,\alpha=1}^{3} m_i^\alpha = 0\,, \tag{49}$$

which is the condition for magnetic neutrality of a system of nine Abelian monopoles. It may be written

$$m_1 + m_2 + m_3 = 0\,. \tag{50}$$

There are three possibilities:

(a) $m_1 = m_2 = m_3 = 0$; (b) $m_1 = m_2 = \pm 3\,,\ m_3 = \mp 6$;

(c) $m_1 = 3\,,\ m_2 = 0\,,\ m_3 = -3$.

$SU(3)_s$ monopoles of charge $+1$ and -2 have topologically identical vortex strings; similarly for monopoles of charge $+2$ and -1. Hence, quarks with monopole configurations (1, 1, 1), (1, 1, -2), (1, -2, -2), and (-2, -2, -2) are allowed. These are topologically distinct from the configurations (2, 2, 2), (2, 2, -1), (2, -1, -1), and (-1, -1, -1). The configurations corresponding to the permitted values of m_i are

(1) $m_i = 0$: $\pm\,(1, 1, -2)$,

(2) $m_i = \pm 3$: $\pm\,(1, 1, 1)$ or $\pm\,(2, 2, -1)$,

(3) $m_i = \pm 6$: $\pm\,(2, 2, 2)$.

They give rise to a variety of possible magnetic charge compositions for the nucleon (and for any baryon with zero E.D.M.). But, in terms of positive (+) and negative ($-$) magnetic charges, it is easily verified that cases (a), (b), and (c) yield four different structures. These are shown in figure 3.10. Whether they are subject to further physical restrictions will not be discussed here.

For future reference, the composition of a proton and a neutron in terms of $\mathcal{P}$ and $\mathcal{N}$ omegons is predicted to be that shown in figure 3.11.

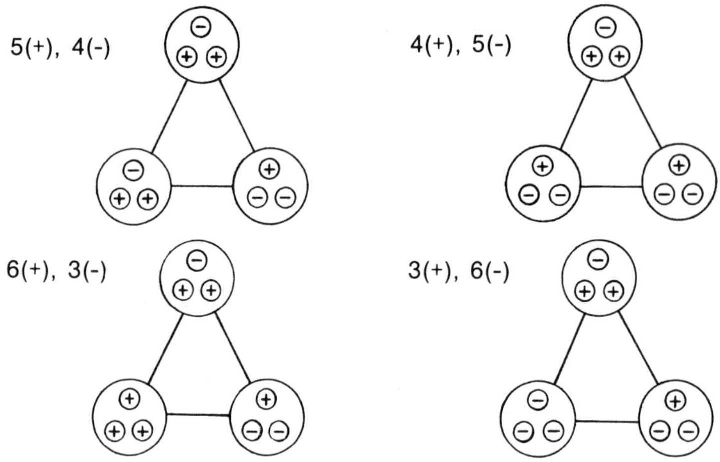

Fig. 3.10

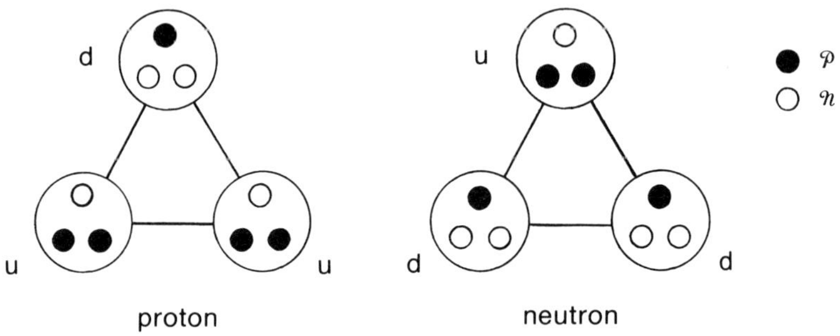

Fig. 3.11

REFERENCES

1. S. M. Phillips, Phys. Letters, **84B,** 133 (1979) (reproduced here).
2. H. C. Tze and Z. F. Ezawa, Phys. Rev., **D14,** 1006 (1976).
3. Z. F. Ezawa and H. C. Tze, Nucl. Phys., **B100,** 1 (1975).
4. Ibid.
5. E. Corrigan and D. Olive, Nucl. Phys., **B110,** 237 (1976).
6. Phillips, ref. 1.
7. Ezawa and Tze, ref. 3.
8. G. E. Harrison et al., Phys. Rev. Letters, **22,** 1263 (1969); J. K. Baird et al., Phys. Rev., **179,** 1285 (1969).

CHAPTER 4

Two Hypotheses concerning Micro-Psi

And as imagination bodies forth
The form of things unknown, the poet's pen
Turns them to shapes, and gives to airy nothing
A local habitation and a name.

Shakespeare

U.P.A.'s are identified as the omegon constituents of quarks. This suggests a formative principle that determines the composition of an M.P.A. Its predictions are tested quantitatively, using micro-psi data, and the agreement is shown to have high statistical significance.

CONTACT BETWEEN PHYSICS AND MICRO-PSI

In 1895, Annie Besant and C. W. Leadbeater began their comprehensive investigation of the atomic structure of the elements. After their initial examination of the gases in the atmosphere, they reported[1] the observation of an egg-shaped object containing two interlaced, triangular arrays of bodies, each body being made up of three U.P.A.'s. They counted eighteen U.P.A.'s in the object. They also observed two other objects, one containing 261, the other 290, U.P.A.'s. From the relative sizes (1:14.50:16.11) of their populations and from the proximity of these ratios to the chemical atomic weights of hydrogen, nitrogen, and oxygen, respectively, the investigators inferred that they had examined individual atoms of these elements. Their later study of various chemical compounds containing hydrogen, nitrogen, and oxygen confirmed this, so it seemed. Further work revealed that no simpler object with less than eighteen U.P.A.'s ever appeared, whatever the element that they examined, and this strengthened their belief that the egg-shaped body was an atom of the lightest element—hydrogen. In the hydrogen M.P.A. (fig. 4.1), some of the groups of three U.P.A.'s (called "Hydrogen Triplets") were found to consist of two (+) U.P.A.'s and one (−) U.P.A., while others consisted of one (+) and two (−) U.P.A.'s. One triangular array of Hydrogen Triplets (called a "Hydrogen Triangle") was made up of five (+) and four (−) U.P.A.'s. The other array had four (+) and five (−) U.P.A.'s. The M.P.A. was made up of nine (+) and nine (−) U.P.A.'s. Another form of the M.P.A. with ten (+) and eight (−) U.P.A.'s was discovered in 1932 (both varieties are fully discussed in chap. 7). The reader should now compare the structures of the Hydrogen Triangles shown in figure 4.1 with the Omegon Model prediction of the magnetic charge composition of a nucleon, shown in figure 3.10. Their evident similarity suggests that a Hydrogen Triangle is a

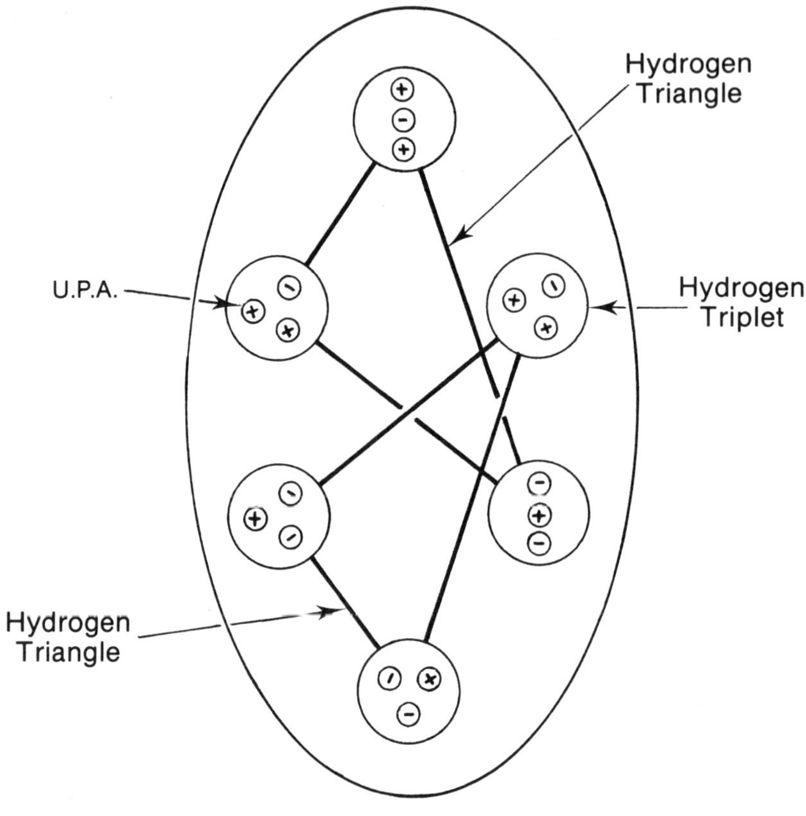

Fig. 4.1. The hydrogen M.P.A.

proton, a Hydrogen Triplet is a u or d quark, and a U.P.A. is a $\mathcal{P}$ or $\mathcal{N}$ omegon (a more detailed analysis is given in chaps. 5 and 7). The implication of this identification, however, is that a hydrogen M.P.A. is not one proton (as the investigators' claim to be able to observe hydrogen atoms would have led one to expect) but *two* protons. Moreover, the Hydrogen Triangles were reported to be overlapping, not separated in space. This implies that, if the Hydrogen Triangles depicted in figure 4.1 are protons, they must be close enough to be within range of their nuclear forces. But it is well known that the diproton is unstable as a bound state because the nucleon-nucleon interaction is too weak to permit two protons binding together in the spin singlet state allowed by the Exclusion Principle. However, the hydrogen M.P.A. was reported to be stable, no spontaneous dissociation of the M.P.A. into free Hydrogen Triangles ever having been reported by Besant and Leadbeater. It is obvious that the M.P.A. does not represent two protons merely bound together by their short-range nuclear forces. Also, it cannot be a hydrogen molecule because the two protons in the molecule are 100,000 times farther apart than the Hydrogen Triangles as depicted in figure 4.1, which indicates a distance apart about equal to the "size" of a proton, that is, to the expectation value of the radius of the three-quark bound state. If the observers had examined just one end of a hydrogen molecule, they would have reported seeing only one Hydrogen Triangle instead of two. But it is obvious that two are present in the M.P.A., for they are quite distinctive. The possibility is raised, therefore, that M.P.A.'s of elements are *not* nuclei but physical systems created by the perturbative effect on nuclei of micro-psi observation. The problem of the stability of the hydrogen M.P.A. is related to the question of how it is formed, and these issues will be discussed in chapter 7.

The resemblance between the quark/omegon structure of a proton predicted by the Omegon Model and the form of a Hydrogen Triangle suggests the following hypothesis:

HYPOTHESIS 1: *The "ultimate physical atom" (U.P.A.) is either member of the omegon isospin doublet* ($\mathcal{P}$, $\mathfrak{N}$) *belonging to the (10, 9) fundamental representation of the unified gauge group* $SU(10)_f \times SU(9)_c$ *for omegons.*

According to this, the U.P.A. has nine colour-shade states that form the fundamental representation of $SU(9)_c$. It is an $SU(9)_c$ magnetic monopole, and the (+) and (−) chiral forms differ in their having opposite magnetic polarity (discussed in detail in chap. 5).

As mentioned earlier in the cases of hydrogen, nitrogen, and oxygen, when the investigators compared the U.P.A. populations of the M.P.A.'s of various elements, they discovered that the numbers were approximately proportional to the appropriate atomic weights (taking the atomic weight of hydrogen as 1). For example, helium (with an atomic weight of 4) had an M.P.A. with 72 U.P.A.'s, four times the number for hydrogen, and carbon (with an atomic weight of 12) had an M.P.A. containing 216 U.P.A.'s, twelve times the number for hydrogen. Of the 111 different M.P.A.'s recorded by Besant and Leadbeater, 39 were found to have U.P.A. populations that were exact integer multiples of 18. This mathematical relation served as a means of identifying the element whose atoms they thought they had observed: they had merely to count all the U.P.A.'s in the M.P.A., divide by 18, and then compare the resultant number with listed chemical atomic weights. The method appeared to be trustworthy because of the consistent success that they had in deducing the identity of elements in materials known to contain them prior to micro-psi examination. As the hydrogen M.P.A. has the omegon content of two hydrogen nuclei, consistency with this relation of proportionality requires that the M.P.A. of any element should have the omegon content of two nuclei of that element. Accordingly, the following additional hypothesis is proposed:

HYPOTHESIS 2: *The M.P.A. of an element is a multi-omegon system that is compounded from the omegons making up the nucleons in* TWO *nuclei of that element.*

This is a preliminary, partial statement of a process of M.P.A. formation that will be described in chapter 7. It is a sufficient statement for the present task of comparing the micro-psi data with the mathematical consequences of Hypotheses 1 and 2, which are now stated: according to these hypotheses, since both protons and neutrons contain nine omegons, the number of U.P.A.'s predicted to be in an M.P.A. formed from two nuclides of an element with mass numbers A_1 and A_2 is

$$N(A_1, A_2) = 9A_1 + 9A_2 = 9(A_1 + A_2) .$$

If the nuclides are identical (as, it will be shown, was commonly the case for the M.P.A.'s that were recorded by Besant and Leadbeater) and have mass number A, the number of U.P.A.'s is predicted to be

$$N(A) = 18A ,$$

that is, the population of U.P.A.'s is proportional to the mass number of the nuclide selected for micro-psi observation. This explains why, as the investigators discovered, the relation of proportionality to the atomic weight of the element is only approximate. It must be remembered that the concepts of mass number, nucleons, and atomic nuclei were unknown to Besant and Leadbeater, who completed much of their work before Rutherford proposed in 1911 the nuclear model of the atom. They may be excused, therefore, for thinking that their discovery of an approximate correlation between populations of U.P.A.'s and atomic weights was of fundamental significance. It is obvious, however, that the atomic weight of an element, whether based on the physical or on the chemical scale of units, cannot be a physically meaningful parameter relating M.P.A.'s to atomic nuclei. This is because the atomic weight of an element is not a parameter that refers to a particular nucleus but, instead, is an arithmetic average of the atomic masses of all its isotopes, weighted according to their relative, natural abundances. The approximate nature of the linear proportionality between populations of U.P.A.'s and atomic weights is due to the fact that the mass numbers of common, stable isotopes are generally within range of the atomic weight by one or two units. Although comparison of number weights with atomic weights was incorrect, it was accurate in practice as a method of assigning elements to M.P.A.'s. The investigators did not depend solely on this method, for they found other correlates that could help to identify the element whose atoms they believed they had observed. These enabled them to check the accuracy of their labelling of M.P.A.'s. An important correlation was that between the form of an M.P.A. and the position of the element in the periodic table. This will be discussed in chapter 7.

A TEST OF HYPOTHESES 1 AND 2

The mathematical model described above will now be tested, using as data the population of the M.P.A.'s given in reference 5 of chapter 1. Table 4.1 lists all the 111 M.P.A.'s recorded by Besant and Leadbeater, together with their counted U.P.A. populations (N), predicted omegon populations (n), and errors of counting ($e = N - n$). Various criteria determine the selection of the most appropriate nuclide for a given M.P.A. These are specified below (not in any order of priority):

1. *Abundance.* The most abundant isotope of an element is normally chosen, since the atoms of this isotope are the most likely to be selected (the selection of atoms by the micro-psi observer is assumed to be a random process). If this choice conflicts with other criteria, the next most common isotope satisfying all the criteria is selected.
2. *Stability.* Only stable isotopes are considered, except, of course, in the case of naturally radioactive isotopes. All choices refer to the ground state of the nuclide.
3. *Plausibility of error.* The size of the counting error implied by a certain choice must be credible in relation to the symmetry of the M.P.A. The reason for this is as follows: instead of counting every U.P.A. in an M.P.A., the investigators restricted themselves to counting only the U.P.A.'s present in one funnel, and they *calculated* how many U.P.A.'s were in all the funnels by multiplying this count by the number of funnels. They applied a similar procedure to M.P.A.'s having spikes, bars, or arms. They regarded this as a reliable expedient because they had noted, as the result of many observations, that all funnels, spikes, and so on, in a given M.P.A. had identical contents (with the exception of a few elements, which they treated differently). But this procedure magnifies artificially any error of counting made during the examination of a funnel, spike, or other feature, so that the total error predicted for

TABLE 4.1

COMPARISON OF PREDICTED OMEGON AND COUNTED U.P.A. POPULATIONS OF THE M.P.A.'S OF THE ELEMENTS

Element	Nuclide	U.P.A. Number N	Omegon Number n	Error $N-n$
Hydrogen	H^1	18	18	0
Deuterium ("Adyarium")	H^2	36	36	0
Helium ("Occultum")	He^3	54	54	0
	He^4	72	72	0
Lithium	Li^7	127	126	+ 1
Beryllium	Be^9	164	162	+ 2
Boron	B^{11}	200	198	+ 2
Carbon	C^{12}	216	216	0
Nitrogen	N^{14} $+N^{15}$	261	261	0
Oxygen	O^{16}	290	288	+ 2
	O^{17}	310	306	+ 4
Fluorine	F^{19}	340	342	− 2
Neon	Ne^{20}	360	360	0
Neon ("meta")	Ne^{22}	402	396	+ 6
Sodium	Na^{23}	418	414	+ 4
Magnesium	Mg^{24}	432	432	0
Aluminium	Al^{27}	486	486	0
Silicon	Si^{28}	520	504	+16
Phosphorus	P^{31}	558	558	0
Sulphur	S^{32}	576	576	0
Chlorine	Cl^{35}	639	630	+ 9
	Cl^{37}	667	666	+ 1
Argon ("proto")	Ar^{36} $+Ar^{40}$	672	684	−12
	Ar^{40}	714	720	− 6
Argon ("meta")	Ar^{38} $+Ar^{40}$	756	702	+54
Potassium	K^{39}	701	702	− 1
Calcium	Ca^{40}	720	720	0
Scandium	Sc^{45}	792	810	−18
Titanium	Ti^{48}	864	864	0
Vanadium	V^{51}	918	918	0
Chromium	Cr^{52}	936	936	0
Manganese	Mn^{55}	992	990	+ 2
Iron	Fe^{56}	1,008	1,008	0
Cobalt	Co^{59}	1,036	1,062	−26
Nickel	Ni^{60}	1,064	1,080	−16
Copper	Cu^{63}	1,139	1,134	+ 5
Zinc	Zn^{64} $+Zn^{66}$	1,170	1,170	0
Gallium	Ga^{69} $+Ga^{71}$	1,260	1,260	0
Germanium	Ge^{72}	1,300	1,296	+ 4
Arsenic	As^{75}	1,350	1,350	0
Selenium	Se^{78} $+Se^{80}$	1,422	1,422	0
Bromine	Br^{79}	1,439	1,422	+17
Krypton	Kr^{82}	1,464	1,476	−12
Krypton ("meta")	Kr^{84}	1,506	1,512	− 6
Rubidium	Rb^{85}	1,530	1,530	0
Strontium	Sr^{88}	1,568	1,584	−16
Yttrium	Y^{89}	1,606	1,602	+ 4
Zirconium	Zr^{90}	1,624	1,620	+ 4
Niobium	Nb^{93}	1,719	1,674	+45
Molybdenum	Mo^{97}	1,746	1,746	0
Technetium	Tc^{99}	1,802	1,782	+20
Ruthenium	Ru^{102}	1,848	1,836	+12
Rhodium	Rh^{103}	1,876	1,854	+22
Palladium	Pd^{106}	1,904	1,908	− 4
Silver	Ag^{107} $+Ag^{109}$	1,945	1,944	+ 1

TABLE 4.1—*Continued*

Element	Nuclide	U.P.A. Number N	Omegon Number n	Error $N-n$
Cadmium	Cd^{112}	2,016	2,016	0
Indium	$In^{113}+In^{115}$	2,052	2,052	0
Tin	Sn^{118}	2,124	2,124	0
Antimony	Sb^{121}	2,169	2,178	− 9
Tellurium	$Te^{130}+Te^{125}$	2,223	2,295	−72
Iodine	I^{127}	2,287	2,286	+ 1
Xenon	Xe^{129}	2,298	2,322	−24
Xenon ("meta")	Xe^{130}	2,340	2,340	0
Caesium	Cs^{133}	2,376	2,394	−18
Barium	Ba^{136}	2,455	2,448	+ 7
Lanthanum	La^{139}	2,482	2,502	−20
Cerium	Ce^{140}	2,511	2,520	− 9
Praseodymium	Pr^{141}	2,527	2,538	−11
Neodymium	Nd^{143}	2,575	2,574	+ 1
Promethium ("Illinium")	Pm^{147}	2,640	2,646	− 6
Promethium ("meta")	Pm^{151}	2,736	2,718	+18
Samarium	Sm^{154}	2,794	2,772	+22
Europium	Eu^{153}	2,843	2,754	+89
Gadolinium	Gd^{160}	2,880	2,880	0
Terbium	Tb^{159}	2,916	2,862	+54
Dysprosium	Dy^{164}	2,979	2,952	+27
Holmium	Ho^{165}	3,004	2,970	+34
Erbium	Er^{168}	3,029	3,024	+ 5
Thulium	Tm^{169}	3,096	3,042	+54
Ytterbium	Yb^{174}	3,131	3,132	− 1
Lutecium	Lu^{175}	3,171	3,150	+21
Hafnium	Hf^{178}	3,211	3,204	+ 7
Tantalum	Ta^{181}	3,279	3,258	+21
Tungsten	W^{183}	3,299	3,294	+ 5
Rhenium	Re^{187}	3,368	3,366	+ 2
Osmium	$Os^{190}+Os^{192}$	3,430	3,438	− 8
Iridium	Ir^{193}	3,458	3,474	−16
Platinum ("A")	Pt^{194}	3,486	3,492	− 6
Platinum ("B")	Pt^{196}	3,514	3,528	−14
Gold	Au^{197}	3,546	3,546	0
Mercury ("A")	Hg^{199}	3,576	3,582	− 6
Mercury ("B")	Hg^{200}	3,600	3,600	0
Thallium	Tl^{205}	3,678	3,690	−12
Lead	Pb^{207}	3,727	3,726	+ 1
Bismuth	Bi^{209}	3,753	3,762	− 9
Polonium	Po^{210}	3,789	3,780	+ 9
Astatine ("85")	At^{219}	3,978	3,942	+36
Emanation	Em^{222}	3,990	3,996	− 6
Emanation ("meta")	Em^{220}	4,032	3,960	+72
Francium ("87")	Fr^{223}	4,006	4,014	− 8
Radium	Ra^{226}	4,087	4,068	+19
Actinium	Ac^{228}	4,140	4,104	+36
Thorium	Th^{232}	4,187	4,176	+11
Protactinium	Pa^{234}	4,227	4,212	+15
Uranium	U^{238}	4,267	4,284	−17
Hybrid M.P.A.'s				
"X"	$Ru^{102}+Os^{192}$	2,646	2,646	0
"Y"	$Rh^{103}+Ir^{191}$	2,674	2,664	+10
"Z"	$Pd^{108}+Pt^{194}$	2,702	2,718	−16
"Z isotope"	$Pd^{106}+Pt^{194}$	2,716	2,700	+16
"Kalon"	$Xe^{124}+Em^{222}$	3,054	3,114	−60
"Kalon" ("meta")	$Xe^{124}+Em^{220}$	3,096	3,096	0

all these features must contain their number as a factor. The total error for the M.P.A. must be compatible with its symmetry, since it may be either partly or wholly compounded from errors of counting made for one funnel, spike, or other feature. If it is incompatible, the nuclide must be rejected as an unrealistic choice. For example, a nuclide that implies an error $e = 3$ is not a plausible choice for an M.P.A. having four funnels, because the overcounting cannot be the same for each funnel, as the counting procedure requires. In a number of cases, criterion 3 is found to conflict with criterion 1. This necessitates a realistic selection of less abundant isotopes. In most cases, little subjective judgement is needed in order to assess and to compare the plausibilities of errors associated with various choices of nuclides.

These criteria serve to eliminate ambiguity of choice for the majority of M.P.A.'s. Ambiguity remains only for a very few elements (such as tellurium) that possess many stable isotopes. These cannot be eliminated by a careful application of the criteria of choice, and for them, the question of the most appropriate nuclide is a matter of personal, arbitrary choice, since no candidate alone satisfies all the criteria better than the rest.

In a few cases, a zero error is found for a radioactive isotope whose mass number is exactly halfway in value between the mass numbers of two stable nuclides. For example, exact agreement between predicted and counted populations is found for the radioactive isotope Zn^{65} of zinc, whose mass number is mid-way between those of the common, stable isotopes Zn^{64} and Zn^{66}. It is suggested that these cases arise because the M.P.A. was formed from two different, stable isotopes of the element with mass numbers A_1 and A_2 that have an arithmetic mean that happens to coincide with the mass number of a radioactive isotope of the element, that is,

$$N(A_1, A_2) = 9(A_1 + A_2) = 18A_3 ,$$

where A_3, the arithmetic mean of A_1 and A_2, is the mass number of a radioactive isotope. Other examples are gallium, selenium, and indium. In other cases where it proves necessary to consider two different nuclides, choice of a single nuclide is incompatible with the three criteria described above (nitrogen is an exception that is discussed in chap. 7).

The error column in table 4.1 shows that the level of agreement between predicted and counted populations is excellent, discrepancies being generally less than 1 percent. The few anomalous cases, such as europium and terbium, with relatively large errors are readily explainable and are not as serious as they might appear to be on first glance. For example, the M.P.A. of europium, with an error of +89, belongs to the Tetrahedron Group, and each of its four funnels is composed of a group of twelve identical bodies and two groups each of nine similar bodies. A mere miscounting of the U.P.A.'s inside one of these bodies by 1 or 2 would lead to a total error of 48 or 96 for the first type of body and an error of 36 or 72 for the second type. Similarly, the M.P.A. of terbium, with an error of +54, belongs to the Octahedron Group, and each of its eight funnels is made up partly of a group of four identical bodies. If one of these were miscounted by merely 1 or 2, the total error would be 32 or 64. It is apparent, then, that the errors shown in table 4.1 represent the accumulated or total errors, not the errors of counting actually made by the investigators during their examination of M.P.A.'s. The latter errors were artificially magnified by their counting procedure, namely, that of multiplying the number of U.P.A.'s within a group by the number of such groups inside a funnel and that of multiplying the number of U.P.A.'s inside a funnel by the number of funnels displayed by the M.P.A. (similarly for groups inside spikes, bars, etc.). The *actual* discrepancies between prediction and observation are generally much smaller than the numbers in the error column.

TABLE 4.2

Nuclide	Number of U.P.A.'s (N)	Number of Omegons (n)	Error per Arm [$(N-n)/6$]
Ne^{20}	360	360	0
"Meta": Ne^{22}	402	396	+1
"Proto": $Ar^{36}+Ar^{40}$	672	684	−2
Ar^{40}	714	720	−1
"Meta": $Ar^{38}+Ar^{40}$	756	702	+9
Kr^{82}	1,464	1,476	−2
"Meta": Kr^{84}	1,506	1,512	−1
Xe^{129}	2,298	2,322	−4
"Meta": Xe^{130}	2,340	2,340	0
Em^{222}	3,990	3,996	−1
"Meta": Em^{220}	4,032	3,960	+12

A remarkable illustration of this error-compounding effect is shown in table 4.2, which compares the predicted and counted populations of the M.P.A.'s of the inert gas elements. These belong to the Star Group (described in chap. 1), and all possess the same central core consisting of a dodecahedral array of identical bodies. The M.P.A.'s differ in the number of U.P.A.'s in the six arms of the star, all of which are similar, for a given M.P.A. The symmetry of the core precludes the possibility of observational error here, and so any error predicted for the M.P.A. must be due to miscounting the particles inside the arms. Only a single arm was examined by the investigators in detail, and this means that the predicted total error must be an integer multiple of 6. Table 4.2 shows that the error per arm is always an integer, as required. It is remarkable that nuclides of inert gas elements can be found that provide errors that are only integer multiples of 6, especially in view of the fact that one arm of the M.P.A.'s of these elements can contain hundreds of U.P.A.'s. If the relation

$$N(A) = 18A$$

were not exact, errors other than integer multiples of 6 could have been expected to be revealed by the correct relation, and these would have been irreconcilable with the sixfold symmetry of the M.P.A.'s. That this is not the case provides strong support for the model.

In graph 4.1 the errors predicted by the relation

$$N(A) = 18A$$

are plotted against the atomic numbers of the elements. The most important feature to note is the random scatter of points about the zero-error axis. The change in the sign of the error from one element to the next is random, the scatter of points being biased neither towards positive nor towards negative values. This is what would be expected if errors are interpreted in terms of randomly made mistakes of counting by the two investigators. With most M.P.A.'s containing several thousand U.P.A.'s, it was inevitable that miscounting took place. A fundamental deficiency in the model would have appeared in the graph as a systematic variation in plotted errors with increasing atomic number, with no even dispersion of points about the zero-error axis. Instead, the variation is random. The absolute magnitude of the error increases slightly with increasing atomic number, and this feature is consistent with miscounting becoming more probable the more U.P.A.'s that had to be counted.

Graph 4.2 shows the variation of the frequencies of errors predicted according to the relation given above. The relatively few errors numerically larger than 22 are omitted for the sake of clarity of representation. The histogram has the approximate form of a Gaussian or normal distribution curve with a large, narrow peak at zero error. It shows that the predicted errors are distributed in a way that is consistent with their having a random occurrence, that is, they are due to chance mistakes of counting further exaggerated by the method of counting used by the investigators. Graph 4.3 shows the variation of frequencies with the absolute values of the errors. It, too, conforms to a normal distribution curve.

Graph 4.4 shows the variation of the number of U.P.A.'s (Y) counted in the M.P.A. with the mass number (X) of the most appropriate nuclide. The linear relationship between X and Y is very apparent. This is confirmed analytically by submitting the data to a statistical test. This will be carried out in the next section.

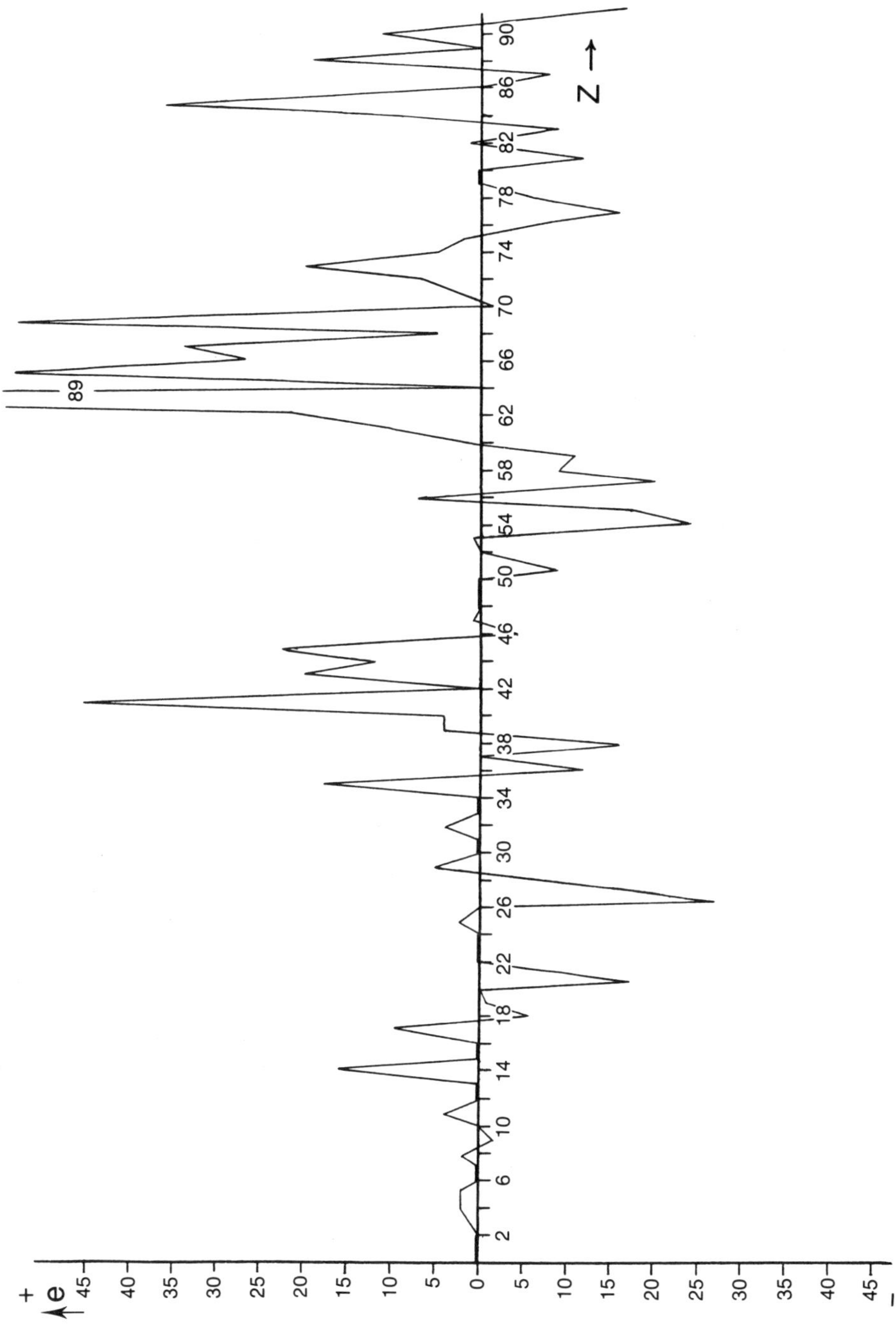

Graph 4.1. Graph of error (e) versus atomic number (Z)

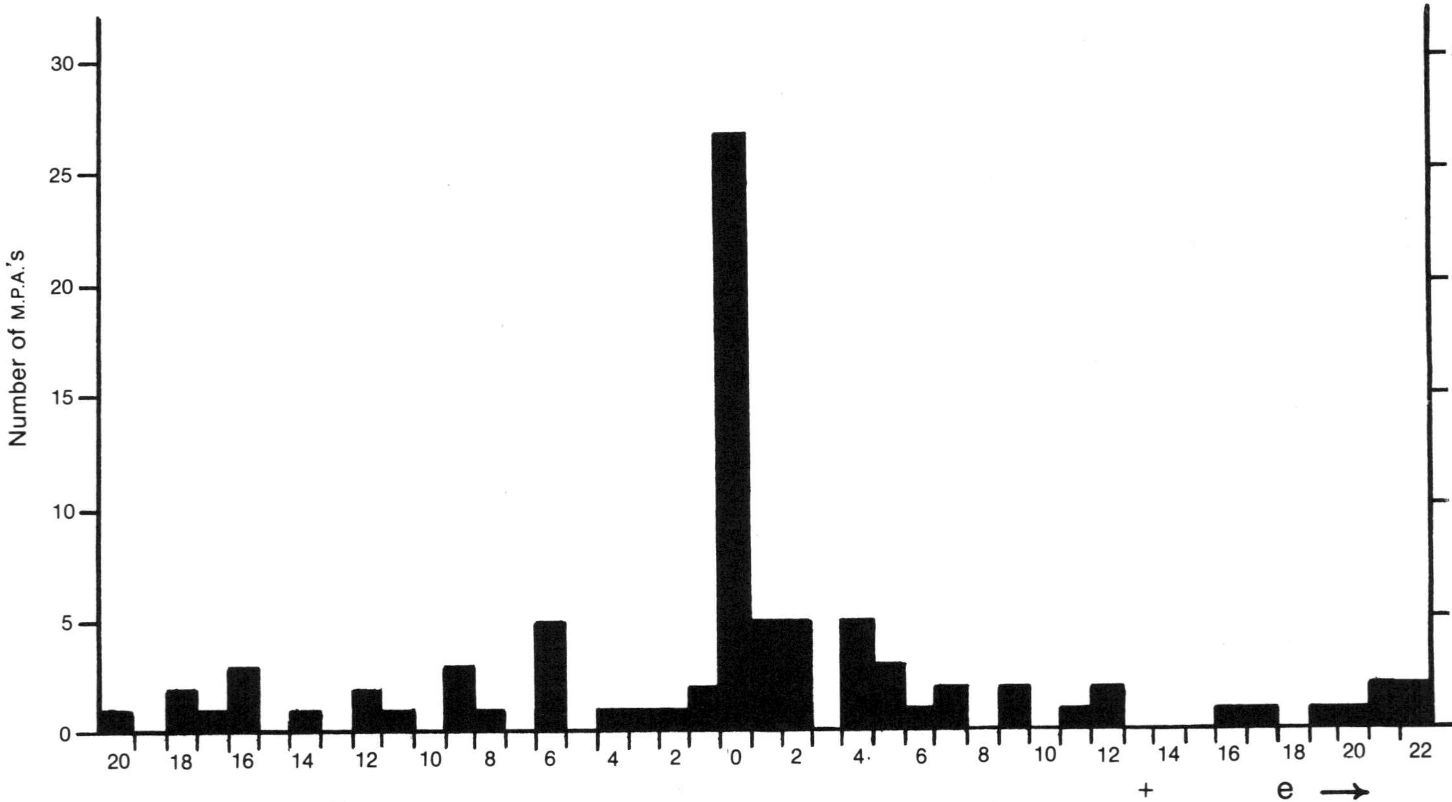

Graph 4.2. Histogram of number of M.P.A.'s versus error (*e*)

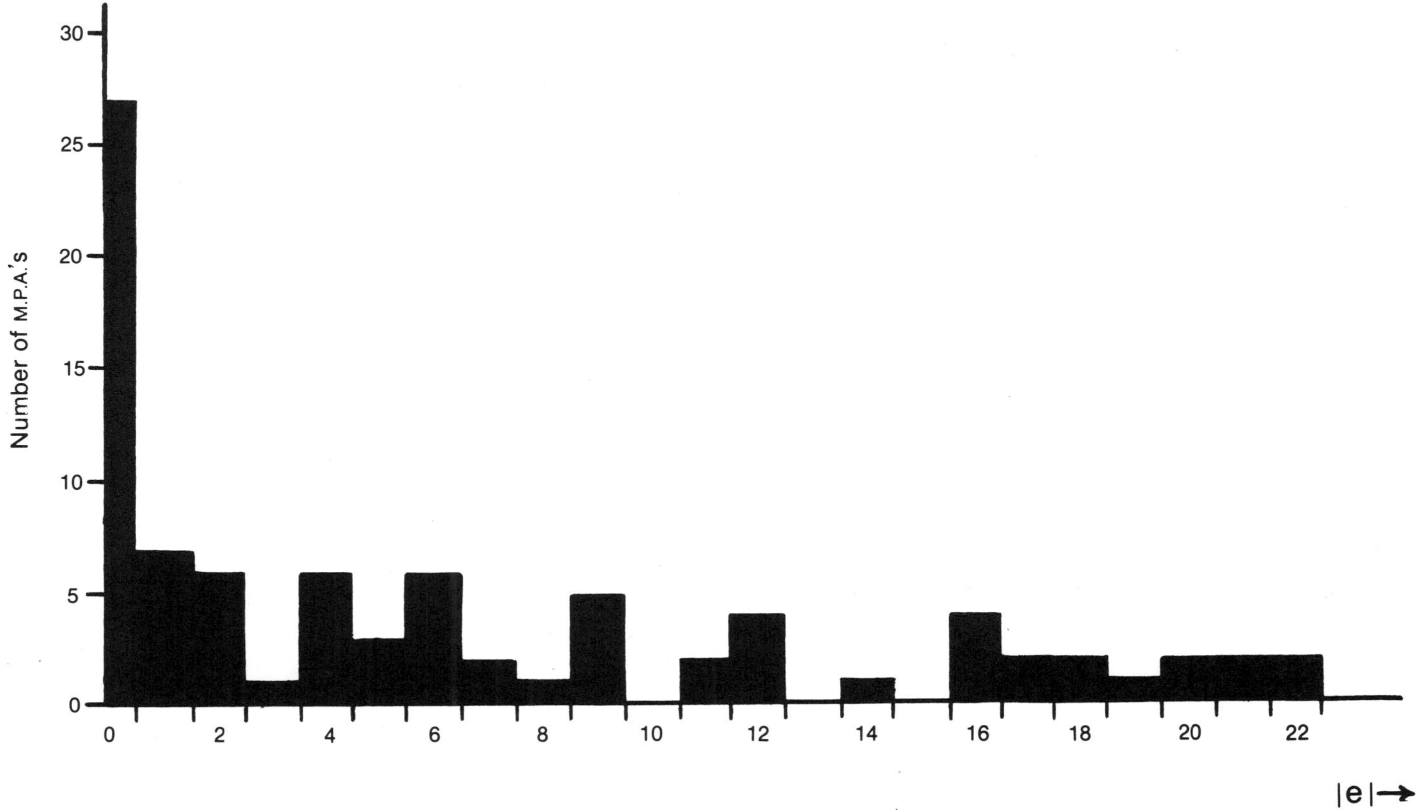

Graph 4.3. Histogram of number of M.P.A.'s versus $|e|$

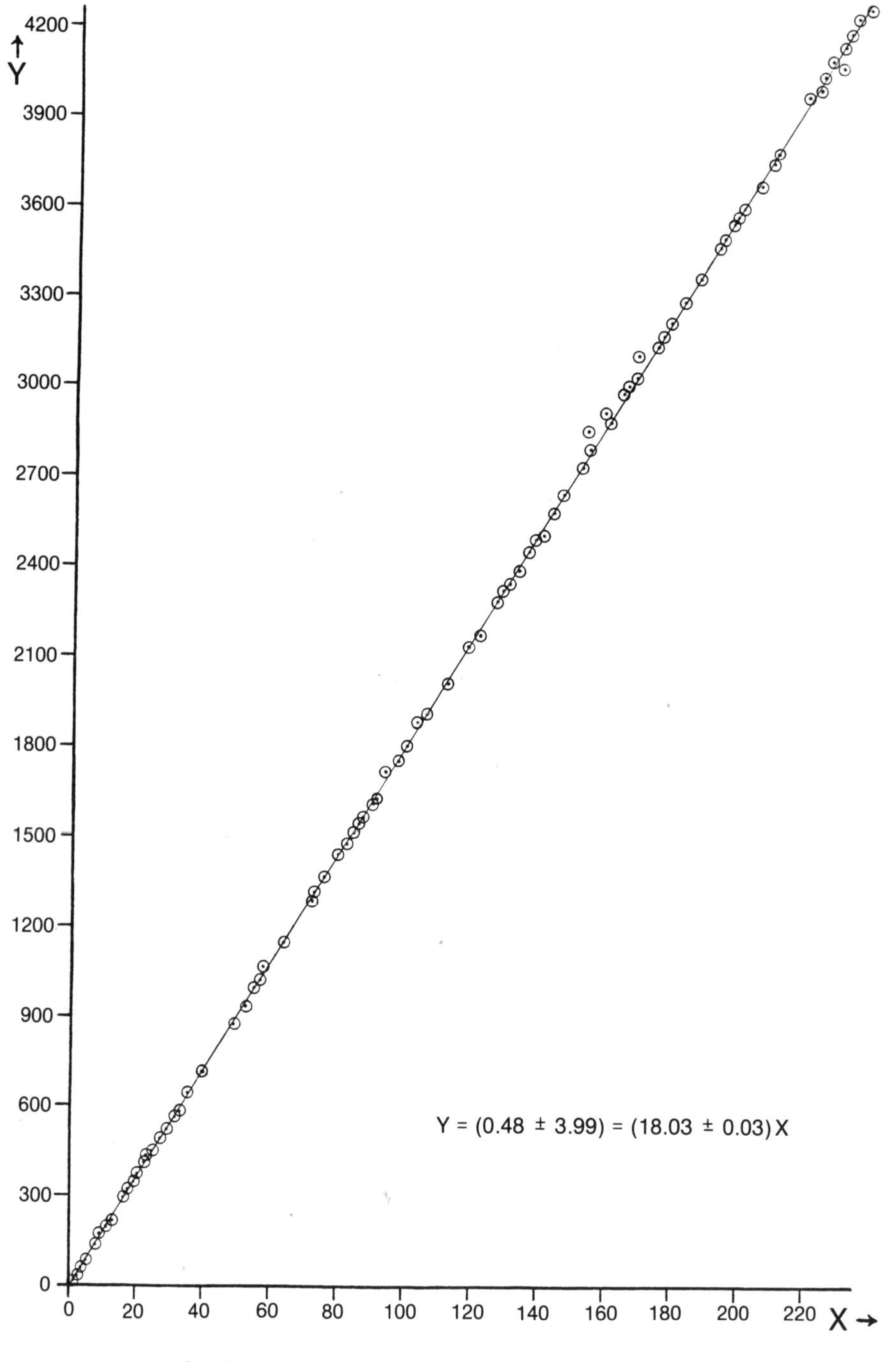

Graph 4.4. M.P.A. population (Y) versus mass number (X)

A TEST OF STATISTICAL SIGNIFICANCE

Let Y be the population of U.P.A.'s in the M.P.A. of a nuclide of mass number X. Graph 4.4 of Y versus X is that of a straight line. A linear relationship between Y and X is expected. Accordingly, the following mathematical model is expected to relate the number Y_i of U.P.A.'s in an M.P.A. to the mass number X_i of the nuclide selected for micro-psi examination:

$$Y_i = \alpha + \beta X_i + e_i ,$$

where e_i is the random error of counting and α and β are constants to be determined. In the particular theory under consideration, $\alpha = 0$ and $\beta = 18$. These two hypotheses will be tested by using the technique of linear regression. The method of least squares is strictly valid only if the errors are normally distributed, independent, and of constant variance. Graph 4.2 confirms their normal distribution. The errors are independent, since graph 4.5 of error versus X shows a random scatter of points. Graph 4.6 of error versus $18X$ also displays a random scatter of points, and so the errors have constant variance. Therefore, all three criteria for the validity of the method of least squares are satisfied. Estimated values of α ($\hat{\alpha}$) and β ($\hat{\beta}$) are obtained from the theoretical expressions

$$\hat{\alpha} = \bar{Y} - \hat{\beta}\bar{X} , \quad \hat{\beta} = S_{XY}/S_{X^2} ,$$

where

$$\bar{Y} = \frac{1}{n}\sum_{i=1}^{n} Y_i , \quad \bar{X} = \frac{1}{n}\sum_{i=1}^{n} X_i ,$$

$$S_{X^2} = \sum_{i=1}^{n} X_i^2 - n\bar{X}^2 , \quad S_{XY} = \sum_{i=1}^{n} X_i Y_i - n\bar{X}\bar{Y} .$$

The best estimate of the error variance is

$$S^2 = [1/(n-2)][S_{Y^2} - (S_{XY})^2/S_{X^2}] ,$$

where

$$S_{Y^2} = \sum_{i=1}^{n} Y_i^2 - n\bar{Y}^2 .$$

The estimates of the standard errors of $\hat{\alpha}$ and $\hat{\beta}$ are

$$\sigma_\alpha = S(1/n + \bar{X}^2/S_{X^2})^{1/2} \quad \text{and} \quad \sigma_\beta = S/(S_{X^2})^{1/2} .$$

The t-test statistic for the hypothesis $\alpha = 0$ is

$$t_\alpha = \hat{\alpha}/\sigma_\alpha .$$

The t-test statistic for the hypothesis $\beta = 18$ is

$$t_\beta = (\hat{\beta} - 18)/\sigma_\beta .$$

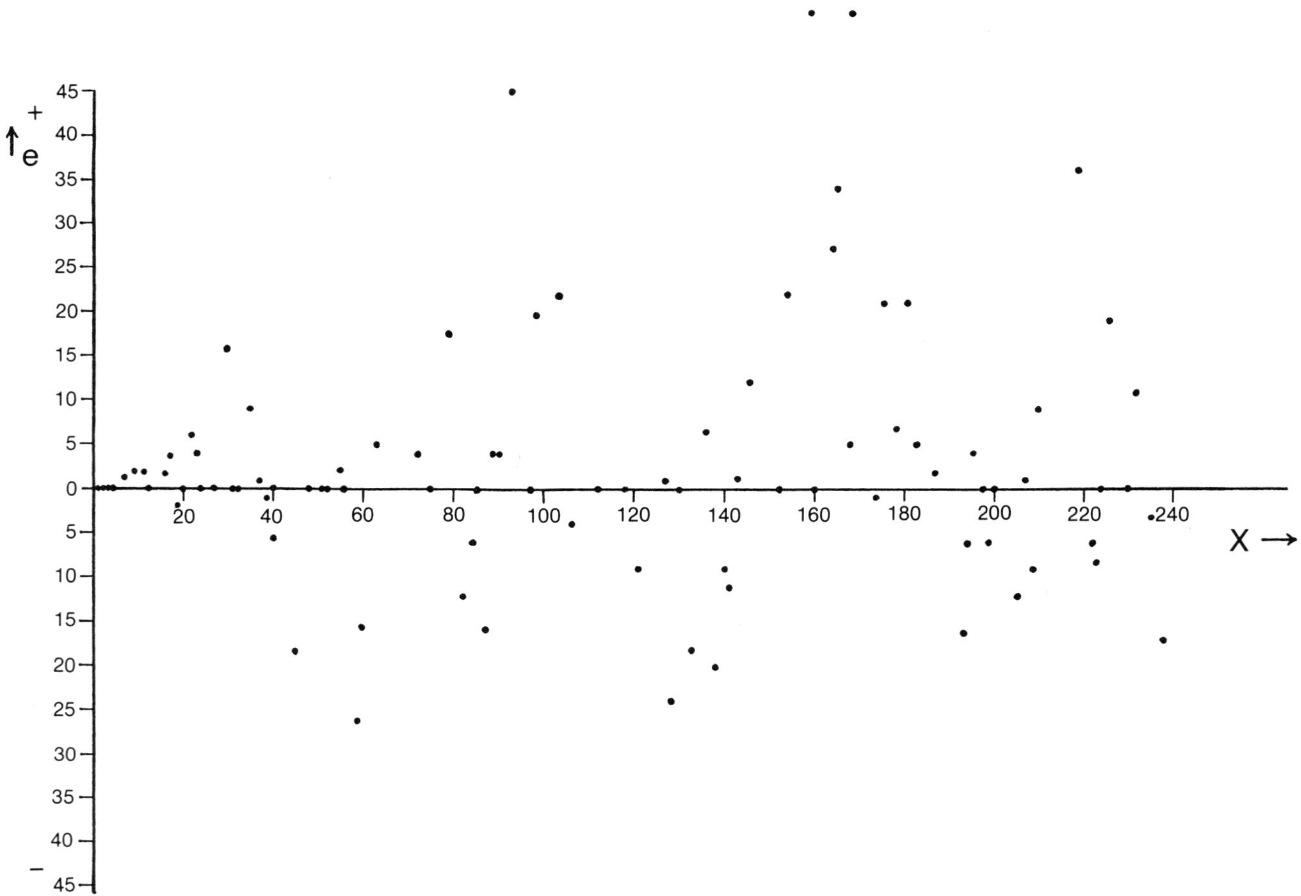

Graph 4.5. Graph of error (e) versus X

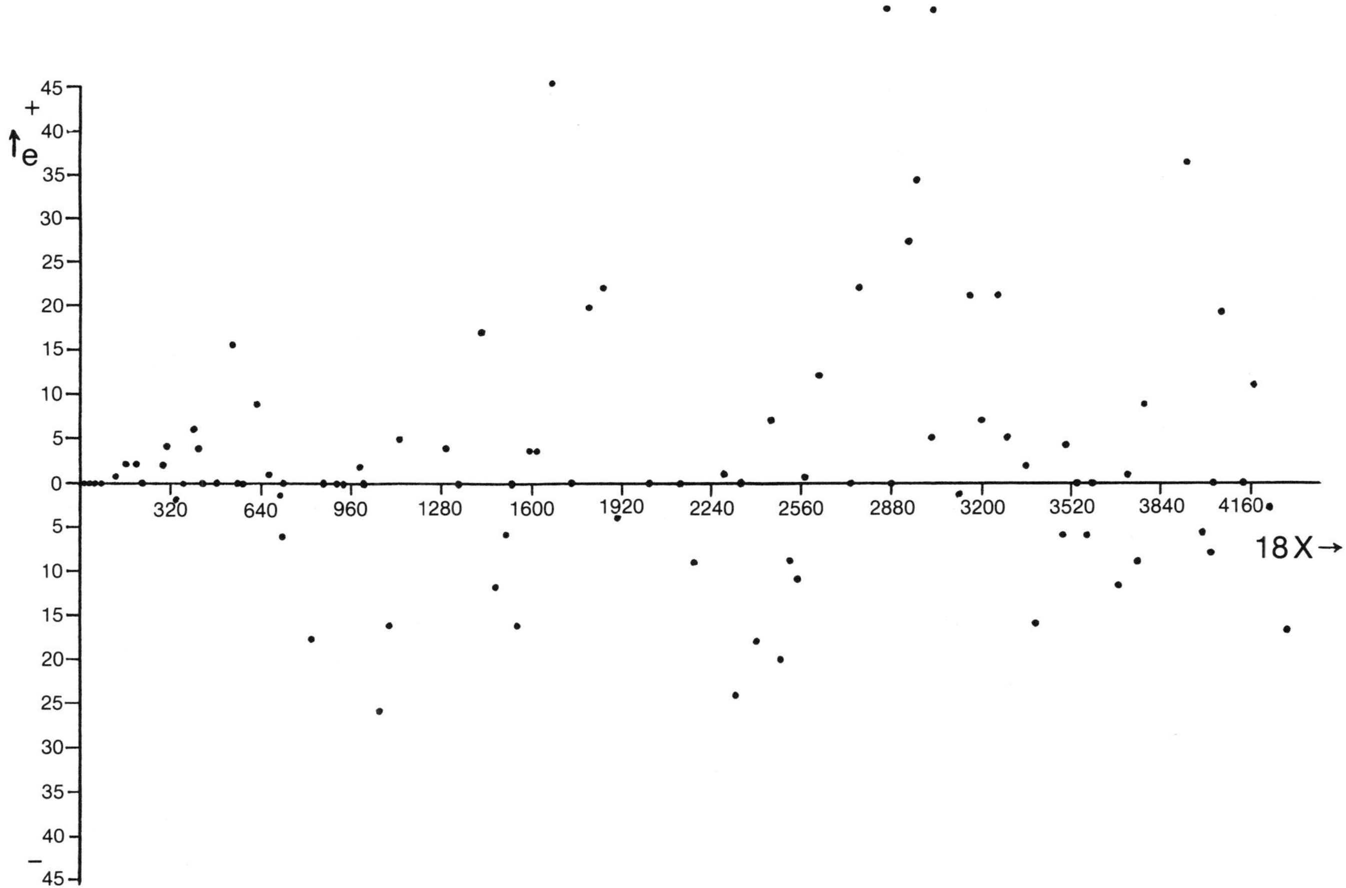

Graph 4.6 Graph of error (e) versus $18X$

Both statistics have $n - 2$ degrees of freedom. Excluded from the test are the "hybrid M.P.A.'s" (to be discussed shortly), and M.P.A.'s listed in table 4.1 as being formed from two different nuclides, since they are not examples of the model under test. This leaves a sample of size $n = 95$. The results are

$$\hat{\alpha} = 0.48\,, \qquad \sigma_{\alpha} = 3.99\,,$$
$$\hat{\beta} = 18.03\,, \qquad \sigma_{\beta} = 0.03\,\cdot$$
$$t_{\alpha} = 0.1191 \text{ (statistical probability } = 0.9052)\,,$$
$$t_{\beta} = 0.8755 \text{ (statistical probability } = 0.3814)\,.$$

The data are best fitted by the relation

$$Y = (0.48 \pm 3.99) + (18.03 \pm 0.03)X$$

and support the hypotheses $\alpha = 0$ and $\beta = 18$ at an acceptable level of confidence ($p > 0.90$ and 0.38, respectively). The data support strongly the prediction

$$N(A) = 18A$$

and, therefore, Hypotheses 1 and 2.

It should be noted that this test does not prove the absolute validity of the model $Y = 18X$. In general, the identities of the nuclides that actually formed the M.P.A.'s cannot be known with certainty independently of their counted populations. This is because the investigators supplied no information concerning the materials that they examined with micro-psi vision. Consequently, the data are fitted to nuclides whose identities are conjectured and are not known with certainty. All that the test demonstrates (and all that is claimed here for it) is that the model based on Hypotheses 1 and 2 *can* fit the data satisfactorily when the most appropriate nuclides are chosen according to a set of criteria. It does not demonstrate that the model actually fits a sample of 95 M.P.A.'s whose identification is certain. This, of course, cannot be shown because of the intrinsic, unavoidable uncertainty of their identification. Only the capability of the model to provide a statistically significant fit to the data is shown. The absolute validity of the model may be inferred from this demonstration only insofar as the choices listed in table 4.1 represent the nuclides that in reality formed the M.P.A.'s. The three criteria used in the construction of table 4.1 to select the most appropriate nuclides so severely limit possible candidates that they make this identification highly probable, if not certain. To this extent the successful fitting of the data may be regarded as a confirmation of the absolute validity of the model.

The identification of M.P.A.'s of elements with isotopes having 100 percent terrestrial abundance is free of ambiguity, since it is certain that the investigators would have examined only the single, naturally occurring isotope of these elements. These (and only these) provide a strictly valid test of the model. Using a sample of 23 such elements, the best estimate of β is given by $\hat{\beta} = 18.042 \pm 0.044$. The hypothesis $\beta = 18$ is acceptable, this value differing from the "best fit" hypothesis by less than 1 standard deviation. The absolute validity of the model is supported by the data.

In chapter 2 the possibility that the work of Besant and Leadbeater was fraudulent was discussed and rejected as implausible. This question will now be re-examined. Let us assume hypothetically that Besant and Leadbeater did not possess genuine micro-psi ability and that, intent on falsely demonstrating that they could exercise a faculty of extra-sensory perception, they fabricated the populations of fictitious M.P.A.'s so as to obtain agreement between their calculated number weights and chemical atomic weights, which were listed in scientific reference works available to the general public at the time. It was only this correlation that would have impressed contemporary scientists, because most of Besant's and Leadbeater's descriptions of atoms and molecules at best bore no relation to classical physics and chemistry and at worst conflicted with them very seriously. It is, therefore, central to the issue of the scientific significance of the investigations. Let us suppose, then, that Besant and Leadbeater decided (for some reason) to construct the simplest M.P.A.—the M.P.A. of hydrogen—out of eighteen U.P.A.'s and that they made contact with chemistry by assigning to other elements M.P.A.'s with populations of U.P.A.'s giving number weights approximately equal to the known atomic weights. According to this suggestion, they might have made the agreement only approximate in order to create the impression that their work was just as prone to error as scientific experiments were, so helping to enhance the plausibility of their claims. Table 4.3 is the list of elements that were claimed to have been examined up to 1908 (it appears on p. 20 of *Occult Chemistry* [2d ed., 1919]). The asterisks affixed to some elements are intended to indicate that these had not been discovered by orthodox science until then. The second column is the population (Y), the third column is the number weight (population/18) that Besant and Leadbeater were supposed to have calculated from the second column, and the fourth lists atomic weights (X) taken (according to the text) from "the latest list of atomic weights, the 'International List' of 1905, given in Erdmann's 'Lehrbuch der Unorganischen Chemie.' " Suppose that they used the relation

$$Y = \beta X$$

(taking $\beta = 18$) in order to fabricate the figures in the second column that they claimed to have counted. If this were true, the data would be expected to support the hypothesis $\beta = 18$ at an acceptable level of confidence. On

TABLE 4.3

Element	U.P.A. Population (*Y*)	Number Weight (*Y*/18)	Atomic Weight (*X*)
Hydrogen	18	1	1
*Occultum	54	3	...
Helium	72	4	3.94
Lithium	127	7.06	6.98
Beryllium	164	9.11	9.01
Boron	200	11.11	10.86
Carbon	216	12	11.91
Nitrogen	261	14.50	14.01
Oxygen	290	16.11	15.879
Fluorine	340	18.88	18.90
Neon	360	20	19.9
*Meta-Neon	402	22.33	...
Sodium	418	23.22	22.88
Magnesium	432	24	24.18
Aluminium	486	27	26.91
Silicon	520	28.88	28.18
Phosphorus	558	31	30.77
Sulphur	576	32	31.82
Chlorine	639	35.50	35.473
Potassium	701	38.944	38.85
Argon	714	39.66	39.60
Calcium	720	40	39.74
*Meta-Argon	756	42	...
Scandium	792	44	43.78
Titanium	864	48	47.74
Vanadium	918	51	50.84
Chromium	936	52	51.74
Manganese	992	55.11	54.57
Iron	1,008	56	55.47
Cobalt	1,036	57.55	57.7
Nickel	1,064	59.11	58.30
Copper	1,139	63.277	63.12
Zinc	1,170	65	64.91
Gallium	1,260	70	69.50
Germanium	1,300	72.22	71.93
Arsenic	1,350	75	74.45
Selenium	1,422	79	78.58
Bromine	1,439	79.944	79.953
Krypton	1,464	81.33	81.20
*Meta-Krypton	1,506	83.66	...
Rubidium	1,530	85	84.85
Strontium	1,568	87.11	86.95
Yttrium	1,606	89.22	88.34
Zirconium	1,624	90.22	89.85
Niobium	1,719	95.50	93.25
Molybdenum	1,746	97	95.26
Ruthenium	1,848	102.66	100.91
Rhodium	1,876	104.22	102.23
Palladium	1,904	105.77	105.74
Silver	1,945	108.055	107.93
Cadmium	2,016	112	111.60
Indium	2,052	114	114.05
Tin	2,124	118	118.10
Antimony	2,169	120.50	119.34
Tellurium	2,223	123.50	126.64
Iodine	2,287	127.055	126.01
Xenon	2,298	127.66	127.10

* See text.

TABLE 4.3—*Continued*

Element	U.P.A. Population (Y)	Number Weight ($Y/18$)	Atomic Weight (X)
*Meta-Xenon	2,340	130	...
*Kalon	3,054	169.66	...
*Meta-Kalon	3,096	172	...
Osmium	3,430	190.55	189.55
Iridium	3,458	192.11	191.56
Platinum A	3,486	193.66	193.34
*Platinum B	3,514	195.22	...
Gold	3,546	197	195.74

repeating the analysis for table 4.3, one obtains the best estimate $\hat{\beta} = 18.055 \pm 0.014$. This differs from the expected value $\beta = 18$ by about 4 standard deviations, and the hypothesis is, therefore, unacceptable ($p < 10^{-4}$). It is concluded that there is no statistical support for the suggestion that Besant and Leadbeater fabricated the population data from lists of atomic weights.

ELEMENTS "X," "Y," "Z," AND "KALON"

As a result of their examination of the three triads of transition elements

iron	cobalt	nickel
ruthenium	rhodium	palladium
osmium	iridium	platinum

belonging to Group VIII of the periodic table, Besant and Leadbeater reported[2] in 1907 their observation of three M.P.A.'s that were distinct, both in population of U.P.A.'s and in their constituent groups, from the M.P.A.'s of these transition elements, although they had the form characteristic of the Bars Group to which the latter belonged. Because the three M.P.A.'s could not be accommodated in the periodic table, the investigators had to label them "X," "Y," and "Z," believing that they were atoms of new elements not at that time discovered by science. Their belief was strengthened by their discovery that, when the M.P.A.'s of the known elements were arranged according to their external shapes, they fell into groups that were arranged in the same way that elements were classified in the periodic table proposed by Sir William Crookes, the famous chemist. X, Y, and Z also fitted this scheme of classification, and Besant and Leadbeater even formulated a version of Crookes's table that provided natural positions for these "missing elements."[3] But it is well known that this table is incorrect, and atomic theory provides no room for a missing group of transition elements in the modern periodic table. No element can exist that corresponds to any of these M.P.A.'s. This provides, of course, an irrefutable argument against the interpretation of an M.P.A. as an atom or atomic nucleus. What can X, Y, and Z be? By comparing the populations listed in table 4.1, it can be seen that the differences between their populations and those of the M.P.A.'s of the three triads of elements are too large for them to be merely isotopic variations of the latter. It is proposed here that they are "hybrid M.P.A.'s," formed from nuclei of two *different* triad elements at the time when compounds of these elements were examined by the investigators. Such a possibility raises the question of why, out of the numerous recorded examples, only three M.P.A.'s should have been formed in this way (apart from the M.P.A. of "Kalon," which will be discussed shortly). Answers to this question can be speculated, but they will not be discussed here.

The following evidence supports such an interpretation: it can be seen from table 4.1 that the populations of the M.P.A.'s of X, Y, and Z are almost exactly the arithmetic mean of those of the M.P.A.'s, respectively, of ruthenium and osmium, rhodium and iridium, and palladium and platinum—just what would be expected if they had, indeed, been formed from nuclei of each pair of elements. Ruthenium has the stable isotopes Ru^{102} and Ru^{104}, and osmium has the stable isotopes Os^{190} and Os^{192}. The M.P.A. formed from either Ru^{102} and Os^{192} or Ru^{104} and Os^{190} (the former pair being favoured because both nuclides are more common) would contain 2,646 omegons, and this is exactly the number of U.P.A.'s counted in the M.P.A. of X. Close agreement can be found for the other hybrid M.P.A.'s (see table 4.1).

Also observed in 1907 was an M.P.A. that, according to its external form, belonged to the Star Group (comprising the inert gases) but cannot be accommodated in the modern periodic table as a new inert gas. Its population of 3,054 U.P.A.'s placed it between xenon and emanation in terms of number weight (and, therefore, effective atomic weight). It was reported to be rare: "Its rarity was then described by saying that there might be one in the atmosphere of an ordinary-sized room."[4] Besant and Leadbeater believed that they had spotted with micro-psi vision an atom of an inert gas unknown to science then, calling the new element "Kalon." It hardly needs to be pointed out that, if an M.P.A. were an atom or atomic nucleus, atomic physics would forbid the existence of one atom or nucleus of kalon even in the entire universe, let alone in an "ordinary-sized room." Its population of 3,054 U.P.A.'s is almost exactly the arithmetic mean of the populations of the xenon and emanation M.P.A.'s (2,298 and 3,990, respectively), implying an effective mass number of about 169 or 170. But it cannot be due to an isotope of either of these elements because neither has isotopes with such mass numbers. It is proposed here that the M.P.A. of kalon is another hybrid M.P.A., like those of X, Y, and Z, formed from nuclei of xenon and emanation during the investigation of the inert gases. The low abundance of these elements in the earth's atmosphere would explain the reported rarity of observations of the M.P.A.'s.

A possible candidate for kalon is the pair of nuclides Xe^{124} and Em^{222} (radon). An "isotopic variation" of the M.P.A. ("meta-Kalon") with 3,096 U.P.A.'s is probably due to the pair of nuclides Xe^{124} and Em^{220} (thoron). The predicted error for kalon is -60; for meta-kalon, it is zero. The former is an integer multiple of 6. Kalon also illustrates the error-compounding effect discussed earlier for the inert gases, namely, that their M.P.A.'s must have errors of counting that are integer multiples of 6 because the investigators counted U.P.A.'s in only one arm of the star-shaped M.P.A.

In conclusion, U.P.A.'s have been identified as $\mathcal{P}$ and $\mathcal{N}$ omegons—the spin-1/2 constituents of u and d quarks predicted by the Omegon Model. The predicted populations of M.P.A.'s agree with the counted numbers to a statistically significant extent. A generalization of Hypothesis 2 explains why several M.P.A.'s should have been reported, despite the fact that none of them can correspond to any known element.

Details of the dynamical transformation of pairs of nuclei into M.P.A.'s are discussed in chapter 7.

REFERENCES

1. Lucifer, November 1895 (London: Theosophical Publishing House).
2. Memorandum concerning "three new Interperiodics X, Y & Z," published with other reports in The Theosophist, January–December 1908.
3. Occult Chemistry, 3d ed., p. 33, fig. 12.
4. Ibid., p. 249.

CHAPTER 5

Micro-Psi Support for the "String Model" of Elementary Particle Physics

There are therefore Agents in Nature able to make the Particles of Bodies stick together by very strong Attractions. And it is the Business of experimental Philosophy to find them out.

Sir Isaac Newton

Small things wax exceeding mighty,
Being cunningly combined;
Furious elephants are fastened
By ropes of grass-blades, twined.

Hitopadesa

MAGNETIC MONOPOLES

Over forty years ago, the English physicist P. A. M. Dirac[1] introduced to physics the concept of "magnetic monopoles" (pointlike magnetic charges), analogous to electric monopoles (pointlike electric charges) such as the electron, which has no size or structure that can yet be revealed by high-energy scattering experiments. He pointed out that the possible electric charges (Q) and magnetic charges (g) that particles could carry are not independent of each other. The laws of quantum mechanics require that they be related by the equation

$$Qg = \tfrac{1}{2}n\hbar c \quad (n = 0, \pm 1, \pm 2, \ldots),$$

where $\hbar = h/2\pi$ (h is Planck's constant) and c is the speed of light. The existence of magnetic monopoles of charge $g = g_0 = \hbar c/2e$ implies that the electric charge Q must be an integer multiple of the unit of electron charge:

$$Q = ne.$$

This was a well-known fact at the time when Dirac discussed magnetic monopoles. These particles, however, were undetected, hypothetical objects, and so his simple explanation of why atomic particles carried only integer multiples of the charge of an electron remained speculative.

Over the years, experimental search for monopoles has yielded mostly negative results. Efforts to generate monopoles in large particle accelerators[2] have been unsuccessful. Cosmic-ray observations[3] have failed to find them, and searches for monopoles trapped inside magnetic material such as ocean-bottom samples[4] and moon rocks[5] have not been fruitful. But magnetic monopoles might exist for another reason: they restore the so-called "dual" symmetry to Maxwell's equations for the electromagnetic field. Nature appears to have given monopoles no role to play in the construction of the universe, for the latter seems to be built in an asymmetric fashion only from electrically charged particles. The success of the Quark Model in explaining many facts of hadron physics had suggested that quarks were not merely mathematical entities but were also real constituents of hadrons. But they, too, have remained elusive to the experimental physicist. For example, fractionally charged particles have been sought[6] by mass spectrographic techniques in materials that have undergone little geochemical ageing (e.g., lunar dust and meteorites) or in materials such as sea water in which fractionally charged atoms may have been accumulating. These experiments determine the upper limits: 2×10^{-22} per nucleon for lunar soil and 3×10^{-24} per atom for atoms with either $-e/3$ or $+2e/3$ quarks attached. One of the few positive results of recent years was that of G. S. La Rue et al.,[7] who reported possible observation of fractional charges in superconducting niobium spheres levitated in magnetic fields. This result, however, was not confirmed by Schiffer et al.,[8] who gave an upper limit of 10^{-22} per nucleon for $e/3$ charges emitted by niobium, tungsten, and iron filaments heated in an electrostatic potential. Noting the elusive character of both monopoles and quarks, Schwinger[9] suggested that they might be one and the same particle. He proposed that hadrons were made up of particles carrying both electric and magnetic charges ("dyons"). He argued that the absence of free magnetic monopoles in the world at large was due to the superstrong electromagnetic coupling between dyons. The strength of the Coulomb force between electric charges is determined by the fine-structure constant $\alpha = e^2/\hbar c \sim 1/137$, whereas magnetic charges would have the coupling constant $g_0^2/\hbar c = 1/4\alpha$. This would mean that the force between dyons is several times stronger than the nuclear force holding nucleons together in nuclei. For example, it would require 7 GeV of energy to separate a pair of Dirac monopoles from a distance of 10^{-13} cm to infinity. Free monopoles (quarks) would be very rare because of the enormously energetic processes (natural or technological) required to dissociate magnetically neutral matter into monopoles of opposite charge. But purely magnetostatic binding cannot lead to "asymptotic freedom" for quarks—the property revealed by high-energy inelastic electron-proton scattering experiments whereby the forces between quarks appear to weaken with decreasing distance. Nor does it lead to stable, bound systems for three quark monopoles only and not more.

The lack of production of free quarks in high-energy collisions of particles and the apparent absence of free quarks in matter can be explained in four ways:

1. Quarks do not exist, and the success of the Quark Model is accidental.
2. Unlike atoms in molecules or nucleons inside nuclei, quarks are not discrete objects but a kind of collective excitation of hadronic matter, perhaps analogous to phonons in a crystal.
3. Quarks are discrete particles that are bound by superstrong forces which in principle permit their escape from hadrons but which cannot yet be overcome by the current generation of particle accelerators.
4. Unlike other forces of nature that weaken with increasing distance between particles, the forces binding quarks together remain constant with increasing distance, so that they are permanently trapped inside hadrons.

An example of the fourth proposal is a class of theories that will be referred to collectively as the "String Model." In this chapter, micro-psi observations made about eighty years ago will be shown to support the essential physical ideas on which this theory of strong forces and of quark confinement is based. Its history is reviewed below.

NON-ABELIAN VORTICES AND QUARK CONFINEMENT

In 1933, Meissner and Ochsenfeld discovered that, when a magnetic field was applied to a superconductive material cooled to low enough temperatures for it to exhibit no electrical resistance, the magnetic flux that had passed straight through the specimen when it was at room temperature (fig. 5.1, *a*) was almost completely expelled from its interior (fig. 5.1, *b*). This "Meissner Effect" is now known to arise from the wholesale formation in the superconducting material of large numbers of "Cooper pairs" of electrons, attractively bound in pairs through their mutual interaction with the atomic lattice—an interaction that overcomes their mutual repulsion due to their electric charges. Their motion in the externally applied magnetic field produces a magnetic field inside the material that opposes and effectively cancels it, so that there is no resultant field. Only when it is superconducting is there a sufficient number of Cooper pairs to achieve this cancellation. Since 1962, two classes of superconductors have been recognized—type 1 and type 2. Only the latter is relevant to the present discussion. When a magnetic field is applied to a type 2 superconductor, the normal, conducting regions of the material form an array of filaments of negligible thickness, arranged parallel to the external field and surrounded by the remaining superconducting material (fig. 5.2, *a*). These filaments are extended bundles of flux lines and are, along their entire length, the centre

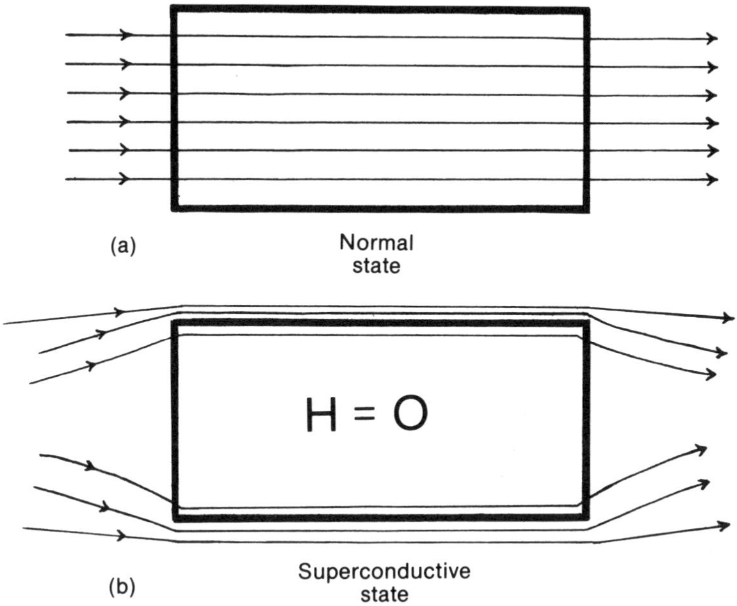

Fig. 5.1

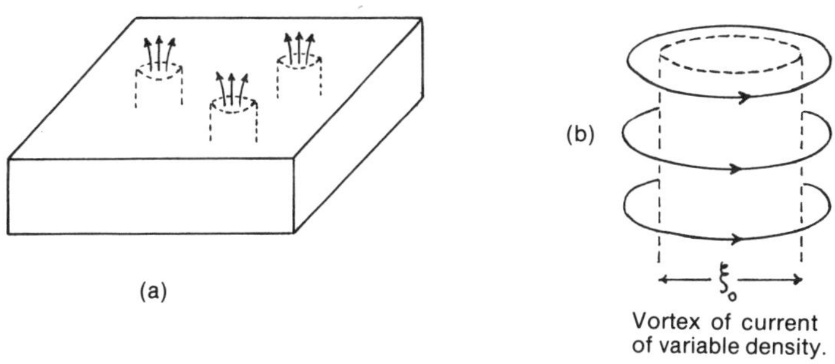

Fig. 5.2

of a whirlpool or vortex of electric current made up of circulating electrons (fig. 5.2, *b*). Inside this cylindrical vortex tube, the magnetic flux is trapped in regions of normal material with non-zero electrical resistance, being expelled from the surrounding superconductive material through the Meissner Effect. Each flux line is one quantum of magnetic flux: $\phi_0 = hc/2e = 2 \times 10^{-7}$ gauss cm^2. The flux in a vortex tube is quantized in terms of this unit. Its core is approximately equal in size to the "coherence length"

ξ_0 and contains most of the flux. But the magnetic field extends into the superconducting region over a distance that is characterized by the "London penetration depth," $\lambda_L = (mc^2/4\pi ne^2)^{1/2}$, where m is the electron mass and n is the density of superconducting electrons. For type 2 superconductors, $\xi_0 \ll \lambda_L$.

The short penetration depth of a magnetic field inside a type 2 superconductor may be interpreted in terms of the photon becoming massive as the result of spontaneous breakdown of Maxwell's field equations for the electromagnetic field (loss of U(1) symmetry in the superconductor). This is an example of the Higgs mechanism in particle physics whereby massless Yang-Mills gauge fields such as the electromagnetic field can acquire mass and so become short-range. In 1973, Nielsen and Olesen[10] pointed out a remarkable parallelism between, on the one hand, the Higgs model of broken gauge symmetry and the London-Ginzburg theory of superconductivity, and, on the other hand, the Dual String Model and flux lines in type 2 superconductors. Their work prompted Nambu[11] to study Dirac monopoles embedded in an Abelian superconducting vacuum, the Higgs field taking over the role of superconducting electrons. He found that pairs of oppositely charged monopoles were permanently bound together. With the identification of quarks as magnetic monopoles and mesons as monopole-antimonopole pairs, this result explained why mesons could not decay into free quarks. In a classical vacuum, the flux lines of a positively charged monopole diverge in all directions (fig. 5.3, a). The flux lines of a monopole immersed in a superconducting Higgs vacuum are confined as quantized bundles within a flux tube or vortex (fig. 5.3, b), being expelled from the surrounding vacuum by the Meissner Effect. The Nielsen-Olesen vortices are in the spinless Higgs meson field, which behaves as a superfluid. Far from the centre of the vortex, the Higgs field density is constant and the

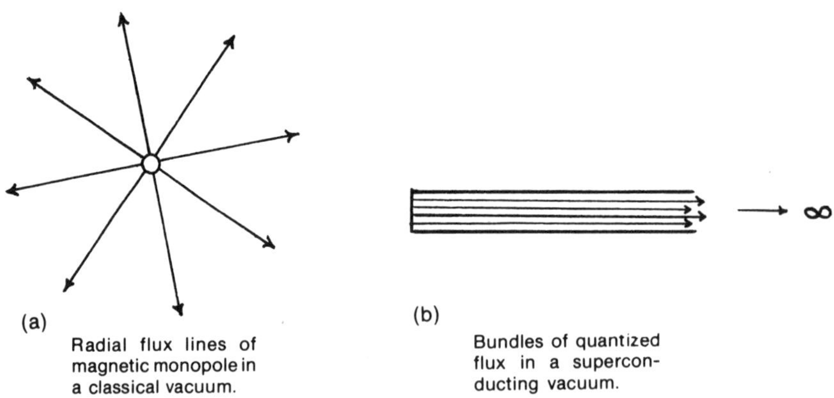

Fig. 5.3

vacuum is superconducting. Closer to the centre, the density progressively decreases until it falls to zero at the centre of the flux tube, where the vacuum is normal and the U(1) symmetry of the electromagnetic field of the monopole is restored (fig. 5.4). Single monopoles are end-points of infinitely

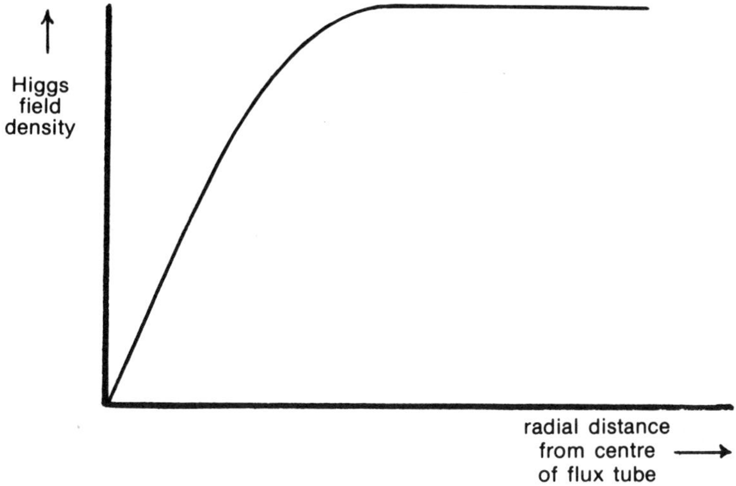

Fig. 5.4

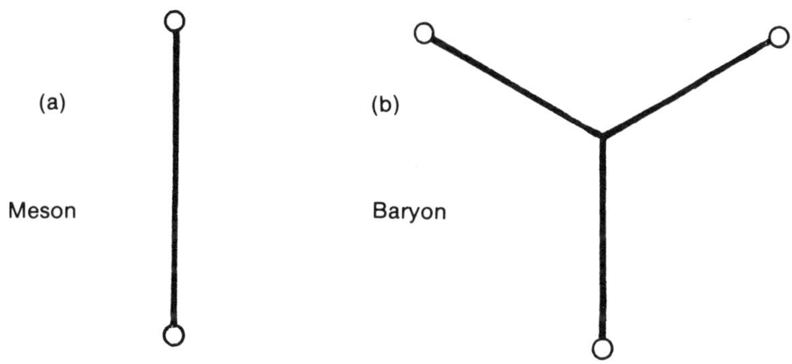

Fig. 5.5

long vortices ("open strings"), while the ends of finite vortices ("closed strings") are monopole-antimonopole pairs, tied together permanently by a bundle of flux lines that emanate from the monopole and terminate on the antimonopole (fig. 5.5, *a*). The flux tube is in equilibrium against the pressure of the surrounding superfluid, and the flux inside it is quantized in units of ϕ_0.

The energy of a flux tube is found to be proportional to its length.[12] Therefore, pairs of monopoles bound at the end of a flux tube cannot be dissociated into free monopoles through absorption of a finite amount of energy. From the energy absorbed in any stretching of a flux tube, a new monopole and antimonopole would materialize. The new monopole would replace the one removed from the pair, and the new antimonopole would join with the displaced monopole, forming another pair, that is, another meson. Quark monopoles could never exist alone. This is the String Model mechanism for quark confinement. However, Abelian vortices cannot bind quarks inside baryons. This is because U(1) is an infinitely connected gauge group and vortices with uniquely defined end-point charges $g = ng_0$ ($n = 1, 2, 3, \ldots$) are physically distinct, so that they do not guarantee baryons to be made up of only three quarks, as hadron spectroscopy indicates. As discussed in chapter 3, the triply connected gauge group $SU(3)_c/Z_3$ is needed to restrict possible bound states of quarks to groups of three, for then Nielsen-Olesen vortices contain magnetic flux defined modulo $3\phi_0$, permitting only three quark monopoles to be bound by them in a Y-shaped configuration (fig. 5.5, *b*). Such Y-shaped strings with quarks at their ends have been proposed for baryons by Artu[13] and by Mandelstaam.[14]

It was proposed in chapter 3 that quarks are Y-shaped strings with omegons as end-points. Micro-psi observations are shown later in this chapter to support this composite picture of quarks.

MICRO-PSI OBSERVATION OF NIELSEN-OLESEN VORTICES

At the end of chapter 1, it was stated that Annie Besant and C. W. Leadbeater claimed that matter is built up from two kinds of U.P.A.'s—a "positive" and a "negative" variety whose spiral forms are mirror images of each other. In chapter 4 they were identified as SU(9) monopoles of opposite magnetic polarity. The investigators observed a large variety of clusters of U.P.A.'s in the M.P.A.'s of the elements, some examples of which are shown in figure 5.6. But they rarely reported more than nine U.P.A.'s in a single cluster. This is significant because, according to the theory presented in chapter 3, the superconducting Higgs vacuum (which implements spontaneous breakdown of the SU(9) gauge symmetry of the strong interaction between omegons) can support a stable system of only nine magnetic monopoles, bound by Nielsen-Olesen vortices, although other vacua can support fewer monopoles if the latter have fewer colour degrees of freedom (as will be illustrated shortly). In most cases, "bright lines" or "streams of light" were seen both to enter and to leave each U.P.A. (fig. 5.7). Less common cases where lines appeared to terminate on U.P.A.'s will be discussed later. These lines were regarded by the investigators as "lines of force": "Force pours into the heart-shaped depression at the top of the Anu, and issues from the point, and is changed in character by its passage."[15] This was noticed in both (+) and (−) U.P.A.'s.

It is proposed that "lines of force" are non-Abelian Nielsen-Olesen

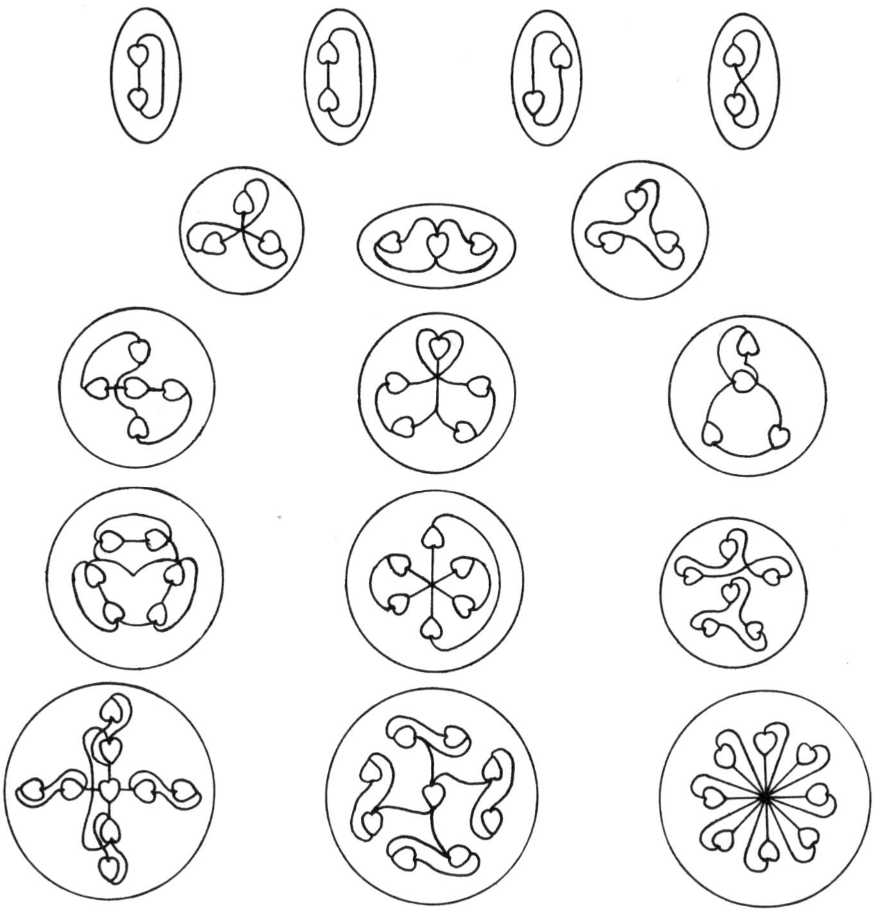

Fig. 5.6. Examples of bound states of U.P.A.'s

Fig. 5.7

vortices carrying quantized flux. Then the above observation may be understood as follows: as well as being the end-point of a single vortex, an SU(N) monopole may be the joint end-point of two such vortices. Because the quantum of magnetic flux in an SU(N) vortex is defined modulo N (in terms of the Dirac unit), a vortex carrying b units of flux in one direction is physically identical with a vortex carrying $N - b$ units in the opposite direction ($b < N$). The intuitive meaning of this is clear: an SU(N) vortex carrying N units of flux (or integer multiples thereof) is equivalent to no vortex at all, that is, to the ground state of the vacuum. If N units of flux are "switched on" in the direction opposite to that of the flux quantum b, the resulting vortex is physically the same as before, but, since b quanta of flux are now nullified, it is equivalent to a vortex carrying the remaining $N - b$ flux quanta in the opposite direction (fig. 5.8). An SU(N) monopole of charge $g = m$ is topologically identical with a monopole of charge $g = m - N$ (and to one of charge $g = m + N$), because magnetic charge is defined only by modulo N. This means that an SU(N) monopole of charge $g = m > 0$ that is the joint end-point of two vortices carrying away flux quanta of a and b ($m = a + b$) is also the joint end-point of two vortices carrying flux quanta of b and $N - a$, respectively, away from and towards the monopole (fig. 5.9, a). Similarly, a monopole of charge $g = -m < 0$, on which two vortices carrying flux quanta of a and b terminate, is equivalent to the joint end-point of two vortices carrying flux quanta of a and $N - b$, respectively, towards and away from it (fig. 5.9, b). Because flux

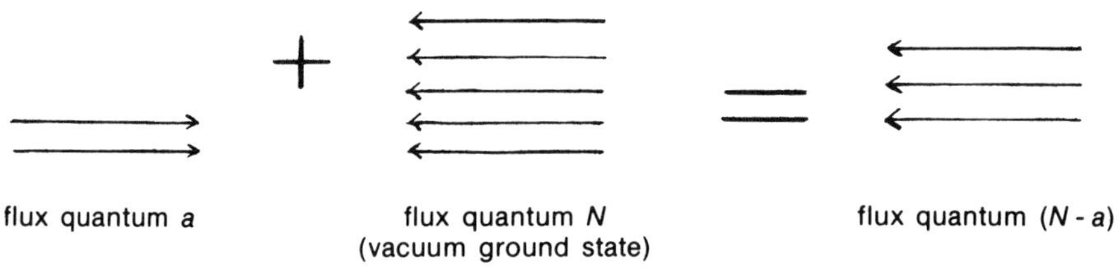

Fig. 5.8

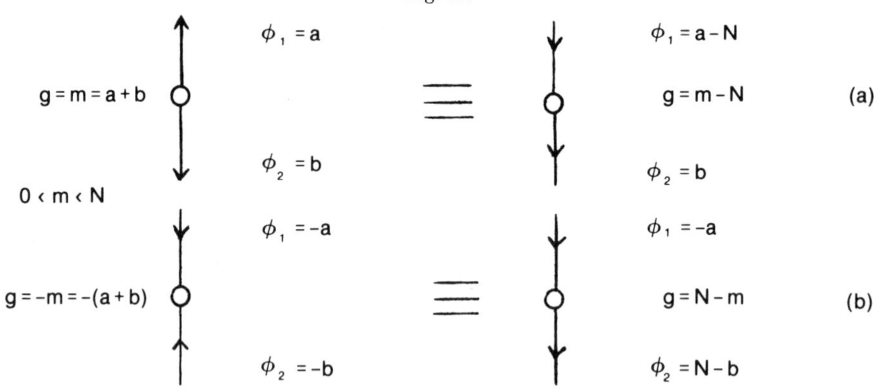

Fig. 5.9

quanta and monopole moments are defined only by modulo N, flux direction is not uniquely defined. However, the directions of the lines of force between U.P.A.'s were specified by the investigators, who noted that the lines entered one end of the U.P.A. and passed out of the opposite end. To facilitate analysis of their observations, the following convention is adopted: flux lines terminate at the "depression" end of a U.P.A. and commence at the "pointed" end.

In conclusion, the "force" referred to above by the investigators is a bundle of non-Abelian gauge flux lines confined mostly along the cores of Nielsen-Olesen vortices of the superconducting Higgs vacuum. The "change in character" of this force as it enters and leaves a U.P.A. is the difference in flux densities of the two vortices joined by the monopole, which acts as a source or sink of flux.

U.P.A.'s AND THEIR STRINGS

Examples of recorded duads of U.P.A.'s bound by lines of force appear in figure 5.10. If U.P.A.'s were always magnetic monopoles embedded in the SU(9) superconducting Higgs vacuum, these duads would not have been observed. This is because such a vacuum permits only groups of nine monopoles, as three clusters of three, to be free systems, according to the theory presented in chapter 3. However, in chapter 7 it will be proposed that one of the physical effects of micro-psi vision on confined constituents of nucleons is to induce local phase transitions of the SU(9) Higgs vacuum to SU(N) vacua ($N < 9$), domains of which trap and confine omegons released from nucleons destabilized by these transitions. In the new vacua, omegons have fewer colour degrees of freedom, although their flavours are unchanged. Duads are bound states of two colour SU(2) monopoles, the strings linking them carrying one flux quantum. Monopoles bound by either one string (fig. 5.10, *a*) or three strings (fig. 5.10, *c*) have magnetic charges of ± 1; those bound by two strings (fig. 5.10, *b*) have charges of ± 2, since single SU(2) strings cannot carry two units of flux. The first three examples are identical with the finite meson strings discussed in current String Model research and are striking evidence for U.P.A.'s being magnetic monopoles.[16] Pairs of U.P.A.'s that point away from the centre of the bound state are $\mathcal{P}$-$\mathcal{P}$ diomegons, pairs both pointing inwards are $\mathcal{N}$-$\mathcal{N}$ diomegons, and pairs that point in opposite directions are $\mathcal{P}$-$\mathcal{N}$ diomegons. The reason for this assignment of isospin states is given shortly in the discussion of the hydrogen M.P.A.

"Lines of force" were usually observed to emanate from and terminate on U.P.A.'s. In a few cases, however, the investigators noted that "where the point of entry and departure is outside the Anu, it is indicated by a dot."[17] Figure 5.11 shows two examples. The vacuum itself is not a real

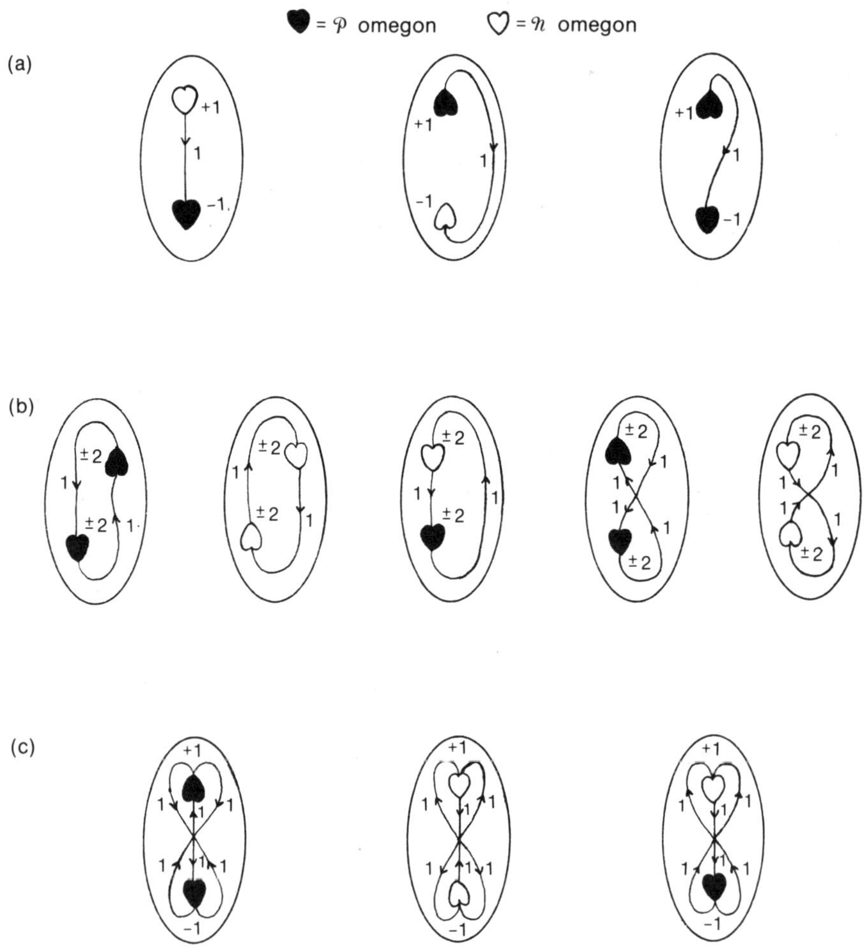

Fig. 5.10. Observed pairs of U.P.A.'s bound by 1, 2, or 3 strings

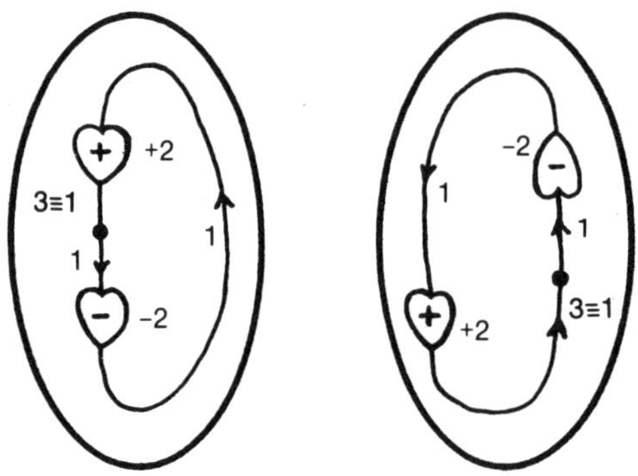

Fig. 5.11

source or sink of magnetic flux. It is suggested that this point is a kind of "branch point" in a string separating sections that carry one and three units of flux. Since the flux in SU(2) vortices is defined by modulo 2, these sections are physically identical. But they appear to the micro-psi observer to have a join or pointlike discontinuity at the point where the flux changes in value by two units.

Examples of free triplets of U.P.A.'s are shown in figure 5.12. They were recorded in the M.P.A. of nearly every element. This abundance is a feature to be expected, since they were identified as quarks at the beginning of chapter 4. The drawings indicate that quarks consist of either triangular or linear clusters of three U.P.A.'s. The first example, that of a Y-shaped string with SU(3) magnetic monopoles at its ends, is identical with a String

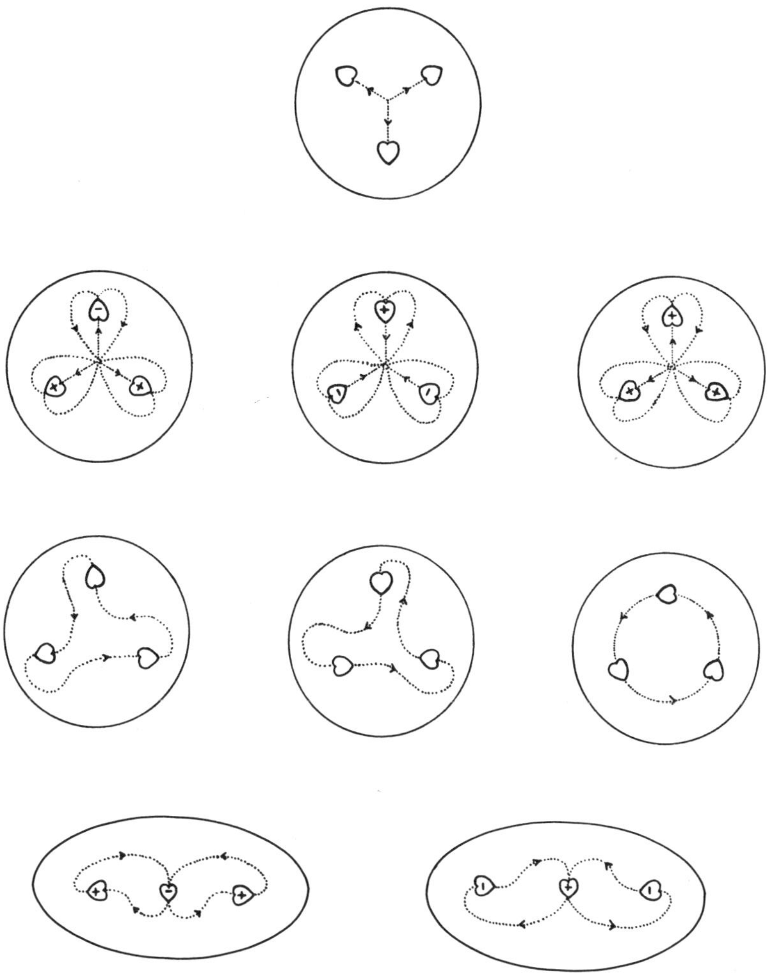

Fig. 5.12. Examples of triplets of U.P.A.'s (quarks), observed in the free (unconfined) state

Model representation of a baryon, with non-composite quarks as end-points. It is one of the strongest pieces of evidence[18] supporting the primary claim made in this book, namely, that the Theosophists Besant and Leadbeater were able to describe the composite character of quarks and protons through their use of micro-psi vision, a siddhi of Yoga. The fourth triplet with three (+) U.P.A.'s indicates that (+) and (−) U.P.A.'s really do differ in their magnetic polarity and not in their electric polarity. According to the Omegon Model, two omegons with a positive electric charge and one with a negative charge make up a u quark, while two with a negative charge and one with a positive charge make up a d quark. A triplet with three (+) U.P.A.'s could not be a u or a d quark if (+) U.P.A.'s carried only positive electric charges. Furthermore, it could not be a $\mathcal{P}$-$\mathcal{P}$-$\mathcal{P}$ bound state with spin 3/2, since it was observed in a Hydrogen Triangle, that is, a proton. Only a definition modulo 3 of the magnetic charge of a monopole with the global gauge group $SU(3)/Z_3$ allows such a bound system of three *positive* magnetic charges. Like diomegons, free (and coloured) quarks cannot exist in the normal vacuum. But, in the vacuum state locally created during the "slowing-down" stage prior to micro-psi observation (for details see chap. 7), they can lose their colour and become free.

Figure 5.13 shows a selection of groups of four or more U.P.A.'s. As configurations of magnetic monopoles bound by SU(N) Nielsen-Olesen vortices, they demonstrate the variable colour-shade valency exerted by omegons in M.P.A.'s, a property that will be discussed further in chapter 7.

In figure 5.14 are shown examples of groups where one or more U.P.A.'s are end-points of single strings. These were not often observed by the investigators. The first one (seen in the M.P.A. of the O^{17} isotope of oxygen) is noteworthy in that it shows four U.P.A.'s as monopole end-points of single SU(4) strings. They have magnetic charges of +2 and −2.

THE HYDROGEN M.P.A.

Hydrogen was one of the first elements examined by Besant and Leadbeater in 1895. In the same year, they published[19] diagrams of its M.P.A., showing the lines of force between the Hydrogen Triplets. These are reproduced in figure 5.15*a*. The isospin states and magnetic charges of the U.P.A.'s and the string flux quanta are assigned in figure 5.15*b*. Re-examined in 1932, the hydrogen M.P.A. was depicted as in figure 5.16, which also shows the groups observed at successive stages of the disintegration of the M.P.A. First, it dissociates into its constituent "protons"* (stage 1); then each free "proton" breaks up into a diquark and a free quark (stage 2); at stage 3, each diquark splits up into two free quarks; finally, all quarks

* By "proton" is meant a proton subject to micro-psi observation, not one in its natural state.

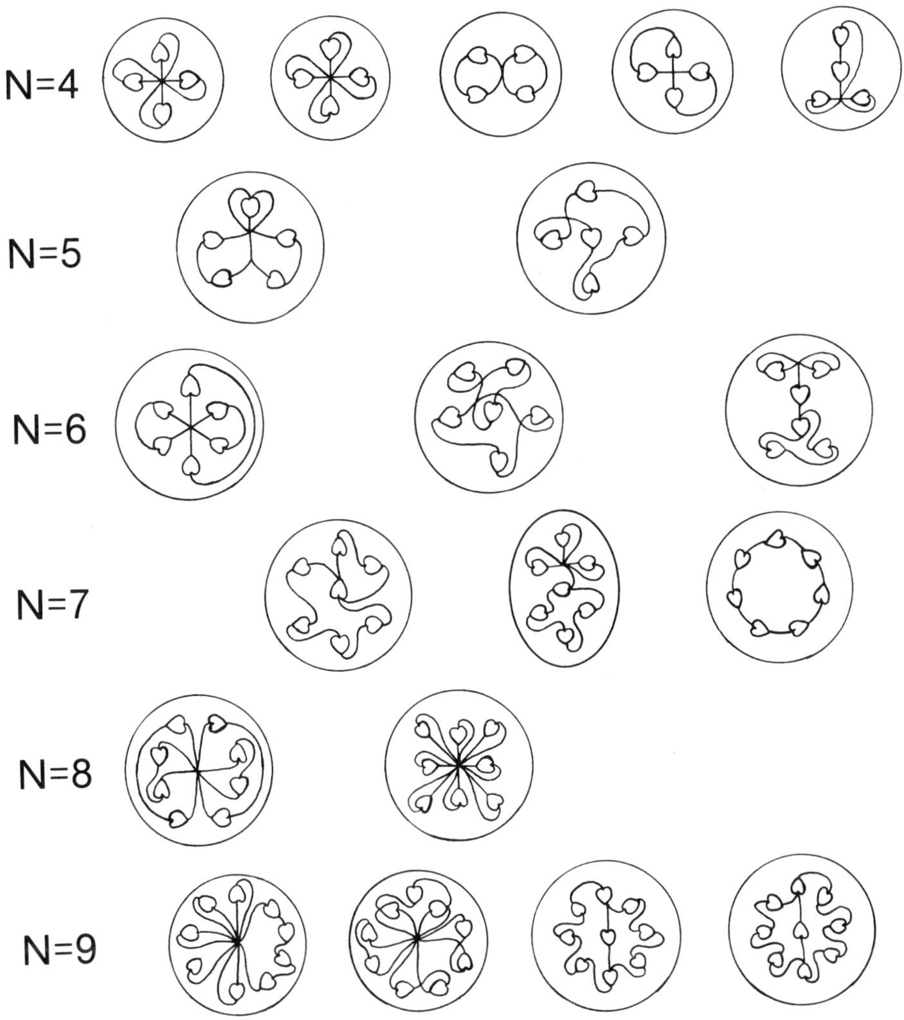

Fig. 5.13. Examples of micro-psi observation of groups of N U.P.A.'s ($4 \leq N \leq 9$), bound by "lines of force."

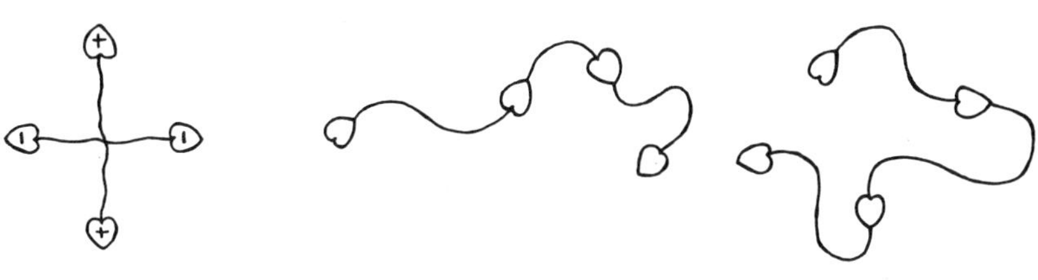

Fig. 5.14

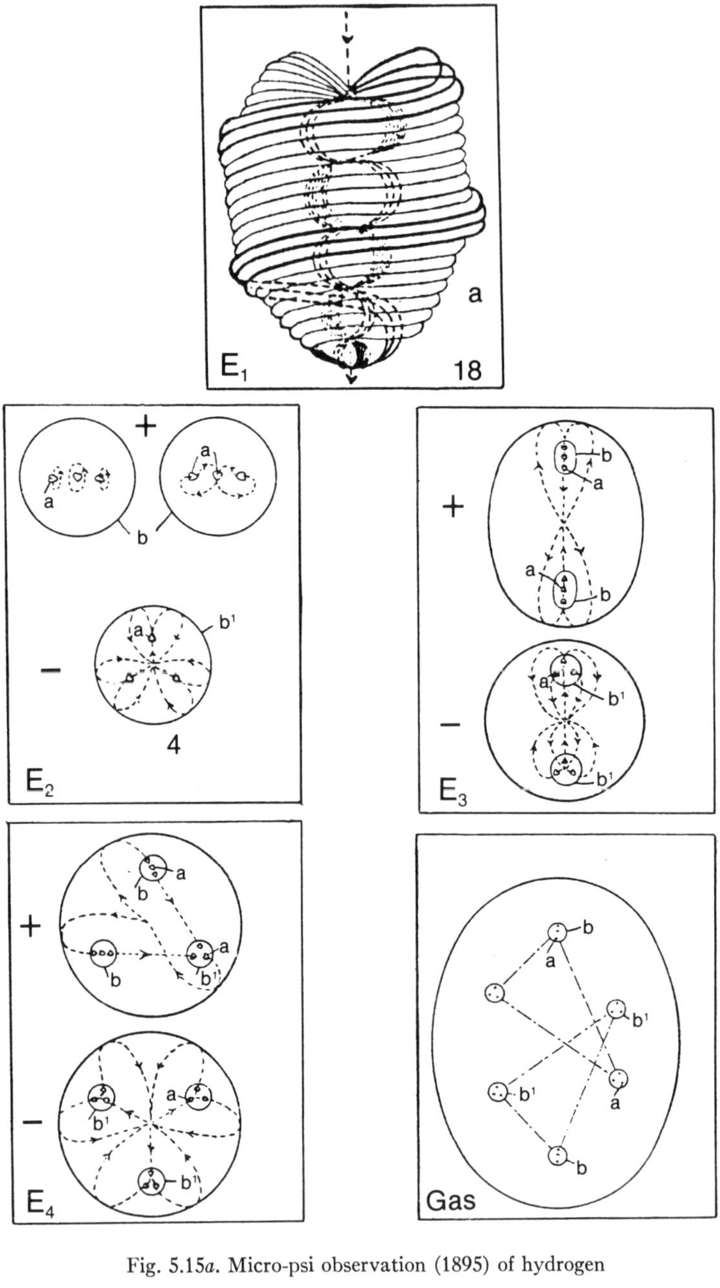

Fig. 5.15*a*. Micro-psi observation (1895) of hydrogen

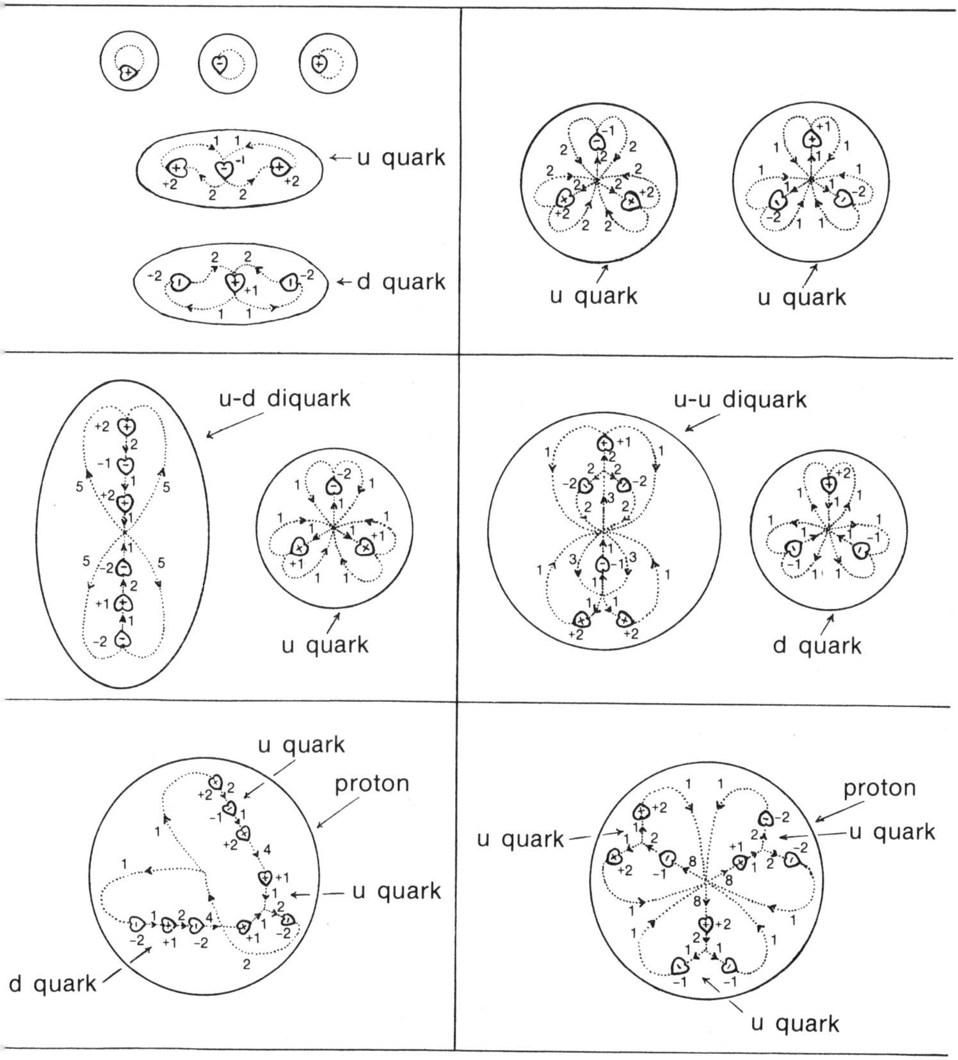

Fig. 5.15*b*. Configuration of gluon flux lines inside proton

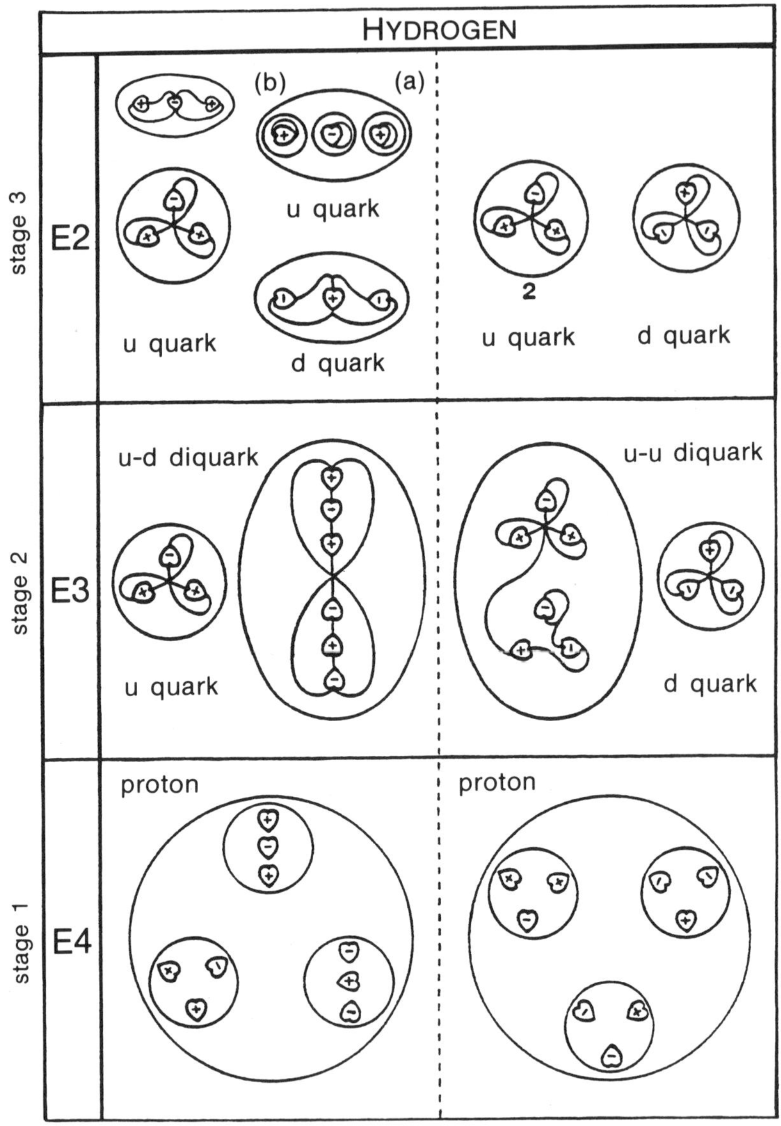

Fig. 5.16. Disintegration of the hydrogen M.P.A.

break up into free omegons (stage 4). Why the M.P.A. should contain two protons is explained in chapter 7.

The investigators claimed that they could distinguish between what they called "positive" and "negative" groups. They portrayed this distinction in diagrams by means of the following convention: "Speaking generally, positive groups are marked by the points of Anu being turned outward and negative groups by the points being turned inward towards each other and the centre of the group."[20] They did not elucidate what they meant by "positive" and "negative" but remarked only that "combinations of three or more Anu are positive, negative or neutral, according to the internal molecular arrangement; the neutral are relatively stable, the positive and negative are continually in search of their respective opposites, with a view to establishing a relatively permanent union."[21] The following consideration shows that the terms signify the sign of the total electric charge of a group of U.P.A.'s: the Hydrogen Triangle in figure 5.16 with four (+) and five (−) U.P.A.'s has two triplets with their U.P.A.'s pointing outward even though one has two (+) U.P.A.'s and one (−) U.P.A. and the other has two (−) U.P.A.'s and one (+) U.P.A. A "positive" triplet, therefore, cannot be one that has more (+) U.P.A.'s than (−) U.P.A.'s. But, if the terms refer to electric charge, then the Hydrogen Triangle should contain two positively charged triplets and one negatively charged triplet. The reader is reminded that the Quark Model predicts that a proton consists of two positively charged u quarks and one negatively charged d quark. Figure 5.16 is consistent with two features of the Quark Model: it shows the presence of three bodies in what has been identified in chapter 4 as a proton —a Hydrogen Triangle. It also indicates that two bodies are "positive" and one is "negative."

Unfortunately, similar considerations cannot be applied to the other Hydrogen Triangle because diagram *a* in figure 5.16 does not indicate clearly whether the linear triplet is positive or negative. But since, for this Hydrogen Triangle, the triangular triplet with its U.P.A. pointing outward must be a u quark and the linear triplet with its outside U.P.A. pointing inward must be a d quark, it may be inferred that *a* is a u quark, or, more accurately speaking, a u quark depicted *after* its disintegration into free omegons. It would appear that the diagram is one showing a linear Hydrogen Triplet (*b*) that was accidentally broken up by the investigators while they observed it. This is because, according to the investigators' own testimony on the appearance of free U.P.A.'s (given shortly), the U.P.A.'s are depicted in diagram *a* as being in the free state.

Figure 5.15*b* shows two different string configurations for the proton. In one of them, three lines of force connect each triplet to the centre of their orbital motion. It is interpreted as follows: quarks are bound to one another in baryons by SU(9) Nielsen-Olesen vortices extending between their constituent omegons. The total flux emanating from the junction of the nine vortices is 0 (mod 9). Quarks are clusters of three SU(9) magnetic monopoles that are internally bound by Y-shaped SU(3) vortices carrying either one or two flux quanta. The flux emanating from the junction of a Y-shaped string is 0 (mod 3). Note that the trigonal symmetry of the sets of three SU(9) strings is identical with the symmetry of arrangement of the SU(3) strings between omegons in a free quark. This is because each quark contributes a net flux of three quanta to the junction of the nine SU(9) strings binding omegons together in a baryon, while each omegon contributes a net flux of one quantum to the junction of the nine SU(3) strings binding them in a free quark. The internal Y-shaped string structure of a bound quark is identical with current models of baryons as three SU(3) quark monopoles bound by Y-shaped strings.

It must be emphasized here that all drawings of groups of particles have only qualitative significance and are not even approximately correct in scale. This is because the micro-psi observer employs widely varying powers of magnification in order to discern and to examine objects of different size. Hence, he finds accurate comparison and depiction of their relative sizes to be quite impossible tasks: "It should be specifically noted that the diagrams are not drawn to scale, as such drawings would be impossible in the given space. The dot representing the Anu is enormously too large compared with the enclosures, which are absurdly too small; a scale drawing would mean an almost invisible dot on a sheet of many yards square."[22] Also: "It must be remembered that the bodies shown diagrammatically in no way indicate relative size; as a body is raised from one substate to the one immediately above it, it is enormously magnified for the purpose of investigation."[23] Still, Hydrogen Triangles were described as very large compared with their constituents, a feature that is consistent with experimental evidence for relatively small or pointlike quarks inside protons.[24]

DISINTEGRATION OF GROUPS OF U.P.A.'S

Besant and Leadbeater claimed not only that they could observe atoms but that they could also disintegrate them into their constituent bodies, and these in turn into smaller groups, until everything was finally broken up into free U.P.A.'s (or at least they could manipulate micro-psi images in this way). They stated: "The first thing that happens on removing a gaseous atom from its 'hole' or encircling 'wall' is that the contained bodies are set free, and, evidently released from tremendous pressure, assume spherical or ovoid forms, the Anu within each re-arranging themselves, more or less, within the new 'hole' or 'wall.' "[25] An example of this which is important theoretically is the disintegration of Hydrogen Triplets. These are very often described as being capable of disintegration into duads of U.P.A.'s and free U.P.A.'s (fig. 5.17). The most interesting feature to note is that "positive" triplets with their U.P.A.'s shown pointing outward always form "positive" duads and that "negative" triplets with inwardly pointing U.P.A.'s always break up into "negative" duads and a free U.P.A. This implies, according to the interpretation made earlier, that a positively charged triplet contains at least two U.P.A.'s that are positively charged and that a negatively charged triplet has at least two U.P.A.'s that are

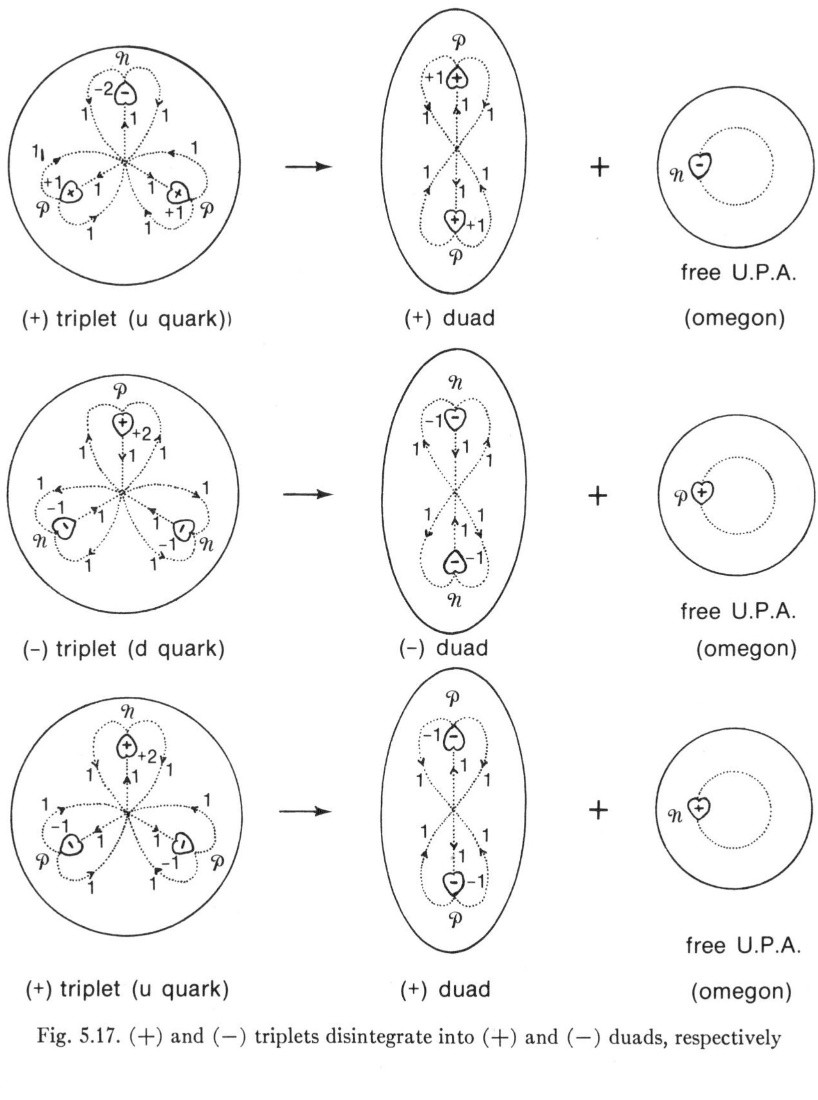

Fig. 5.17. (+) and (−) triplets disintegrate into (+) and (−) duads, respectively

negatively charged, as a pair. This conclusion agrees with the charge signs of omegons in a u and in a d quark:

$$\text{u: } (5/9, 5/9, -4/9)\ ; \quad \text{d: } (-4/9, -4/9, 5/9)\ .$$

Many examples of group disintegration will be given in chapter 7.

Another interesting feature exhibited by nearly all groups is that they are described as enclosed in what the investigators called a "hole" in space. This was referred to in the last quotation. It will be discussed further in chapter 7, but the conclusions of that discussion are summarized here: a "hole" containing a cluster of N U.P.A.'s is a vacuum excitation consisting of a finite-sized, inhomogeneous Higgs vacuum domain with SU(N) Nielsen-Olesen vortices. The enclosing surface (or encircling "wall") is the phase boundary between it and the circumambient SU(N') Higgs vacuum ($N \neq N' \leq 9$). Groups with different numbers of U.P.A.'s are embedded in different vacua that co-exist in M.P.A.'s as bag-like objects trapping the omegon magnetic monopoles within them. Figure 5.6 shows examples of these "holes" with their "walls" clearly marked.

How a human being with his micro-psi vision focused on particles can break them up into smaller units is a question that will not be discussed here. Since omegons are probably permanently confined by their strings (this has not yet been rigorously proved), the dissociation cannot be achieved in a mechanical way. Instead, the principle probably at work is this: through the causal link established by the observer between himself and the object he examines, he induces in an SU(N) Higgs vacuum domain a phase transition to an SU(n) vacuum, where $n < N$. This nullifies all string bonds between monopoles *momentarily*, so that they separate. Then, owing to the formation of new strings between them, they reassemble in just those bound states that are allowed by their magnetic charge. For example, a quark is separated from a proton by destroying the SU(9) vortices linking it to other quarks, rather than by mechanically breaking its bonds (which may be impossible). A quark made up of magnetic charges of 1, 1, and -2 embedded in an SU(3) Higgs field domain can be broken up into a free omegon of charge -2 and a diomegon, consisting of two monopoles with charge 1 (mod 2) bound by vortices in an SU(2) Higgs vacuum (see the first example in figure 5.17). Dissociation of a duad of U.P.A.'s into two free U.P.A.'s is achieved by the observer inducing a phase transition of the SU(2) vacuum domain into an Abelian superconducting vacuum, in which omegons lose their colour-shades and cease to function as non-Abelian monopoles. The spherical "hole" that encloses a free U.P.A. (see fig. 5.17) is a U(1) Higgs vacuum domain, its "walls" being the phase boundary between this domain and the ambient SU(9) Higgs field. When in such a free state, an omegon is colour-shadeless and emits no flux, since it no longer has a magnetic charge. As a result, the String Model idealization of a free, isolated magnetic monopole as the end-point of an infinitely long vortex is not apparent to the observer, who describes, instead, a free U.P.A. as shown in figure 5.17. The dipolar magnetic field of the spinning, electrically charged omegon is expelled from the superconducting vacuum, in which it is embedded, and squeezed into a loop.

It should be emphasized that the observer has to restrain the natural motion of the particles in his field of vision sufficiently to be able to describe them. This restraint is imposed during active observation at all stages of disintegration of groups of particles. But its effect is not one of distorting the natural bound-state configurations of strings and monopoles. Micro-psi faithfully reproduces these, although the diagrams are not necessarily accurate in terms of scale and relative sizes.

In conclusion, micro-psi observations provide indisputable support for the String Model. They also indicate that quarks are clusters of three magnetic monopoles. Further observational support is discussed in the next chapter.

REFERENCES

1. P. A. M. Dirac, Proc. Roy. Soc. (London), **A133,** 60 (1931); Phys. Rev., **74,** 817 (1948).
2. H. Bradner and W. M. Isbell, Phys. Rev., **114,** 603 (1959); E. M. Purcell et al., Phys. Rev., **129,** 2326 (1963); E. Amaldi et al., Nuovo Cimento, **28,** 773 (1970); I. I. Gurevich et al., Phys. Letters, **31B,** 394 (1970); R. A. Carrigan et al., Phys. Rev., **D8,** 3717 (1973).
3. W. V. R. Malkus, Phys. Rev., **83,** 899 (1951); W. C. Carithers et al., Phys. Rev., **149,** 1070 (1966); M. A. Ruderman and D. Zwanziger, Phys. Rev. Letters, **22,** 146 (1969); W. Z. Osborne, Phys. Rev. Letters, **24,** 1441 (1970).
4. H. H. Kolm, Sc. J., **4,** no. 9 (1968); H. H. Kolm et al., Phys. Rev., **D4,** 1285 (1971).
5. P. H. Eberhard et al., Phys. Rev., **D4,** 3260 (1971).
6. C. M. Stevens et al., Phys. Rev. Letters, **17,** 60 (1966); Phys. Rev., **D14,** 716 (1976).
7. G. S. La Rue et al., Phys. Rev. Letters, **38,** 1011 (1977); **42,** 142 (1979).
8. J. P. Schiffer et al., Phys. Rev., **D17,** 2241 (1978).
9. J. Schwinger, Phys. Rev., **173,** 1536 (1968); Science, **165,** 757 (1969).
10. H. B. Nielsen and P. Olesen, Nucl. Phys., **B61,** 45 (1973).
11. Y. Nambu, Phys. Rev., **D10,** 4262 (1974).
12. H. W. Wyld and R. T. Cutler, Phys. Rev., **D14,** 1648 (1976).
13. X. Artu, Nucl. Phys., **B85,** 442 (1975).
14. S. Mandelstaam, Phys. Letters, **53B,** 478 (1975).
15. Occult Chemistry, 3d ed., p. 14.
16. They appear on pp. 25 and 96 of Occult Chemistry, 2d ed.
17. Occult Chemistry, 3d ed., p. 24.
18. It appears on p. 96 of Occult Chemistry, 2d ed.
19. Lucifer, November 1895 (London: Theosophical Publishing House).

20. Occult Chemistry, 3d ed., p. 26.
21. Ibid.
22. Ibid., p. 36.
23. Ibid., p. 10.
24. E. D. Bloom et al., in: High Energy Physics, Proceedings of the 15th International Conference on High Energy Physics, Kiev, 1970, edited by V. Shelest (Kiev, U.S.S.R.: Navkova Dumko, 1972).
25. Occult Chemistry, 3d ed., p. 10.

CHAPTER 6

The Testimony of the Micro-Psi Observer

There is a kind of circular motion. Oh yes, there is, a circular motion along the little bit of the beam which is within my sight.

Observation by Geoffrey Hodson of electron Zitterbewegung in a cathode-ray tube, December 7, 1959

This chapter presents the micro-psi observer's own account of what he sees when he directs his magnifying vision to the atomic world. It is intended, first, to illustrate the vivid, dynamic quality of micro-psi imagery; second, to show how apt his descriptions are in relation to the String Model of elementary particle physics; and, third, to show that his testimony is consistent with his claim to be able to magnify objects of atomic size or less.

ZITTERBEWEGUNG

The observation of this phenomenon demonstrates the true magnifying power of micro-psi. In 1930 the physicist Schrödinger[1] pointed out that the Dirac theory of a free electron implies that, superimposed on the observable linear motion of an electron, there is a circular motion about the direction of its spin with a radius equal to its (reduced) Compton wavelength: $\lambda\!\!\!^{-}_C = \hbar/mc = 3.86 \times 10^{-11}$ cm. This means that, while the average velocity of an electron is less than c (the speed of light), its instantaneous velocity is always $\pm c$. This highly oscillatory motion was called *Zitterbewegung* by Schrödinger. It has a frequency given by

$$\omega = 2E/\hbar \,,$$

where

$$E = +(c^2p^2 + m^2c^4)^{1/2}$$

is the energy of the free electron. Therefore, $\omega \geq 2c/\lambda\!\!\!^{-}_C \simeq 10^{21}$ per second, the lower limit being for non-relativistic motion ($E \simeq mc^2$). The resulting trajectory of the electron is a helix of radius $\lambda\!\!\!^{-}_C$ (fig. 6.1), instead of the straight line predicted by classical mechanics. The attempt to localize an electron in space and limit the direction of its spin introduces negative-energy Fourier components into its wavepacket, and these interfere with the positive-energy plane-wave components. The effect is that each plane wave contributes a circular motion to the electron of radius λ_C and frequency ω in a plane perpendicular to its direction of spin, and only the total contribution of all the plane waves vanishes for a spherically symmetric wave-packet. Although not directly detectable experimentally, Zitterbewegung produces a current that can be shown[2] to account for the Dirac magnetic moment of the electron.

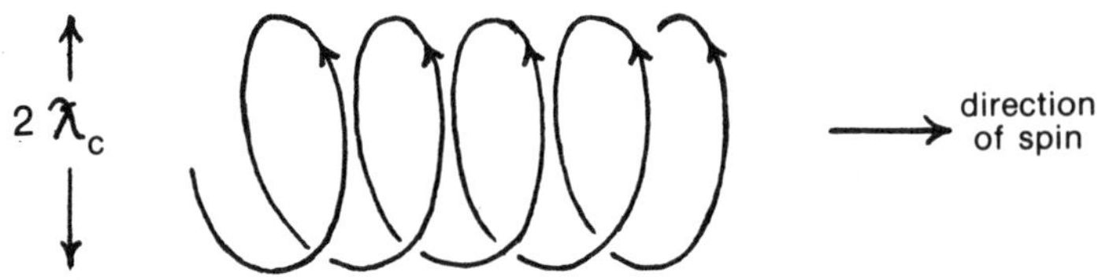

Fig. 6.1

Below are listed comments made by a micro-psi observer[3] during his examination of an electron beam in a cathode-ray tube:

1. "There is a kind of circular motion. Oh yes, there is, a circular motion along the little bit of the beam which is within my sight. I can't get away from that. It's just as if the pencil of beam continued its steady condition but there is associated with it a distinct thickening and thinning going on and it's a wavelike condition."
2. "Oh yes, there's that outside circular movement, helical or spiral movement again."
3. "Yes, again I have to record this circular movement. Round the beam. It's very noticeable. It's very like a series of wave-crests."
4. "It is spiral and it does consist of an enlargement which is moving like an eel around it."
5. "It consists of a flow, a spiral flow along the beam."
6. "Oh, it is so marked, this helical movement, it's almost like a writhing that goes on, quite independent of the pencil which isn't affected by it."
7. "It's very like a series of wave-crests. It consists of particles in a . . . , in a . . . , orbital but not in themselves—round, round the pencil."
8. "I can see the brilliantly lighted pencil of rushing particles going along as the central beam, very well. That's there. But I'm focusing, trying to get a clear image of this other axial phenomenon. Radial phenomenon. Oh yes, it's particles. . . . Oh it is particles . . . going in a circular movement to produce this illusion, as if there were. . . . You could say as if you had a wire wound round with another wire wound round it. And this second wire round it consists of rapidly moving, circularly moving particles. But they don't go along with the wire, the big wire. They go on going round and round and round, as if the wire, the main one went through it."

On the same occasion, electrons themselves were apparently caught sight of in the vacuum of the cathode-ray tube: "It's intermittent. And they're elongated particles. Oh, I got one up by accident, they're spinning. It's a spinning particle. Definitely spinning particle."[4] On another occasion, during examination of a 5 μA D.C. current in a copper wire, streams of lighted points were observed to "move in a spiral fashion from disc to disc or crest to crest of pulsed wave."[5] Finally, another observation of an electron: during examination of U.P.A.'s in a graphite rod through which an electric current passed, a micro-psi observer noticed that a rushing stream of lighted points passed by them. They were much smaller than U.P.A.'s, and some of them appeared to be scattered or deflected during their travel down the rod. This suggests that they were conduction electrons being scattered by carbon atoms as they travelled through the lattice. On close examination of one of them, it was remarked: "I don't say it's an anu, but it looks, it's got a double spiral movement in it. It's got a—I can't give you any relative sizes—it has got a double spiral movement in it, it's got a—perhaps I had better describe it. The first impression was like the, somewhat like the sweet chestnut with its cover round it. That is to say, it hasn't got a smooth surface, it's radiating itself, it is radiating rays and lines of force all round itself and these spike out for a distance of about one, two, three—about a sixth or eighth of the diameter, sideways diameter, of the object. It's—er—they're simply shooting forces out from within themselves all around the sides in addition to their obviously spirally, winding, writhing movement, very noticeable which I've never seen quite to the same extent before, perhaps because I haven't been able to observe it, is this spiked effect. But don't think of anything like a chestnut covering, they are very much closer than that. It's an extremely fine radiation that is going off all round. It's a very strong one too. It gives the impression of rigidity. I've got one. Well, it is shaped like an anu."[6] An electron has the same appearance as a U.P.A., although it looks much smaller. Since experiments have established that the electron behaves in its electromagnetic interaction with other charged particles as a structureless, pointlike particle at least down to distances approaching 10^{-15} cm, this description of an electron with electric lines of force emanating from it indicates that micro-psi can discern objects smaller than 10^{-15} cm.

U.P.A.'S AS SPINNING MAGNETIC MONOPOLES

The two chiral forms of a U.P.A. are shown in figure 6.2. They are mirror images of each other. The ten whorls or currents flow in parallel, spiral-shaped, closed, continuous curves. Three of these appear much "brighter" and "thicker" to the micro-psi observer than the other seven. Of the former,

Annie Besant and C. W. Leadbeater remarked: "In the three whorls flow currents of different electricities."[7] Noting that the constituents of hadrons strongly interact by exchanging gluons, thus changing their "colour," according to the theory of quantum chromodynamics and to its generalization in the Omegon Model, the following statement is curious: "Force pours into the heart-shaped depression at the top of the Anu, and issues from the point, and is changed in character by its passage; further, force

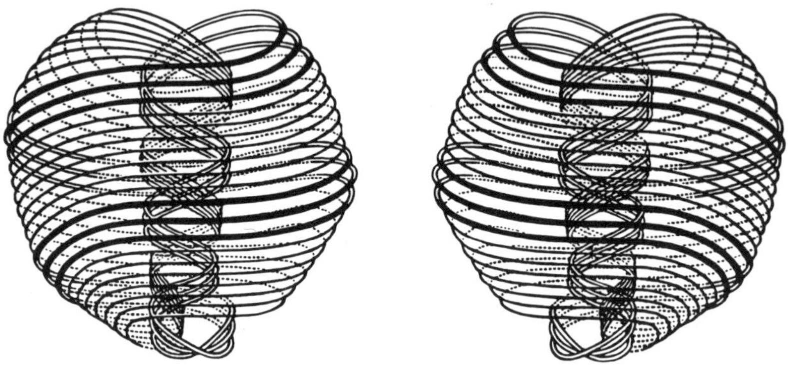

Fig. 6.2. The two chiral types of U.P.A.'s

U.P.A.'s	OMEGONS
1. (+) variety is one "from which force comes out"; (−) variety is one "through which it disappears."	1. Positive and negative magnetic monopoles are, respectively, sources and sinks of flux lines.
2. "Force pours into the heart-shaped depression at the top of the Anu, and issues from the point."	2. Magnetic monopoles can be joint endpoints of two or more strings.
3. Are bound to one another by a "very thin line of lighted force."	3. Are confined by flux lines in vortices of the Higgs field.
4. "The changing shades of colour that flash out from the rapidly revolving and vibrating Anu depend on the several activities of the spirals; . . . with the change of activity from one spiral to another the colour changes."	4. Change their nine colour-shades by emitting or by absorbing gluons.*
5. "It turns incessantly upon its own axis, spinning like a top."	5. Are spin-1/2 fermions.
6. "An electric current brought to bear upon the Anu checks their proper motions, i.e., renders them slower; the Anu exposed to it arrange themselves in parallel lines."	6. As Dirac magnetic monopoles, they possess an electric dipole moment that is orientable in an external electric field.
7. Bound groups are surrounded by a "sphere-wall"; "its 'wall' is the pressed back 'space.' "	7. Are trapped in "bags," "bubbles," or "domains" of the Higgs field superfluid of hadronic vacuum.

* No literal interpretation of "shades of colour" in terms of omegon colour-shade states is implied.

rushes through every spiral and every spirilla,* and the *changing shades of colour* that flash out from the rapidly revolving and vibrating Anu depend on the several activities of the spirals; sometimes one, sometimes another is thrown into more energetic action, and with the change of activity from one spiral to another the *colour changes*."[8] Is this an observation of the dynamical changes accompanying the random absorption or emission of gluons by omegons? One should also ask whether it is coincidental that the number of whorls in a U.P.A. is equal to the number of flavour states of omegons. If not, then each whorl may be the dynamical signature of a potential flavour state.

The reader should note that figure 6.2, which is a copy, made especially for the third edition of *Occult Chemistry*, of a diagram appearing in earlier editions, displays an error of draughtmanship. It shows the two outer, thicker whorls crossing over into each other at the bottom of the U.P.A. This implies that nine whorls constitute the U.P.A., not ten, as Besant and Leadbeater reported. Actually, the three thicker whorls do cross over one another, but they follow separate, closed paths in space (see fig. 5.15*a*—the original diagram). Figure 6.2 was made after the deaths of the investigators and so was not checked.

Besant and Leadbeater reported that U.P.A.'s possessed three intrinsic motions: "The Anu has—as observed so far—three proper motions, i.e., motions of its own, independent of any imposed upon it from outside. It turns incessantly upon its own axis, spinning like a top; it describes a small circle with its axis, as though the axis of the spinning top moved in a small circle; it has a regular pulsation, a contraction and expansion, like the pulsation of the heart."[9] The properties of spin and spin precession were, of course, not known by science to belong to atomic particles at the time (1908) when this observation was first published. Another observer described the spin and pulsation of the U.P.A. thus:

1. "There it is, it was vertical, the one I've got now is on its side. That's interesting. If I look at it, one end of it, yes, the enlarged end. Yes, there it is, it's coming up clearly. Yes, it's the anu. I've got onto one anu. . . . It's on its side as it happens, it's made of spirallae, . . . and if I go to the large end and look down, I can . . . the impression of gyration is clockwise in this particular one. The spinning is immense, the shape isn't maintained all the time. It swells and it goes down, like a centrifugal force makes it swell out, then something else makes it swell in. It's pulsing. It's . . . it's lighted, very brilliantly lighted."[10]
2. "The anu is definitely lighted and I can discern degrees of luminosity according to the spirallae of which it is composed and the . . . I don't

* Each of the ten whorls consists of 1,680 coils or "spirillae" of a closed helix.

see much colour except perhaps yellow, very, very pale, but it's lighted anyway, it looks like a lighted cage spinning very rapidly, egg-shaped or heart-shaped, too flat for egg-shaped—heart-shaped. Now I perceive that it is spinning very rapidly, and the particular one that I am looking at now is spinning. Supposing I were looking down at the top, from the top into the heart, if it were vertical it would be spinning this way, clockwise, this particular one, very rapidly indeed, almost to produce a blur."[11]

3. "It is pulsing in . . . like a heart beat but vertically as well as . . . what's the word, generally. It expands and contracts, that's one pulse, and of course at the same tempo it slightly stretches itself up and down so that there is a change in shape which rather confused me because I happened first of all to get it in an elongated phase of pulse. It looked more like—what—I must be careful—like a carrot."[12]

The use of this vivid simile here is remarkable, for the same one was used, eleven years later, by another micro-psi observer who had no knowledge of the earlier observer's work and who never communicated with him: "The anu appears more like a carrot in form than a complete sphere, the 'top end' alone appearing more spherical."[13]

Spin precession of a U.P.A. about the direction of an external magnetic field was witnessed once during micro-psi examination of the space between the plates of a tuning condenser. When a magnet was brought near, the following effect was noticed: "Well, the vertical, spinning phenomenon remains. Oh, those are . . . that's very peculiar, the axes are rocking. It's like a top which is wobbling." When the magnet was withdrawn, it was noted: "There they are, they . . . no, they are steady in their spin." Then, when it was brought near again: "Yes, it makes the axis wobble."[14] Such an observation indicates that a U.P.A. has a magnetic dipole moment—a property that is implied by its identification as an omegon (Hypothesis 1, stated in chap. 4).

U.P.A.'s should have electric dipole moments if they are Dirac magnetic monopoles. This is indicated by the following statement: "An electric current brought to bear upon the Anu checks their proper motion, i.e., renders them slower; the Anu exposed to it arrange themselves in parallel lines, and in each line the heart-shaped depression receives the flow, which passes out through the apex into the depression of the next, and so on. The Anu always set themselves to the current"[15] (see fig. 6.3). This alignment of the spin axis of a U.P.A. parallel to an external electric field is consistent with its having an electric dipole moment.

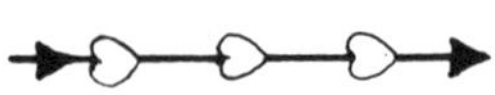

Fig. 6.3

GLUON CLOUDS AND VORTICES

According to the String Model, a meson is a monopole-antimonopole pair bound by a bundle of flux lines that thread a closed Nielsen-Olesen vortex in an SU(3) superconducting Higgs vacuum. Far from the dipole, the vacuum is homogeneous, that is, the vacuum expectation value of the Higgs field is space-time independent and is equal to its value at infinity. Closer to the dipole, the vacuum becomes inhomogeneous, with the density of the Higgs field decreasing steadily until it becomes zero at the centre of the vortex. Along this nodal line of the Higgs field, the colour SU(3) gauge symmetry, spontaneously broken by the non-zero vacuum expectation value of the field, is restored. The magnetic flux between the monopole and antimonopole is channelled into a flux tube whose width is given by the "coherence length" ξ_0. The flux also penetrates the ambient vacuum to an extent measured by the "penetration depth" Λ, which is, therefore, a measure of the width of the massive gluon cloud enveloping the flux tube, that is, $\Lambda \simeq m_V^{-1}$, where m_V is the gluon mass. Such a finite string (fig. 6.4)

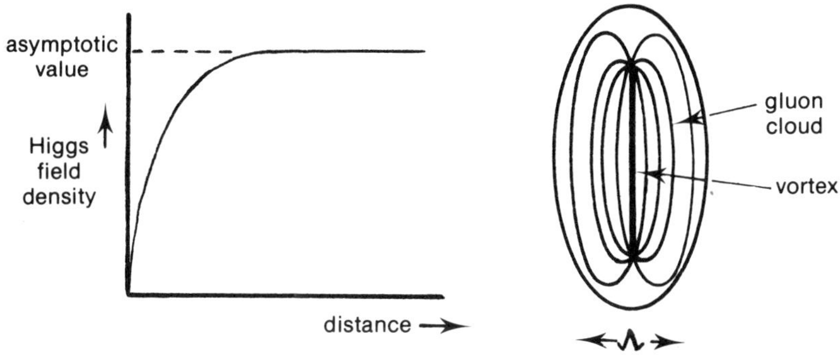

Fig. 6.4

is a localized vacuum excitation in which a domain of inhomogeneous vacuum is surrounded by the homogeneous, superconducting vacuum from which flux is expelled and confined in bundles of width m_V^{-1}. For SU(2) strings, the Higgs field increases in density to its asymptotic value more rapidly, the stronger the coupling between the magnetic charges.[16] For a wide range of coupling constants, the vacuum becomes homogeneous at about a distance Λ from the core of the flux tube. Most of the flux lines are bundled together within this distance.

The micro-psi picture of pairs of U.P.A.'s bears a striking similarity to the String Model account. Some were depicted as in figure 6.5. Joined by one or two "lines of force," the U.P.A.'s were described as being enclosed in a "hole" with an ovoid "wall." The surface of the "wall" appears to micro-psi vision to be transparent and under tension, as if repelled by the forces emanating from the U.P.A. in a way similar to that of air trapped in a bubble in a liquid, the surrounding fluid being held back by the pressure of the air: "A sphere-wall is a temporary effect, caused by one or more Anu in rotation. Just as a stream of air under pressure will make a hole on the surface of water, by pushing back that water, so it is with the groups. As they revolve, the force of their motion drives back the circumambient medium."[17] Of the "wall," the investigators remarked: "The wall belongs to space, not to the atom."[18] It was suggested earlier that a "wall" is a boundary separating different Higgs vacua. In the case of a duad of U.P.A.'s, an SU(2) Higgs vacuum domain enclosing a pair of omegon magnetic monopoles with charge ± 1 is surrounded by the normal vacuum.

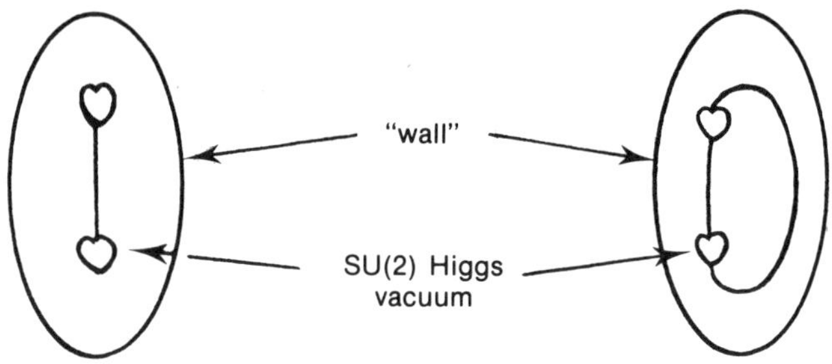

Fig. 6.5

The following observations of "force fields" are interpretable in terms of the clouds of gluons that envelope the strings between monopoles:

1. "It will be seen that the Anu is a sphere, slightly flattened, and there is a depression where the force flows in, causing a heart-like form. Each is surrounded by a field."[19]
2. "Though there were no final conclusions on the matter, it appeared to the investigator as if the sphere-wall was composed of forces radiating from the centre, which after travelling a certain distance, returned to the centre."[20]
3. "The groups show all kinds of possible combinations; the combinations spin, turn head over heels and gyrate in endless ways. Each aggregate is surrounded with an apparent cell-wall, a circle or oval, due to the pressure on the surrounding matter caused by its whirling motion. The surrounding fields strike on each other, and the groups, and rebound, dart hither and thither, for reasons we have not distinguished."[21]

In the second edition of *Occult Chemistry*, there is added the footnote: "That is, the surrounding magnetic fields strike on each other."[22]

The next quotation refers to the magnetic character of U.P.A.'s: "I am looking at one of the major ones* at the moment. It has its own magnetic

* A reference to one of the three "thicker" whorls in a U.P.A.

field, that is to say, that in addition to the force going down this spiral there is a dance or play of energy outside of it, too. It's like a magnet."[23] The "string" character of the lines of force binding U.P.A.'s together is evident from the following remarks:

1. "These particles which looked at first as if they were all jammed together are now seen to be apart. I'm inside and they're connected with each other by a very thin line of lighted force."[24]
2. "All of them are in their own individual rotatory movement which is very rapid. There is a certain extraordinary rigidity in spite of the fact that the whole is composed only of spinning vortices with their lines of inter-connecting force."[25]

Though not specifically examined, these "lines of inter-connecting force" were not reported to have any noticeable thickness. This implies

$$\xi_0 \ll \Lambda\,,$$

which is the condition for a type 2 superconductor. This is just the state of the Higgs field that confines monopoles. The next observation appears to be one of this field: "There is a transparent and to me invisible fluid, along with which they are carried and it is this fluid which appears to have electromagnetic or magnetising qualities which cause the anu affected by it to lock."[26] In this final observation, the circular current of Higgs particles constituting a Nielsen-Olesen vortex seems to have been noticed: "The sight I have of these objects* is, I think, improved from the earlier observations. They're surrounded by a field of spinning particles going round them. The one I've got hold of is like a spinning top—the old-fashioned spinning top, but imagine that with (spinning rapidly) a mist or field round it of at least half its own dimension, of particles spinning in the same direction much smaller than itself. The Anu itself is not only the heart-shaped corrugated form that I have described, it is the centre of a great deal of energy and activity outside it and within it. Outside it, as I have said, there's this rushing flood of particles, the corrugations themselves are alive with energy and some of it is escaping—not all of it, but some of it, and this gives it a tremendously dynamic look. Inside, it's almost like a furnace, it is like a furnace (I don't mean in heat) of boiling activity—organised by the bye, yes, in some form of spiral fashion admittedly, but there's a great deal of activity of free, minuter particles."[27] These particles described here as circulating around a U.P.A. are Higgs mesons whose collective motion constitutes the Nielsen-Olesen vortex or string, the end-points of which are the U.P.A.'s themselves.

* The "objects" were U.P.A.'s.

To summarize: firsthand accounts by micro-psi observers describe U.P.A.'s as "spinning like a top," as having a "regular pulsation" and "changing shades of colour," and as "wobbling" when external magnetic fields are switched on. They "arrange themselves in parallel lines" in external electric fields, have "surrounding magnetic fields," and are connected by a "very thin line of lighted force." These and other remarks, made either by Besant and Leadbeater during the first decade of the twentieth century or by Geoffrey Hodson during the years 1958–59, bear a striking resemblance to current ideas (some only a few years old) about spinning, coloured magnetic monopoles that are confined by strings or bundles of colour flux lines, squeezed into a narrow tube by the ambient, superconducting vacuum. The aptness of their vivid description is a strong argument for the objectivity of micro-psi observation.

REFERENCES

1. E. Schrödinger, Sitzber. Preuss. Akad. Wiss., Physik.-math. Kl., **24,** 418 (1930).
2. K. Huang, Am. J. Phys., **20,** 479 (1952).
3. From unpublished transcripts of investigations by Dr. D. D. Lyness and Geoffrey Hodson (hereinafter G. H.), December 7, 1959; recordings of selected sessions available on cassette tapes.
4. Ibid.
5. G. H., December 2, 1959.
6. G. H., February 11, 1959.
7. Occult Chemistry, 3d ed., p. 14.
8. Ibid.
9. Ibid.
10. G. H., October 27, 1959.
11. G. H., April 26, 1958.
12. G. H., January 25, 1959.
13. Mrs. R. Watson, in letter to Dr. M. G. Hocking, January 14, 1970.
14. G. H., December 15, 1959.
15. Occult Chemistry, 3d ed., p. 15.
16. H. W. Wyld and R. T. Cutler, Phys. Rev., **D14,** 1648 (1976).
17. Occult Chemistry, 3d ed., p. 28.
18. Ibid., p. 15.
19. Ibid., p. 13.
20. Ibid., p. 15.
21. Ibid., p. 26.
22. Occult Chemistry, 2d ed., p. 25.
23. G. H., April 26, 1958.
24. G. H., October 22, 1959.
25. G. H., December 9, 1959.
26. G. H., April 27, 1958.
27. G. H., January 26, 1959.

CHAPTER 7

Micro-Psi Atoms

What is now proved was once only imagin'd.
William Blake

INTRODUCTION

A comprehensive, scientific investigation of micro-psi vision should ask and seek the correct answers to three questions: (1) Do images experienced in this altered state of consciousness refer to contemporary physical objects in real space-time? (2) If so, what are these objects that are accessible to scientific examination and measurement? (3) How does the observer gain such scientifically verifiable information about the microcosm? In this book only the first two questions will be considered. For a scientist or layman who cannot accept the metaphysical answers given by Indian yoga, the last question is the most difficult, partly because of the current lack of experimental data concerning the physical effect (if any) of micro-psi observation on matter but also because micro-psi is an altered state of consciousness (albeit one still of the physical world), and it must encompass both neurological and parapsychological issues as well as the formal question of its cognitive character. In view of this, speculation about its modus operandi would be premature. The third question is also irrelevant to the present task, which is to demonstrate that micro-psi provides valid and accurate knowledge of the microcosm, insofar as it strongly supports the Quark, Omegon, and String Models of particle physics. Because its effects on atoms and nuclei are not experimentally known, they can only be inferred from the available reports. Any supposed insight gained by this procedure is, of course, model-dependent, for it comes from an arbitrary theoretical identification of U.P.A.'s. But the arguments and evidence presented in previous chapters make plausible the hypothesis that U.P.A.'s are omegon constituents of quarks, and this has a number of consequences bearing on the nature of the physical effects of the act of micro-psi observation. These will be discussed shortly.

MICRO-PSI AS AN OBJECT-OBSERVER INTERACTION

The micro-psi observers Annie Besant and C. W. Leadbeater claimed that they could sufficiently retard the internal and external modes of motion of atoms and molecules to be able to observe them indefinitely. They also claimed that objects were unaffected in any other way by this intervention: "The object examined, whether an atom or a compound, is seen exactly as it exists normally, that is to say, it is not under any stress caused by an electric or magnetic field. As each object is in rapid motion, the only force brought to bear on it is a special form of will-power, so as to make its movement slow enough to observe the details."[1] In the language of parapsychology, this apparent exertion of mechanical action at a distance would be an example of "psychokinesis" (P.K.). But even if the observed system was not under any stress, it did not follow that it was identical with the system that existed just prior to its examination. It was an arbitrary (though natural) assumption on the part of the investigators that M.P.A.'s and atoms were the same things, for they had nothing with which to compare M.P.A.'s. Many of their reports reveal that they could be more than mere passive spectators. The following quotations vividly illustrate the dynamic interplay that can occur between the observer and the images on which he concentrates:

1. In an apparent synthesis of a single micro-psi molecule (M.P.M.) of methyl alcohol from an M.P.M. of methyl chloride by his replacement of a chlorine M.P.A. by an OH group, Leadbeater remarked: "I stick him on instead of the Chlorine. But when I remove the will from him he does not stay: he pops out. I do not appear to be able to get him to stick together. Leadbeater also noted: "The Oxygen departs as soon as one removes one's will from it."[2]
2. In his examination of the M.P.M. of stannic oxide, Leadbeater said: "If four Hydroxy OH groups are tried, we get $Sn(OH)_4$ but this is unstable and remains so long as the will holds them. If the will is released, SnO_2 is formed and the remaining Oxygen atoms go off with the Hydrogen forming $2H_2O$."[3]
3. Leadbeater commented on the perturbative effect of his ability to slow the motion of the images that he watched: "The molecule is spinning. You have to hold it still and then you have to be careful that you do not spoil its shape. I am always afraid of disturbing the things because I must stop their motion in order to give an idea of them."[4]
4. During his examination of an M.P.M. of carbon dioxide, Leadbeater was told: "Get one of these CO_2 and remove one Oxygen and then see what happens to the other funnels." He replied: "But, see here, you can't

> remove the funnels. The funnels stay behind. You can pull out the Oxygen, but the funnels stay behind and they go and join the rest of the outfit. They go and join the rest and the whole seems to me to break up. I can't hold it together. If I withdraw one Oxygen, the other slips away."[5]

These remarks indicate not that there is some kind of mental manipulation of visual hallucinations but rather that a causal connection can exist between the observer and the object that he cognitively experiences while in his altered state of consciousness. No speculation is offered concerning the nature of this causal link that leads to mutual interaction between the observer and the object under his observation. But the quotations above, taken at their face value, show that micro-psi can disturb the dynamical behaviour and stability of physical objects (whether intentionally or unintentionally). It must be asked: What is the effect of micro-psi on atoms? The clue giving the answer to this all-important question was found in chapter 4, where a statistical correlation was shown to exist between the U.P.A. populations of the M.P.A.'s of elements and the number of omegons in the nucleons constituting two nuclei of the relevant element. This supports Hypothesis 2, which is now stated in full:

HYPOTHESIS 2: *The M.P.A. of an element is a multi-omegon bound system formed from* TWO *nuclei of the element as a result of the ground state of the superconducting Higgs vacuum being perturbed over an atomic-sized region of space by the act of micro-psi observation.*

Before discussing this in more detail, we will summarize certain relevant ideas. Nuclear physics has established that the core or nucleus of an atom is assembled from a distinct particle called the "nucleon." It has two "isospin" states—the electrically charged proton and the slightly more massive neutron, which is neutral. The nucleus, which occupies about 10^{-15} of the volume of the atom, is extremely dense compared with natural or man-made materials, being about 5×10^{12} times as dense as uranium. It is made up of nucleons that are bound together by strong forces whose range extends little beyond the confines of the nucleus. According to the Omegon Model, nucleons consist of nine omegons, bound in three groups of three by string-like excitations of the superconducting Higgs vacuum. The nuclear forces holding the nucleons together arise from the residual coupling between their strings. Experimental high-energy physics seems to indicate that hadrons like the nucleon have constituents that are permanently imprisoned within them. In the Omegon Model, there are two levels of confinement: confinement of colour-shaded omegons in coloured quarks and confinement of quarks in colourless hadrons. Both features should be guaranteed through the Meissner Effect, by which magnetic flux is expelled from the ambient vacuum and channeled in strings that extend between monopoles. These create valence bonds that do not weaken, the farther apart omegons are in quarks and the farther the latter are apart.[6] The 9-fold connectivity of the global gauge group $SU(9)/Z_9$ for the coupling between the nine colour-shade states of omegons implies that the quantum of magnetic flux in vortices is defined modulo 9. This allows nine (and only nine) magnetic monopoles to be bound by vortices in a stable configuration. Saturation of omegon forces for particle states in the SU(9) Higgs vacuum occurs only for quark-antiquark and three-quark bound states. Other multi-quark states do not satisfy the octavalency of omegon string bonds and so cannot exist.

A SPACE-TIME PICTURE OF M.P.A. FORMATION

In chapter 4 "Hydrogen Triangles" were identified as protons. But the investigators rarely reported hydrogen M.P.A.'s or Hydrogen Triangles to be present in the M.P.A.'s of elements. Instead, according to their observations, a large variety of groups of U.P.A.'s are present, among which duads and Hydrogen Triplets are very common. These were identified in chapter 5 as diomegons and quarks, respectively. Unlike nuclei, M.P.A.'s have no single unit from which they are assembled, although the most common particle is the Hydrogen Triplet, often present in multi-quark groups. The general absence of nucleons from M.P.A.'s and the presence, instead, of exotic quark states mean that M.P.A.'s are very different from the nuclear and compound nuclear states (the latter being briefly created in high-energy collisions of nuclei in the physicist's laboratory). The problem of accounting for this difference is related to the more general question of how M.P.A.'s are formed from pairs of nuclei. The existence of exotic states itself provides a clue: in the normal vacuum, omegons possess nine hidden degrees of freedom—their colour-shades. If omegons were embedded in other vacua, some of these degrees of freedom would be suppressed and bound systems of fewer than nine omegons could exist. The processes of formation of M.P.A.'s must also yield other *vacua*.

As a dynamical basis for Hypothesis 2, therefore, it is proposed that, in the initial preparation stage prior to actual micro-psi observation, an effective reduction takes place in the degree of colour-shade symmetry of the vector gluon-mediated interaction between omegons. This lowering of the colour valence of omegons renders nucleons in the nuclei unstable, so that they break up into quarks and omegons. A nucleus is transformed into an amorphous cloud of free particles in chaotic motion. This expands, owing to the net repulsion between the predominantly positively charged groups of particles. As they separate, the monopoles interact once more and new bound states form, the clouds *condensing* into small groups of omegons. But, because the coupling between monopoles embedded in the superconducting Higgs field becomes constant at distances apart much larger than the Compton wavelength of the gluons, the omegons in the separate clouds formed from neighbouring nuclei interact just as strongly as those in the same cloud. Consequently, two such clouds coalesce as they condense and shrink, instead of condensing into separate systems. Omegons in nuclei that do not come under micro-psi observation do not participate in the long-range interaction leading to coalescence because their normal valences are operative in the nuclear state and their power to form string bonds is already at its limit. Once magnetic monopoles are free, the constant long-range force between them leads to binding between monopoles of neighbouring nuclei as well as of the same nucleus, although why only *two* nuclei should be affected (as required by Hypothesis 2) remains problematical. As a result, nuclei that were neighbours before coming under micro-psi observation lose their separate identities and become fused in pairs. Groups aggregate in larger clusters as a result of interaction between their strings and become globally bound in a quasi-nuclear potential that represents their average coupling to all other groups. The original Higgs vacuum with uniform, broken colour SU(9) symmetry becomes a set of SU(N) vacuum domains ($N \leq 9$ is variable), each enclosing a stable group of omegon magnetic monopoles confined in closed string configurations. There is not wholesale recombination of particles into nucleons because these are no longer more stable in the new vacua than other bound states of omegons, so that formation of nucleons is not necessarily the most favoured process. In general, for $m \neq n$, groups of m monopoles and groups of n monopoles cannot be bound by vortex excitations of the same Higgs vacuum. Groups are separated by phase boundaries encircling each vacuum domain. It is proposed that this accounts for the general depiction of groups as enclosed by "walls" (also discussed in chaps. 5 and 6). The investigators remarked: "The wall belongs to space, not to the atom,"[7] that is to say, the surface boundary around a group is not an intrinsic property of the group itself but is a feature, instead, of the vacuum surrounding it: the interfaces separating domains of various vacua are the "walls." Figure 7.1 shows a typical example, present in the M.P.A. of magnesium. SU(2), SU(7), and SU(3) vacuum domains are present, containing a diomegon, a heptad of omegons, and a single colourless quark (one whose constituent omegons behave as SU(3) monopoles instead of as SU(9) monopoles). Numerous other examples

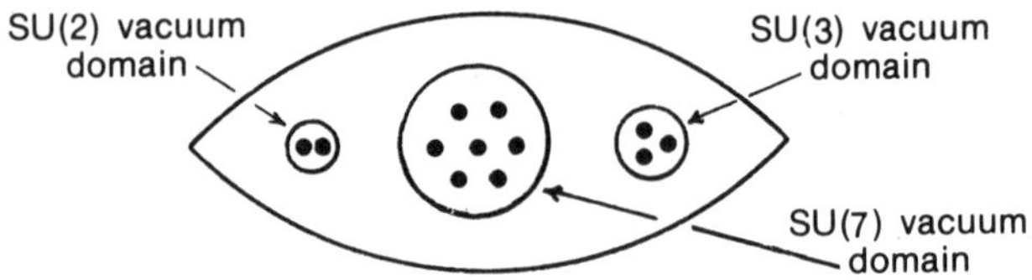

Fig. 7.1

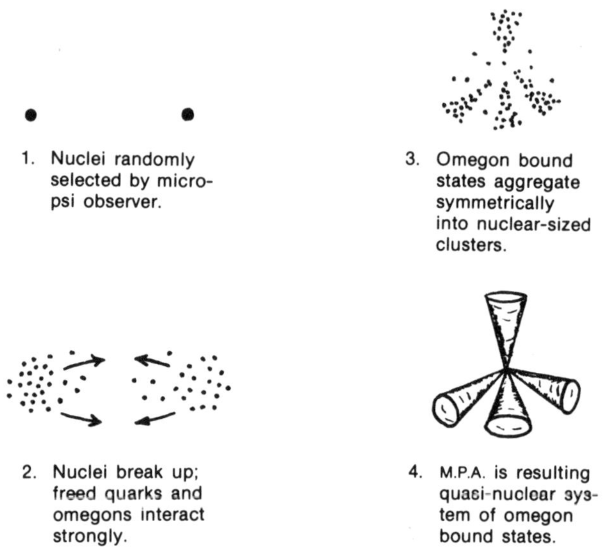

Fig. 7.2

will be met later in this chapter. Figure 7.2 summarizes the above account of the space-time picture of M.P.A. formation.

SHELL STRUCTURE IN M.P.A.'S

M.P.A.'s exhibit an assembly in space of groups of U.P.A.'s that is analogous to the filling up of shells by nucleons in nuclei. First, the similarly shaped M.P.A.'s of a given class may contain the same central core of particles. The Star Group comprising the inert gases typifies this feature, for their M.P.A.'s have the same nucleus of 120 U.P.A.'s (the "Ne120" group; see table 7.1). Second, similarly populated groups may appear several times in an M.P.A., as well as in other M.P.A.'s. This repetition is apparent in the composition of the arms of the M.P.A. of the Star Group and is even more obvious for the Tetrahedron Group elements shown in table 7.2. It is noteworthy here that the common group Ca160 containing 160 U.P.A.'s occurs first for the doubly magic element of calcium, whose Ca^{40} nucleus contains very stable filled shells of twenty protons and twenty

TABLE 7.1

STAR GROUP

Element	Centre	6 Arms of M.P.A.
Neon......	Ne120	6 (Ne22+3Li4+2H3)
Argon......	Ne120	6 (Ne22+N63+Ar14)
Krypton...	Ne120	6 (Ne22+N63+Ar14+mNe15)
Xenon.....	Ne120	6 (Ne22+N63+Ar14+mNe15+2N110+Xe14+Xe15)
Kalon......	Ne120	6 (2Ne22+2N63+2Ar14+2mNe15+2N110+Xe14+Xe15+Ka12)
Radon.....	Ne120	6 (3mNe22+2N63+3Ar14+3mNe15+3N110+Xe14+Xe15+I.7)

NOTE.—Chemical symbols refer to the element in whose M.P.A. the group first appears; numbers denote U.P.A. population of group. Prefix "m" denotes "meta" (isotopic) species of the M.P.A.

TABLE 7.2

TETRAHEDRON GROUP A

Element	Centre	4 Funnels
Beryllium...........	Be4	4 (Be10)
Calcium.............	Ca80	4 (Ca160)
Chromium...........	8N6+8Ad6	4 (Ca160+2Cr25)
Strontium...........	Sr96	4 (2Ca160+2Sr24)
Molybdenum........	N2+Sr96	4 (2Ca160+2Mo46)
Barium.............	I.7+Sr96	4 (2Ca160+2Mo46+Ba33+Li63+Ba80)
Neodymium.........	Ce667	4 (2Ca160+2Mo46+Nd65)
Ytterbium...........	Yb651	4 (3Ca160+2Mo46+Yb48)
Tungsten............	Lu819	4 (3Ca160+2Mo46+Yb48)
Radium.............	Lu819	4 (3Ca160+3Mo46) (also spikes)
Uranium............	Lu819	4 (3Ca160+3Mo46) (also spikes)

NOTE.—Chemical symbols refer to the element in whose M.P.A. the group first appears; numbers denote U.P.A. population of group.

neutrons. These are present in the nuclei of the elements following calcium in table 7.2, but not so in the case of the element preceding it—beryllium. This suggests that such a stable nuclear structure leaves its trace in M.P.A.'s in the form of this common structural unit, which must have a corresponding stability, since it exists in all the heavier elements of Tetrahedron Group A. But the appearance of groups that act as building blocks for the M.P.A.'s in a certain Group is not necessarily determined by the nuclear structures of the elements in the Group. This is because units that are common to one Group may also appear in other Groups. The stability of these must be an intrinsic property of multi-omegon bound states that is unconnected with the nuclear structure of the parent nuclei.

A detailed analysis of the structures of M.P.A.'s reveals that a common group may have variants making them analogous to mirror nuclei. A group in one M.P.A. may be made up of A $\mathcal{P}$ and B $\mathcal{N}$ omegons, while the same unit in the M.P.A. of another element may have A $\mathcal{N}$ and B $\mathcal{P}$ omegons, the $\mathcal{P}$ and $\mathcal{N}$ merely exchanging places. The existence of such variations is consistent with the isospin independence of strong interactions mediated by vector gluon exchange between omegons and with the approximate equality of the masses of the $\mathcal{P}$ and $\mathcal{N}$ omegons.

Except for minor differences that are isotopic in origin, the micro-psi observer reports that all M.P.A.'s of a given element are identical. Therefore, the transformation of nuclei into M.P.A.'s must be determined by their ground-state structures, the same final state always resulting because the "initial conditions" (the nuclear ground state and the atomic environment) are the same for all nuclei of the element in the examined substance. The multiplicity of types of constituents in M.P.A.'s makes nuclear models inappropriate to the understanding of their dynamics. No detailed explanation of their structure will be offered, except for the simplest cases. The task of this chapter will be to demonstrate by detailed analysis that M.P.A.'s are formed from two nuclei, so providing incontestable evidence supporting the thesis of this book that micro-psi descriptions refer to quarks and their constituents. This will be achieved by relating the composition of M.P.A.'s in detail to the omegon/quark structure of nucleons and to the nucleon composition of nuclei. The very high degree of consistency that will be found between the observations and theoretical expectation permits only one reasonable conclusion: the yogic siddhi of micro-psi provides accurate information (when correctly analysed) about the ultimate constituents of matter.

SHAPE OF M.P.A.'S

It was pointed out in chapter 1 that the investigators Besant and Leadbeater found seven basic types of M.P.A.'s: the Spike, Dumb-bell, Tetrahedron, Cube, Octahedron, Bars, and Star Groups. Hydrogen, helium, nitrogen, and oxygen had M.P.A.'s that did not belong to any of these classes. They also discovered that elements with a common valence of either n or $8 - n$ were in the same class, when $n = 1, 2, 3$, or 4. The monovalent alkali metals and the halogens are in the Spike and Dumb-bell Groups; elements with valences of 2 or 6 are in the Tetrahedron Group; the Cube Group contains elements with valences of 3 or 5; the Octahedron Group contains elements with a valence of 4; the Bars Group comprises the three triads of Group VIII of the periodic table; finally, the Star Group is made up of the inert gas elements with zero valence (see table 7.3). This relation between the position of an element in the periodic table and the shape of its M.P.A. served to strengthen the belief of the investigators that M.P.A.'s were atoms. Indeed, they used this relation as a means of confirming the identity of the element corresponding to a given M.P.A., this having been inferred by their comparison of the "number weight" (U.P.A. population/18) of the M.P.A. with published chemical atomic weights. It also enabled

TABLE 7.3—MICRO-PSI CLASSIFICATION OF THE ELEMENTS

SPIKE GROUP	DUMB-BELL GROUP	TETRAHEDRON GROUP		CUBE GROUP		OCTAHEDRON GROUP		BARS GROUP	STAR GROUP
		A	B	A	B	A	B		
IA Lithium	IA Sodium	IIA Beryllium	IIA Magnesium	IIIB Boron	IIIB Aluminium	IVB Carbon	IVB Silicon	VIII Iron	0 Neon
VIIB Fluorine	VIIB Chlorine	VIB (Oxygen)	VIB Sulphur	VB (Nitrogen)	VB Phosphorus	IVB Titanium	IVB Germanium	VIII Cobalt	0 Argon
IA Potassium	IB Copper	IIA Calcium	IIB Zinc	IIIA Scandium	IIIB Gallium	IVA Zirconium	IVB Tin	VIII Nickel	0 Krypton
VIIA Manganese	VIIB Bromine	VIA Chromium	VIB Selenium	VA Vanadium	VB Arsenic	Ln Cerium	Ln Terbium	VIII Ruthenium	0 Xenon
IA Rubidium	IB Silver	IIA Strontium	IIB Cadmium	IIIA Yttrium	IIIB Indium	IVB Hafnium	IVB Lead	VIII Rhodium	"Kalon"
VIIA Technetium ("Masurium")	VIIB Iodine	VIA Molybdenum	VIB Tellurium	VA Niobium	VB Antimony	IVA Thorium		VIII Palladium	0 Radon
IA Caesium	Ln Samarium	IIA Barium	Ln Europium	IIIA Lanthanum	Ln Gadolinium			Elements "X," "Y," and "Z"	
Ln Promethium ("Illinium")	Ln Erbium	Ln Neodymium	Ln Holmium	Ln Praseodymium	Ln Dysprosium			VIII Osmium	
Ln Thulium	IB Gold	Ln Ytterbium	IIB Mercury	Ln Lutecium	IIIB Thallium			VIII Iridium	
VIIA Rhenium	VIIB Astatine ("85")	VIA Tungsten	VIB Polonium	VA Tantalum	VB Bismuth			VIII Platinum	
IA Francium ("87")		IIA Radium		IIIA Actinium					
		VIA Uranium		V Protactinium					

them to place correctly in the periodic table certain elements that science had not discovered at the time. For example, in 1909 they called an element "Illinium" whose M.P.A. belonged to the Spike Group,[8] and in 1932 they gave it an atomic number of 61. This was promethium, first produced artificially in 1947. Also in 1932, they reported their observation of the M.P.A.'s of elements "85" and "87."[9] They placed them in the Dumb-bell Group and Spike Group, respectively, along with other elements of Group I and Group VII. These elements were astatine (first produced in 1947) and francium (found in 1939). The M.P.A. of element "43" was first recorded in 1909 and called "Masurium." It was placed in the Spike Group that comprises Group I and Group VII elements, and was described again in 1932,[10] five years before technetium was made in the laboratory. Its terrestrial existence is very unlikely, its isotopes being too short-lived for any remaining primordial technetium to be detectable on earth. But natural technetium, formed by the spontaneous fission of uranium, has been identified and isolated by chemists, and it must be assumed that the investigators examined such fission fragments with micro-psi vision.

The correlation between the position of an element in the periodic table and the shape of its M.P.A. poses the following problem: the investigators found that M.P.A.'s belonged to larger units that they assumed were molecules. In some of these, M.P.A.'s were bound together by forces directed between funnels of neighbouring M.P.A.'s. If a pair of linked funnels represented a single chemical bond, then M.P.A.'s of the Tetrahedron, Cube, and Octahedron Groups with, respectively, 4, 6, and 8 funnels would form 4, 6, and 8 bonds with other M.P.A.'s. But these elements have valences of 2, 3, and 4, respectively, and so the M.P.A. seems to exert a valence of only 1/2 through a single funnel. Even worse, some M.P.A.'s seem to divide into two parts when observed in M.P.M.'s. Both features are, of course, absurdly wrong and indicate that M.P.A.'s are not ordinary atoms. But suppose a funnel were functionally related in some way to one valence electron. Then the number of funnels in the M.P.A.'s of the Tetrahedron, Cube, and Octahedron Group elements would be equal to the number of valence electrons belonging to two atoms. Since Hypothesis 2 states that the nuclei of *two* atoms participate in the formation of an M.P.A., the possibility arises that valence electrons are responsible for the shape of M.P.M.'s belonging to these groups. If valence electrons determined the funnels, then the M.P.A.'s of the inert gases would not display funnels, for their atoms do not contain valence electrons. This is, indeed, the case. But if some connection exists between the shape of M.P.A.'s and similar valence states or electron configurations in the valence shell of the free atom, one must ask: why do elements in Groups IA and VIIA of the periodic table fill the Spike Group and those in Groups IB and VIIB fill the Dumb-bell Group, while sodium in Group IA is in the Dumb-bell Group and fluorine in Group VIIB occupies the Spike Group? Why do the lanthanides, having similar electron configurations in their atoms, occupy every M.P.A. group except the Bars and Star Groups? Since they are commonly trivalent, why do they not all fill the Cube Group? The variable occupation of the 4*f* electron shell would appear to determine widely different forms of the M.P.A.'s of the lanthanides. These questions will be considered shortly.

Given that an M.P.A. is a stable system of groups of omegons, bound in small clusters by vortices and collectively by some spherically symmetric potential, then the directions of the funnels, bars, and so on, are those that minimize the Coulomb energy of the charged bodies within them. For example: four identically charged bodies bound by a central force take up a tetrahedral orientation because of mutual repulsion of their charges (Tetrahedron Group); six charged bodies assume either a face-centred cubic array (Cube Group) or a hexagonal array (Star Group); Coulomb repulsion between eight bodies with identical charges results in an octahedral arrangement (Octahedral Group); finally, for fourteen bodies, the equilibrium array is a combination of face-centred cubic and octahedral arrays (Bars Group). But what determines the number of funnels, bars, or other structural components of M.P.A.'s? This question will now be considered.

TETRAHEDRON, CUBE, AND OCTAHEDRON GROUPS

Figure 7.3 shows the general forms of the M.P.A. Micro-psi observation of atoms of elements with valence n ($n = 2$, 3, or 4) disintegrates their nuclei into clouds of momentarily free omegons in chaotic motion. Omegons from two neighbouring similar nuclei intermix and cluster into stable bound states formed by their strong colour forces. Because of their electromagnetic coupling to the $2n$ valence electrons of the two atoms, the global aggregation

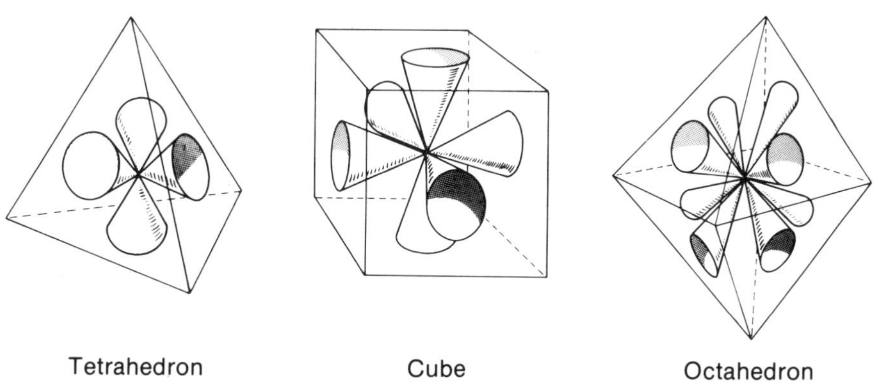

Fig. 7.3

of these groups is not isotropic but takes place along directions in space that are determined by the geometry of the equilibrium configuration of $2n$ classical, negative charges bound to a positively charged central body. It is suggested that non-valence electrons do not influence the shape of an M.P.A. because they occupy orbitals in filled shells centred about each original nucleus and, since they are unshared by atoms, their charge distribution in space does not draw the omegons released from the two nuclei towards one another. Only valence electrons that spend part of their time in orbital motion about other nuclei can aid the aggregation of particles released from the latter. In elements of the Tetrahedron Group, two similar atoms forming the M.P.A. supply four valence electrons, Coulomb attraction to which segregates the omegons freed from their nuclei into four identical clusters of omegon bound states. Because they have the same net electric charge, their mutual orientation in space is tetrahedral. In the Cube Group, two atoms provide six electrons, attraction to which results in a face-centred cubic array of clusters. For the tetravalent elements of the Octahedron Group, two atoms contribute eight valence electrons, and these give rise to an octahedral array of clusters.

As evidence of the role played by the electric charges of particles in determining the shapes of M.P.A.'s, the following result is cited: analysis of M.P.A.'s belonging to these three groups reveals that whether or not they possess central cores can depend on whether the electric charge $2Z$ of the M.P.A. (formed from two nuclei of atomic number Z and mass number A) can be compounded just from the electric charges of the omegons in N identical funnels, that is, whether there exist integers n and m such that

$$\begin{aligned} 18Z &= N(5n - 4m)\,, \\ 18A &= N(n + m) \qquad (N = 4, 6, 8)\,. \end{aligned}$$

If not, the M.P.A. should have a central core, because otherwise the funnels cannot all have the same electric charge. This has been verified for all M.P.A.'s with cores that have been analysed to date. It suggests that the process responsible for M.P.A.'s requires similar funnels to have identical net charges; otherwise, omegon bound states aggregate in central cores that allow the remaining particles to aggregate symmetrically among the funnels so that each has the same charge. The process proposed for formation of M.P.A.'s implies that the contents of a funnel cannot be neutral. This, too, has been verified.

STAR GROUP

The form of the M.P.A. is shown in figure 7.4. The inert gases have atoms with full electron shells and no valence electrons. The above explanation is inapplicable to them. The M.P.A. is formed from the nuclei of two atoms

Fig. 7.4. Star Group M.P.A.

which, although unbonded, are close enough as free atoms at the time of observation to participate in the formation of an M.P.A. Each atom has an outermost set of three p orbitals (p_X, p_Y, p_Z) occupied by pairs of electrons with opposite spins. It is suggested that Coulomb attraction between particles in the omegon clouds and these six pairs of electrons leads to aggregation of particles in the observed hexagonal array of arms of the six-pointed star.

BARS GROUP

The form of the M.P.A. is shown in figure 7.5. The group comprises the three triads of Group VIII elements, which are characterized by variable valence. Their oxidation states vary from two to eight. The number of valence electrons supplied by an atom is variable and cannot be related to the number of bars in the M.P.A., which is fourteen. For iron, cobalt, and nickel, $3d$ and $4s$ electrons are involved in bonding; for ruthenium, rhodium, and palladium, $4d$ and $5s$ electrons participate; for osmium, iridium, and platinum, $5d$ and $6s$ electrons take part. The angular dependence of the five degenerate d orbitals is shown in figure 7.6. The d_{Z^2} and $d_{X^2-Y^2}$ orbitals are directed along the coordinate axes X, Y, and Z, and the d_{XY}, d_{XZ}, and d_{YZ} orbitals are directed along lines that are inclined at 45° to the pairs of axes (X, Y), (X, Z), and (Y, Z), respectively. Consider a cube whose vertical

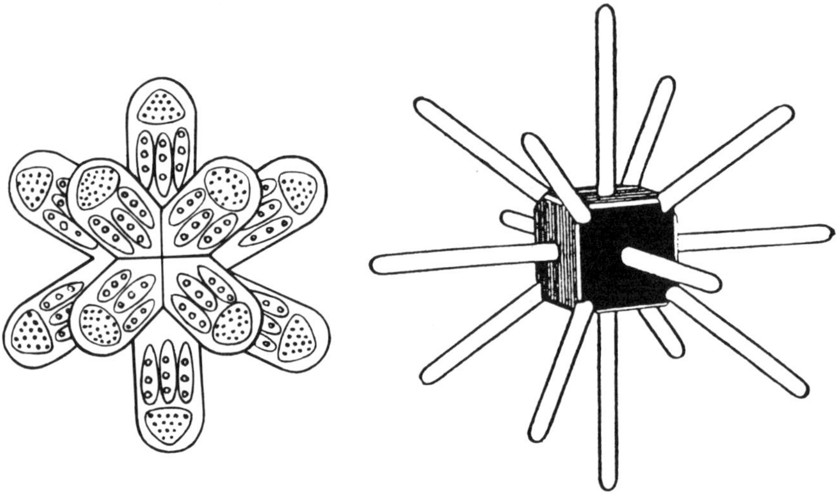

Fig. 7.5

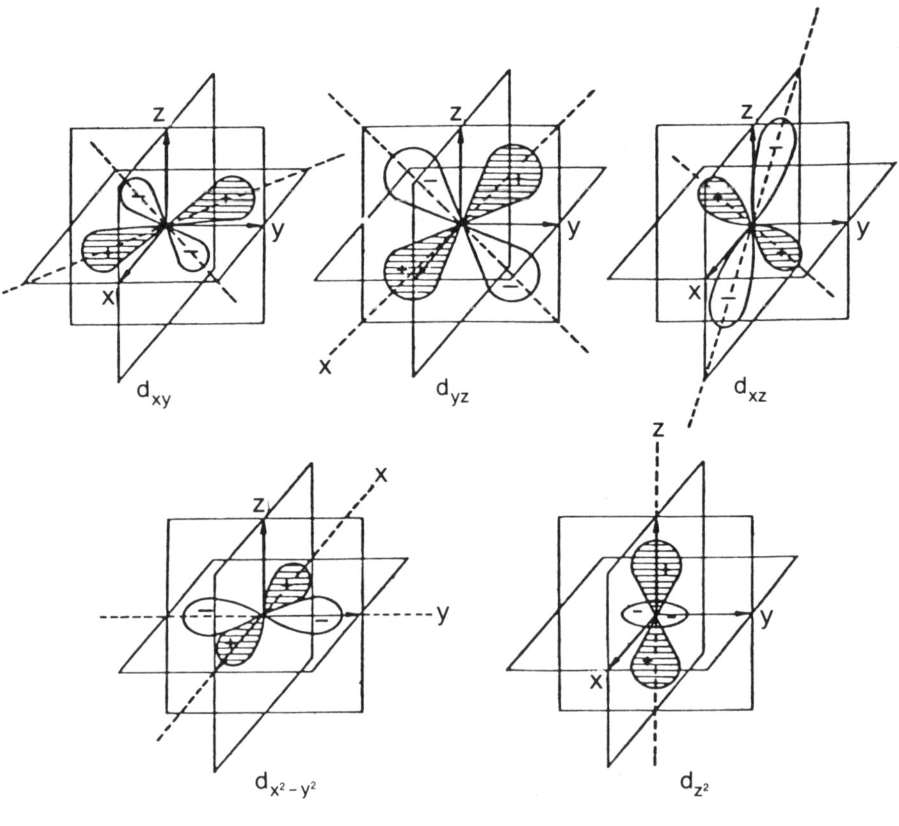

Fig. 7.6. Angular dependence of d_{XY}, d_{YZ}, d_{XZ}, $d_{X^2-Y^2}$, and d_{Z^2} orbitals

diagonal planes $ABA'B'$ and $CDC'D'$ are the YZ and XZ planes of the d orbitals (figure 7.7). By means of the d_{YZ} orbital, the electron density is concentrated along the diagonals AA' and BB'. The d_{XZ} orbital gives a charge density concentrated along the diagonals CC' and DD'. The d_{Z^2} orbital concentrates charge along the Z-axis. The number of electrons occupying these orbitals depends on whether the crystal field is strong or weak. If they are occupied, there are fourteen directions in space along which d electron density is greatest: the eight directions of the body diagonals of a cube and the six face-centred cubic directions. These coincide with the directions of the fourteen bars of the M.P.A. The correspondence would be more significant only if it could be understood how the directions of maximum charge density of d orbitals in the atom determine the form of the M.P.A. formed from *two* atoms. Moreover, why do only Group VIII elements have M.P.A.'s of this type, and not other members of the three transition series that also have partly filled d shells? These problems will not be pursued further.

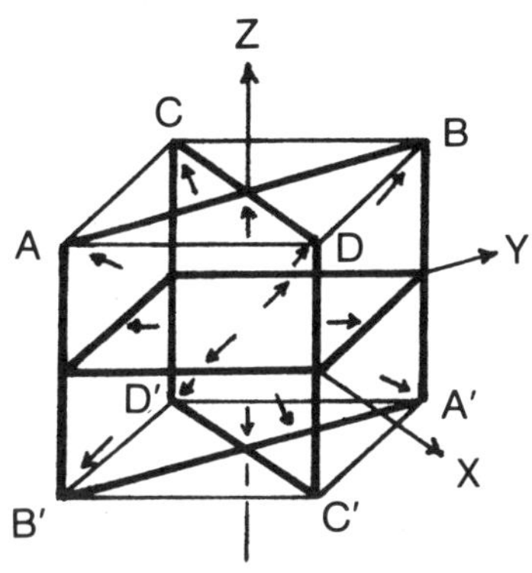

Fig. 7.7. Fourteen axes of symmetry of d_{XY}, d_{YZ}, d_{XZ}, and d_{Z^2} orbitals coincide with directions of fourteen bars of the M.P.A. of Group VIII elements.

SPIKE GROUP

There is no general form of the M.P.A. Figure 7.8 shows that of lithium. The ions of francium, rhenium, promethium, caesium, technetium, and rubidium have an outer octet of electrons. Their M.P.A.'s have sixteen spike-shaped projections, each containing similar groups of particles. This suggests that the outer octets of two ions forming the M.P.A. are responsible for the spikes. However, the ions of fluorine and potassium have outer octets, yet their M.P.A.'s have, respectively, eight and nine spikes. Also manganese has a similar electron configuration to that of technetium and rhenium, yet its M.P.A. has only fourteen spikes. These anomalies will not be discussed further.

Fig. 7.8

DUMB-BELL GROUP

The typical form of an M.P.A. is shown in figure 7.9. It consists of a central connecting rod, at each end of which is a globe. Twelve funnels containing particles project from the centre of the globe, alternate funnels pointing slightly upwards and downwards; that is, there are six pairs. Each funnel has the same composition of particles. The M.P.A. is formed either from two ions or from two covalently bonded atoms in a molecule that have an outer *p* shell occupied by six electrons. It is suggested that these are responsible for the six double funnels at each end of the dumb-bell. Chlorine is chosen as an example: each atom in the molecule is in a distorted tetrahedral sp^3 hybridized valence state, three of the four σ orbitals being filled by lone pairs of electrons and the fourth constituting the σ bond between the two atoms (figure 7.10). A free M.P.A. of chlorine is produced by micro-psi observation of a single molecule. The six non-bonding electrons in sp^3 orbitals directed away from the bond axis of the molecule at each end attract positively charged particles and repel negatively charged ones freed from the disintegrated nuclei. Aggregation of the former takes place along

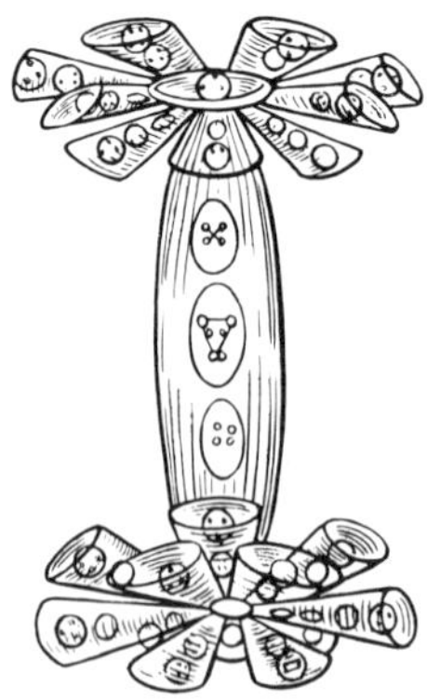

Fig. 7.9

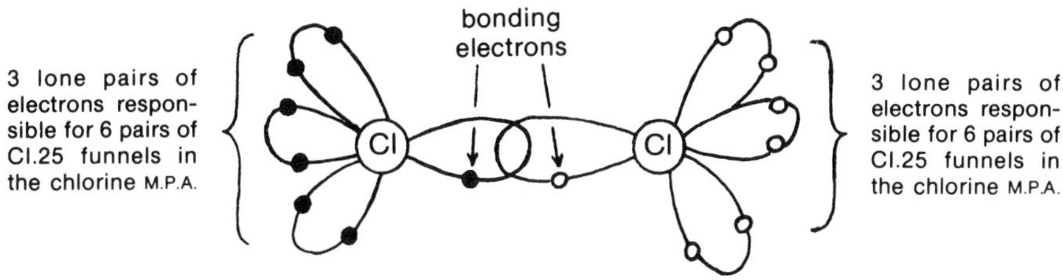

Fig. 7.10. The chlorine molecule

six directions spaced 60° apart, while the latter group along six directions mid-way between the former. In this way, two sets of funnels result, alternating in the polarity of their charges. The feature of positive and negative funnels is confirmed later by the analysis of the M.P.A. of chlorine. It is not clear why, of the alkali metals, only sodium occupies the Dumb-bell Group, although lithium cannot belong to this group because it does not have an outer set of six *p* electrons that can be instrumental in creating an array of six double funnels. It is also not clear why fluorine (which has an electron configuration similar to that of the other halogens) occupies the Spike Group instead of the Dumb-bell Group containing the halogens. Is it because fluorine has a maximum covalency of only four (sharing a maximum of eight electrons with other atoms), while other halogens have a maximum covalency of six (sharing a maximum of twelve electrons)? It may be of significance, then, that the M.P.A. of fluorine has eight spikes.

Discussion of these formative principles is concluded with an account of the lanthanides. Table 7.3 shows that they occupy all groups except the Bars and Star Groups. Since they are trivalent, they might have been expected to occupy the Cube Group containing Group III elements. Their distribution among five groups requires explanation. Table 7.4 shows the valence electron shells of the lanthanide atoms and the number of electrons that they contain. The first feature to note is that, in a given group (apart

TABLE 7.4

VALENCE ELECTRON CONFIGURATION OF LANTHANIDE ATOMS*

SPIKE GROUP	DUMB-BELL GROUP	TETRAHEDRON GROUP		CUBE GROUP		OCTAHEDRON GROUP	
		A	B	A	B	A	B
Pm $4f^5 6s^2$ Tm $4f^{13} 6s^2$	Er $\underline{4f}^{12} 6s^2$ Sm $\underline{4f}^6 6s^2$	Nd $4f^4 \underline{6s}^2$ Yb $4f^{14} \underline{6s}^2$	Ho $4f^{11} \underline{6s}^2$ Eu $4f^7 \underline{6s}^2$	La $4f^0 \underline{5d}^1 \underline{6s}^2$ Pr $\underline{4f}^3 6s^2$ Lu $4f^{14} \underline{5d}^1 \underline{6s}^2$	Gd $4f^7 \underline{5d}^1 \underline{6s}^2$ Dy $\underline{4f}^{10} 6s^2$	Ce $\underline{4f}^2 \underline{6s}^2$	Tb $\underline{4f}^9 \underline{6s}^2$

* Underlined shells contain electrons that determine funnels.

from the Spike and Dumb-bell Groups), the elements can be paired so that one member of the pair differs from the other by its possession of seven extra 4*f* electrons, that is, half the population of the 4*f* shell; second, for elements with this shell either empty, half-full, or full, the number of valence electrons in other shells is exactly half the number of funnels shown by their M.P.A.'s; third, for elements with partially filled 4*f* shells, the number of electrons in valence shells is either the same as or seven more than the number of corresponding electrons in atoms of non-lanthanide elements in the same M.P.A. group (e.g., neodymium has six such electrons, the same as Group VI elements, and cerium has four electrons, the same as carbon). It is proposed, therefore, that

1. Where the 4*f* shell occupancy is very stable (empty, half-full, or full), the other outer electrons are responsible for the shape of M.P.A.'s, the number provided by two atoms being equal to the number of funnels in the M.P.A.
2. Where the 4*f* shell is more than half-filled, only outer electrons and/or 4*f* electrons in addition to the stable half-filled shell are responsible for the funnel structure of the M.P.A.
3. Where the 4*f* shell is less than half-filled, both 4*f* and outer electrons are responsible for funnels.

In the case of the Dumb-bell Group, the six 4*f* electrons of samarium and six of the twelve 4*f* electrons of erbium are responsible for the six double funnels at each end of the dumb-bell. Why promethium and thulium are in the Spike Group is unclear.

"WALLS" OF FUNNELS, BARS, SPIKES, AND STAR ARMS

When the investigators disintegrated M.P.A.'s, they reported that funnels, bars, spikes, and star arms always separated and disappeared, releasing their contents as free bodies. When this happens, the "walls" enclosing groups of particles may change in shape, but individual groups do not break up or suffer any deformation, although their arrangement in clusters of groups may change. Of the "walls," the investigators remarked: "A sphere-wall is a temporary effect, caused by one or more Anu in rotation. Just as a stream of air under pressure will make a hole on the surface of water, by pushing back that water, so is it with the groups. As they revolve, the force of their motion drives back the circumambient medium."[11] In chapters 5 and 6, this "medium" was identified as the Higgs field superfluid permeating space, and a "sphere-wall" was interpreted as a domain boundary separating different Higgs vacua. The shapes of these boundaries are determined by the geometry and motion of the strings confining groups of particles in the domains. An M.P.A. is an aggregate of vacuum domains immersed in the surrounding SU(9) Higgs vacuum of normal space. The global boundary between the normal vacuum and the heterogeneous mixture of domains defines the "walls" of funnels, bars, and so on.

The remainder of this chapter is devoted to the analysis (within the framework of the Quark and Omegon Models) of the M.P.A.'s of all elements from hydrogen to calcium. It is intended to demonstrate by these examples that the detailed descriptions of these systems given by Besant and Leadbeater over half a century ago are consistent with Hypotheses 1 and 2 to a remarkable degree, and, therefore, with the Quark and Omegon Models, the former being supported by a wealth of orthodox scientific evidence. It must be emphasized that the depiction by the investigators of "positive" and "negative" groups of U.P.A.'s as pointing, respectively, outward and inward severely limits the freedom of possible physical interpretation of these groups. This is because the convention employed by Besant and Leadbeater in their published work was shown in chapter 5 to be based on their faculty of being able to distinguish between micro-psi images that refer to particles of opposite electric polarity. It is a notable achievement of the analysis to be given shortly that *all* groups in every M.P.A. that is discussed can be consistently identified in terms of $\mathcal{P}$ and $\mathcal{N}$ omegons and u and d quarks without ever violating this convention—a result that strongly supports the thesis of this book, namely, that Besant and Leadbeater described the existence of quarks in matter.

The following key of identification will be used in the disintegration diagrams:

 = $\mathcal{P}$ omegon (third component of isospin = 1/2; electric charge = 5/9).

 = $\mathcal{N}$ omegon (third component of isospin = −1/2; electric charge = −4/9).

 = $\mathcal{P}$ or $\mathcal{N}$ omegon.

The last symbol denotes U.P.A.'s whose isospin state cannot yet be deduced or found by means of elimination. It also denotes free U.P.A.'s (released through the breakup of duads, triplets, etc.) that are collectively represented in the disintegration diagrams by one U.P.A. with the number of such particles written next to it. These may comprise both $\mathcal{P}$ and $\mathcal{N}$ omegons. All disintegration diagrams are those published in the third edition of *Occult Chemistry* (chap. 1, ref. 5). Only u and d quark labels and the above shades for U.P.A.'s have been added.

HYDROGEN M.P.A.

Micro-psi examination of hydrogen gas reveals a system consisting of two interlaced, triangular arrays of groups of three U.P.A.'s (fig. 7.11). In chapter 4 the triangular array ("Hydrogen Triangle") was identified as a proton, the group of three U.P.A.'s ("Hydrogen Triplet") was identified as a quark, and a U.P.A. was identified as a 𝒫 or 𝒩 omegon. Two forms of the hydrogen M.P.A. were recorded by Besant and Leadbeater: in Hydrogen Variety 1, five (+) and four (−) U.P.A.'s are in one Hydrogen Triangle and four (+) and five (−) U.P.A.'s are in the other, totalling nine (+) and nine (−) U.P.A.'s; in Hydrogen Variety 2, one Hydrogen Triangle contains six (+) and three (−) U.P.A.'s and the other has four (+) and five (−) U.P.A.'s, so that the M.P.A. contains ten (+) and eight (−) U.P.A.'s.

The M.P.A. is formed from the two protons in a molecule of hydrogen (fig. 7.12). The immediate effect of the selection and examination of the molecule by the observer is to destroy the vortices in the superconducting Higgs field inside each proton. This destabilizes the quarks as bound states of three omegons, and so the latter become free, interacting with one another through their electric charges and moving into the space between the former protons. Then vortices reappear, and the omegon magnetic monopoles strongly interact once more, clustering in three groups of three as protons with their string bonds restored. This strong coupling does not weaken with distance,[12] and so it allows monopoles that are initially much farther apart than is typical for particles inside nuclei to recombine into protons. This means that the recombination can take place between omegons belonging to *both* former protons as well as between those belonging to the same former proton. The system thus formed is not a diproton (which is unstable) but two protons with the vortices between their quarks interlocking as a result of this coupling between monopoles sharing the same region of space. Interlocking vortices in the U(1) Higgs model[13] and in the SU(N) model[14] cannot penetrate one another. This is because interlocking and non-interlocking configurations of vortices possess different values of a conserved charge, so that the former cannot evolve into the latter. The M.P.A. represents two protons trapped by their interlocking vortex strings and not two protons bound merely by nuclear forces, although this force was noticed by the investigators: "each of the three groups making one half of Hydrogen are linked to each other across space by lines of attraction."[15] They stated that the two Hydrogen Triangles in the free M.P.A. always appeared to be interlaced. An M.P.A. consisting of separated Hydrogen Triangles was never reported, and spontaneous separation was never observed.

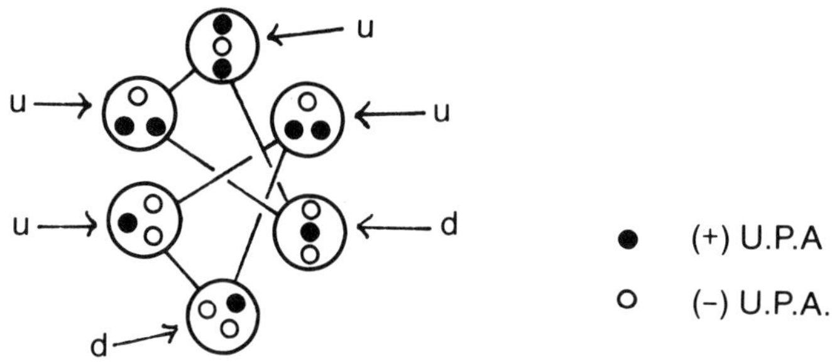

Fig. 7.11

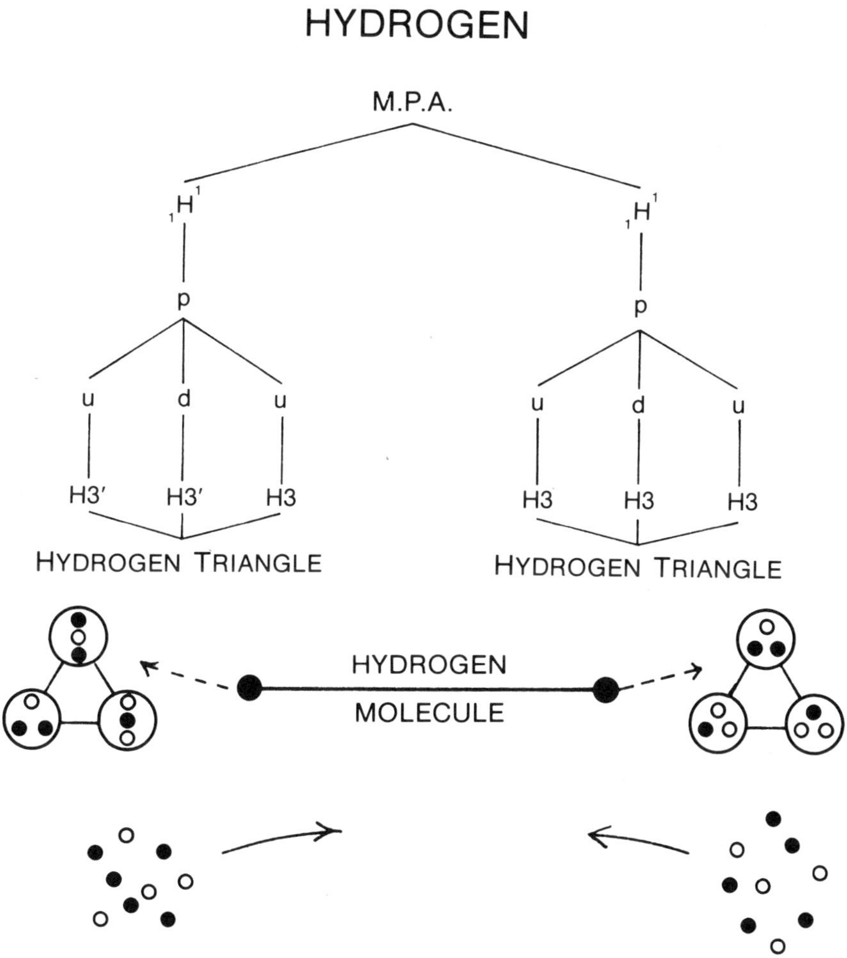

Fig. 7.12

The process described above explains why nothing resembling stable molecules made up of pairs of atoms was observed. This was true for all elements (with diatomic molecules) that were examined in the gaseous state, and it arose because M.P.A.'s are systems formed from both atoms in the molecule.

The products formed at successive stages of the disintegration of the M.P.A. (a process induced by the investigators) are shown in figures 7.13*a* and 7.13*b*. The numbers labelling the stages require explanation: Besant and Leadbeater subscribed to the Theosophical view that matter in the physical universe exists in seven distinct states: the solid, liquid, and gaseous states (of which man has sensory awareness) and the "etheric" state (visible only to the psychic) that has four substates, as different from one another as are solids, liquids, and gases. They assumed that an M.P.A. was a chemical atom, and, not having had at any time during their investigations (1895–1933) a concept of nuclei composed of protons and neutrons, they believed that they were studying the "etheric" state of atoms when they disintegrated M.P.A.'s. Accordingly, they called the various states represented by the bodies released during successive stages of disintegration "ether 4," "ether 3," "ether 2," and "ether 1," the last one being the final state of free U.P.A.'s. In figures 7.13*a* and 7.13*b* and in subsequent disintegration diagrams, these numbers are retained, but only to label the various stages of the breakup of M.P.A.'s and not to denote any qualitative distinction between the hadron states. In the first stage of disintegration (E4), the M.P.A. separates into its two component protons; next, each proton breaks up into a diquark (embedded in an SU(6) Higgs vacuum domain) and a free quark (embedded in an SU(3) vacuum domain) (E3); then the diquark splits up into two free quarks (E2); finally, all quarks break up into free omegons, each embedded in a spherical domain of a superconducting U(1) Higgs vacuum (E1). Note that the convention (discussed in chap. 5) that the U.P.A.'s in "positive" groups point outward and that U.P.A.'s in "negative" groups point inward implies that two triplets in the lower Hydrogen Triangle are positively charged and that the other is negatively charged—exactly the Quark Model prediction. Hydrogen Variety 1 (fig. 7.13*a*), with nine (+) and nine (−) U.P.A.'s, is formed from a hydrogen molecule containing protons made up of five (+) and four (−) U.P.A.'s and four (+) and five (−) U.P.A.'s. Hydrogen Variety 2 (fig. 7.13*b*), with ten (+) and eight (−) U.P.A.'s, is formed from a hydrogen molecule, both of whose protons contain five (+) and four (−) U.P.A.'s (fig. 7.14). The net result is that a $\mathcal{P}$ omegon with magnetic charge +1 is exchanged for a $\mathcal{P}$ omegon with magnetic charge −2.

In Hydrogen Variety 2, the triplet with three (+) U.P.A.'s is a u quark

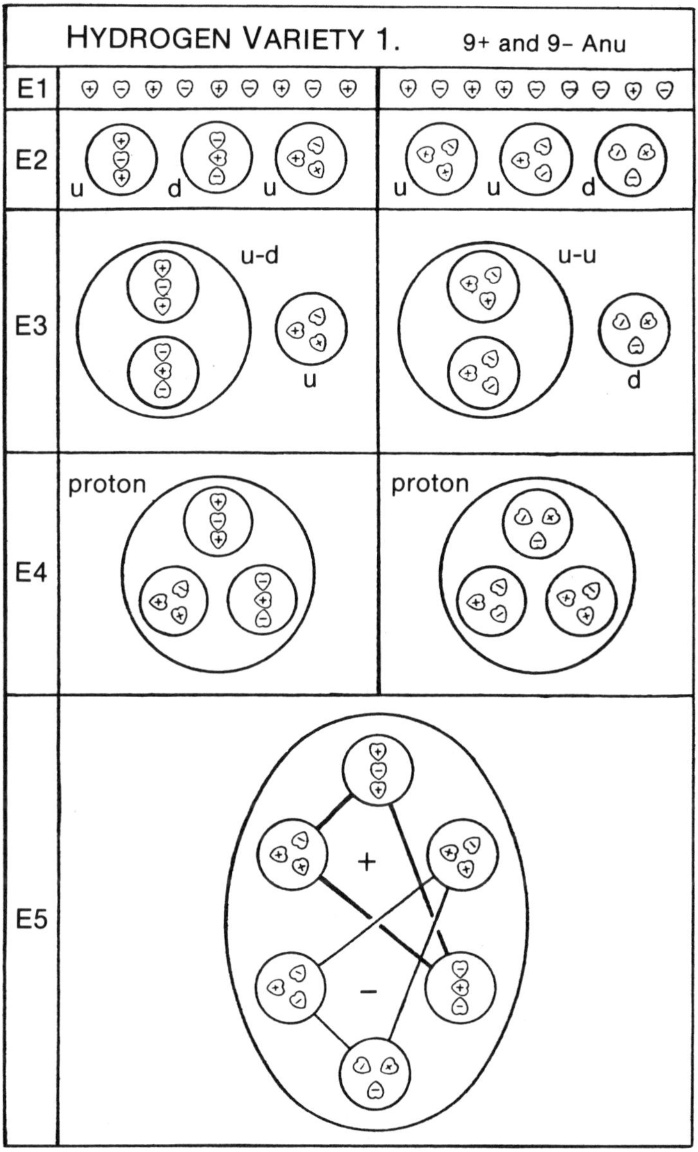
HYDROGEN VARIETY 1.
9+ and 9- Anu
E1
E2
u
d
u
u
u
d
E3
u-d
u
u-u
d
E4
proton
proton
E5
+
–

Fig. 7.13*a*

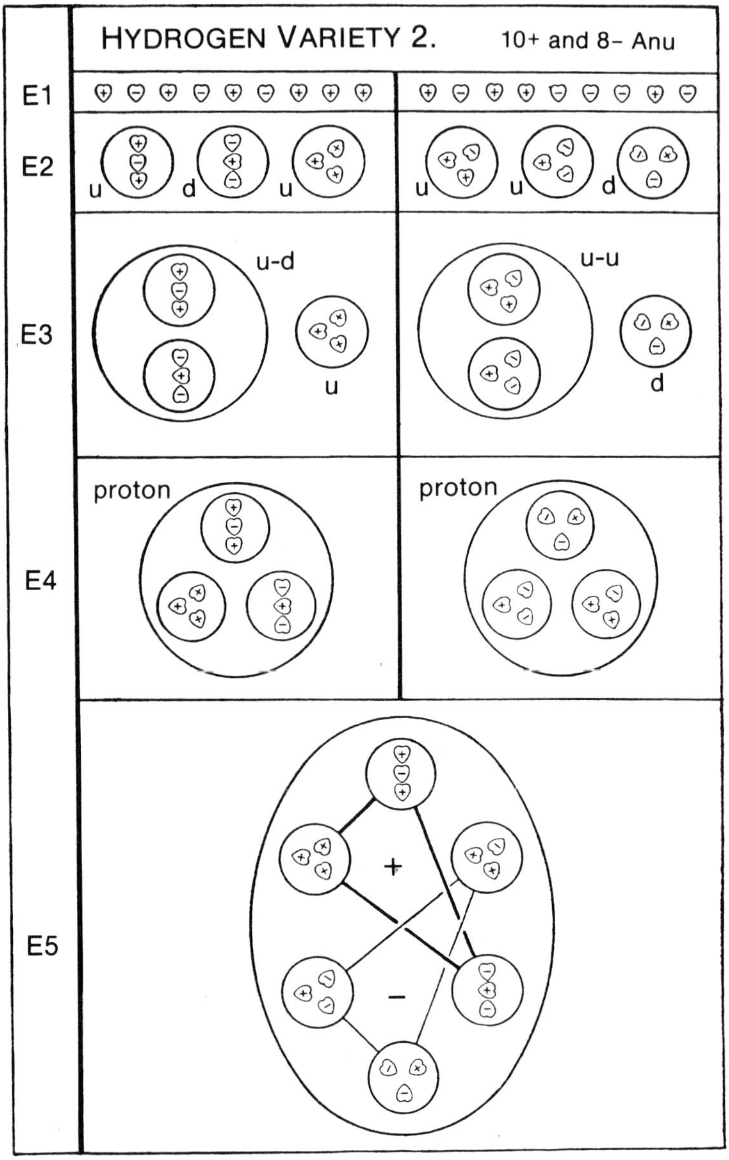
HYDROGEN VARIETY 2.
10+ and 8− Anu
E1
E2
E3
E4
E5
u
d
u
u
u
d
u-d
u
u-u
d
proton
proton
+
−

Fig. 7.13b

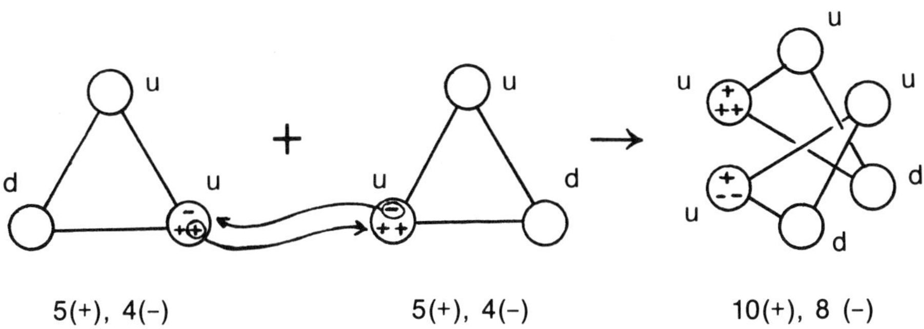

Fig. 7.14. Protons containing identical magnetic charges form Hydrogen Variety 2

made up of three $SU(3)_s$ monopoles, each with magnetic charge $+1$. Such a configuration is stable because of the triple connectivity of the global gauge group $SU(3)_s/Z_3$ governing the quark-forming, superstrong interaction between omegons with different shades of the same colour.

DEUTERIUM ("ADYARIUM") M.P.A.

In 1932, Besant and Leadbeater reported[16] their observation of an object consisting of thirty-six U.P.A.'s, arranged as a tetrahedral array of four H3 triplets (called "Ad12") and a tetrahedral array of four sextets (called "Ad24"). The tetrahedra of triplets and sextets (called "Ad6") interpenetrate, as shown in figure 7.15. The group of six U.P.A.'s making up an Ad6 are arranged in the shape of a "cigar" or elongated hexagon, which revolves rapidly around its longitudinal axis, giving the outward appearance of a pencil sharpened at both ends. The M.P.A. was found in the atmosphere but was regarded as a new element by the investigators (who called it "Adyarium"), even though it had twice the number of U.P.A.'s in the

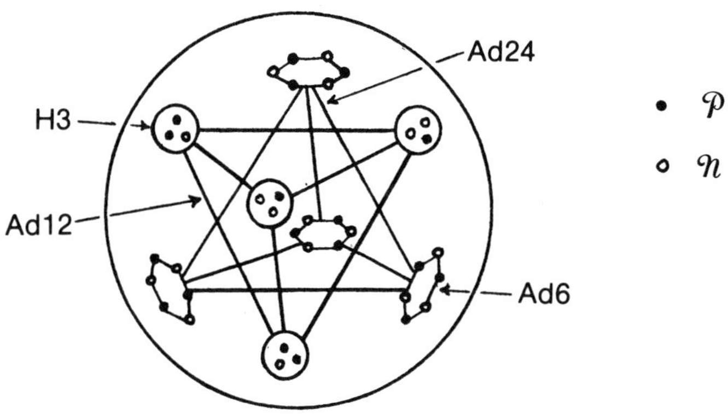

Fig. 7.15. "Adyarium"

hydrogen M.P.A., thus making deuterium a probable source (the isotope had been discovered the previous year by H. Urey and his colleagues). The investigators did not make this identification because they had earlier come to believe (mistakenly, as it turned out) that an object seen during the electrolysis of water was an atom of deuterium.

Adyarium is an M.P.A. of deuterium and is a system that results from micro-psi examination of a molecule of deuterium. The two deuterons split up into six u and six d quarks, which then recombine to form new bound states (fig. 7.16). The Ad12 is a system of two u quarks and two d quarks (each colourless and unconfined, in the sense of quantum chromodynamics), bound together by the strong coupling between their internal strings in a way similar to the binding of two protons and two neutrons in a helium nu-

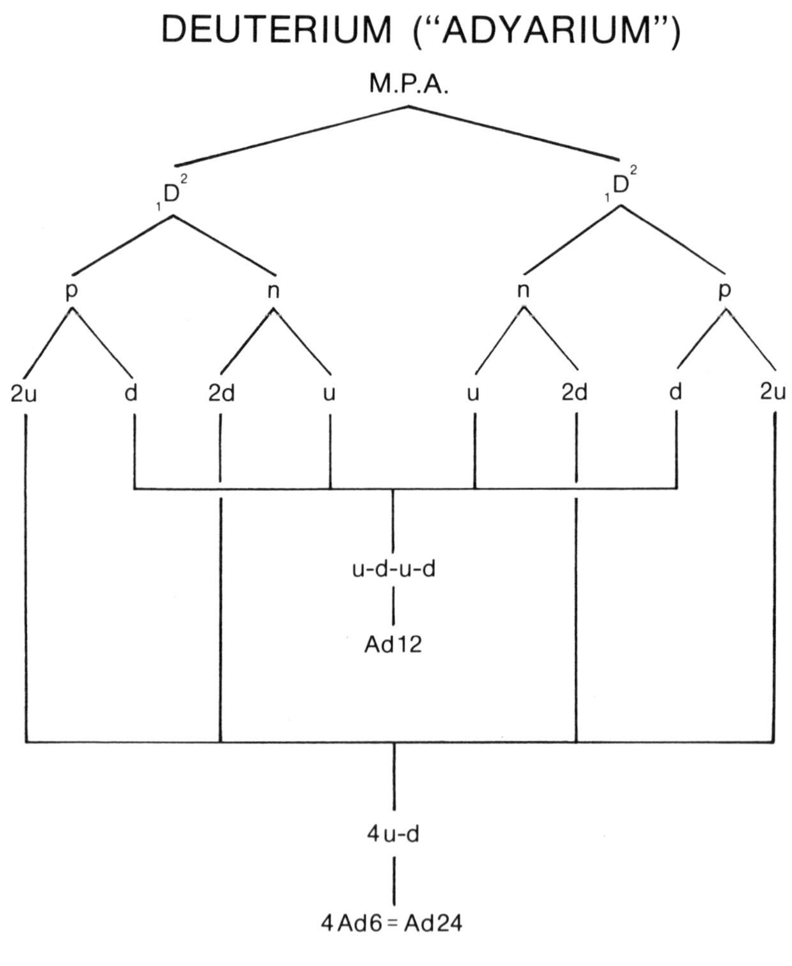

Fig. 7.16

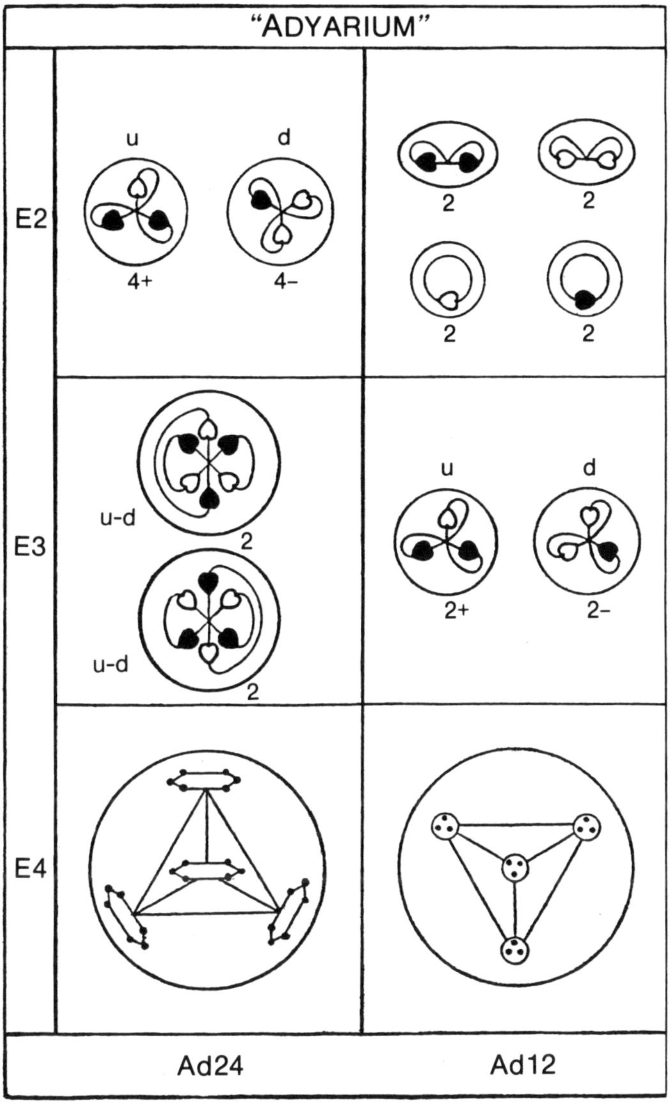

Fig. 7.17

cleus. This system is just the quark counterpart of the latter, and it is the first of many exotic multi-quark states of non-zero triality that will be shown in this chapter to be constituents of M.P.A.'s. It has an electric charge of 2/3 and a spin of 2 (the total nuclear spin of the two deuterons in the molecule of deuterium), the four quarks having parallel spins. Both the spin and isospin wavefunctions of the bound state of four quarks are symmetric, and they occupy a hybridized sp^3 orbital that is described by a space wavefunction that is totally antisymmetric with respect to exchange of their coordinates. This is apparent in figure 7.15 in the tetrahedral arrangement of the H3 triplets. The remaining quarks are present in the M.P.A. as four u-d diquarks. Each diquark (Ad6) is an isospin singlet, colour SU(2) singlet, and spin singlet state. It belongs to the $\underline{21}$ representation of SU(6), which is symmetric in combined spin and unitary spin. The space wavefunction of the two quarks is symmetric. The four identical, spinless bosons have an electric charge of $+1/3$, and their mutual Coulomb repulsion results in the observed tetrahedral arrangement. The Ad24 group is predicted to have four (+) and four (−) triplets. Figure 7.17, showing the products of the disintegration of the M.P.A., confirms this. From figure 7.16 it is predicted that the monopole charges making up the quarks in the Ad12 group are (1, 1, −2) and (−1, −1, 2) (two of each). This is confirmed by figure 7.17, for the H3 triplets break up into two types of duads and a free U.P.A., and the string configuration of either was shown in chapter 6 to be due to charges (1, 1) or (−1, −1). Note that the "positive" and "negative" duads formed from breakup of the H3 triplets in the Ad12 group are $\mathcal{P}$-$\mathcal{P}$ and $\mathcal{N}$-$\mathcal{N}$ diomegons. In figure 7.18 are shown the charges of the magnetic

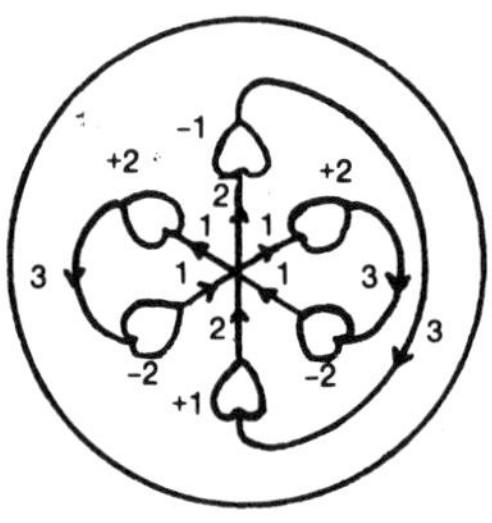

Fig. 7.18. Ad6 group (u-d diquark)

monopoles in the diquark Ad6, together with the flux quanta in their strings. Identification is summarized thus: Ad6 = u-d, Ad24 = 4(u-d), and Ad12 = u-d-u-d.

M.P.A. OF HELIUM ($_2He^3$) ("OCCULTUM")

First observed in 1895 and reported to be rare, the M.P.A. has the form shown in figure 7.19. It was originally thought to be a helium atom, but

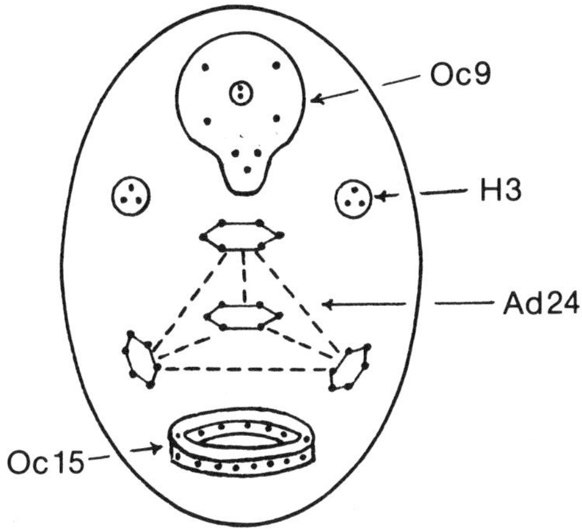

Fig. 7.19. "Occultum"

this view was rejected when the M.P.A. of helium was seen in 1907 to be quite different. It consists of an Ad24 group, two H3 triplets, a ring of fifteen U.P.A.'s (Oc15), and a balloon-shaped body of nine U.P.A.'s (Oc9). It contains fifty-four U.P.A.'s.

Because the concept of isotopes was not available to the investigators in 1907, they could not infer that they had examined the lighter isotope of helium ($_2He^3$). The M.P.A. is formed from two $_2He^3$ nuclei, which provide ten u and eight d quarks (fig. 7.20). Four u-d diquarks constitute the Ad24 (as in the case of "Adyarium"). The two H3 triplets are a u and a d quark. This is confirmed by the disintegration diagram (fig. 7.21), which indicates that they are a (+) and a (−) triplet. The Oc15 is made up of four u quarks and one d quark, bound as fifteen monopoles in a closed, circular configuration of U(1) strings. The Oc9 is made up of omegons in two d quarks and a u quark. Figure 7.21 shows that the products formed by its disintegration consist of two (−) triplets and a duad and free U.P.A. that can make up a (+) triplet, thus confirming this quark composition.

HELIUM M.P.A.

The M.P.A. is shown in figure 7.22. It consists of two Hydrogen Triangles (these are separated and do not overlap, as in the hydrogen M.P.A.), two Ad24 tetrahedra, and two H3 triplets. The Ad24 groups are reported to "revolve round an egg-shaped central body consisting of two H3 spheres, and the triangles spin on their own axes while performing a similar revolution."[17]

The M.P.A. is formed from two $_2He^4$ nuclei (fig. 7.23). These provide a proton and a neutron (the two Hydrogen Triangles), two tetrahedral clusters of four u-d diquarks (Ad24), and two unbound quarks (a d and a u

HELIUM ("OCCULTUM")

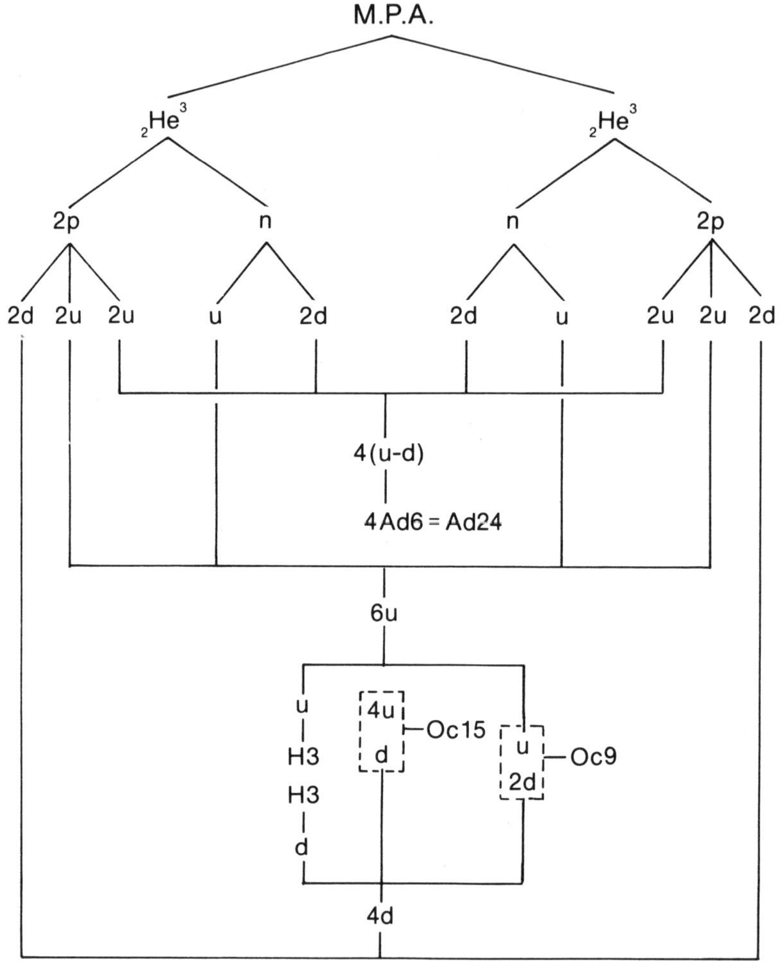

Fig. 7.20

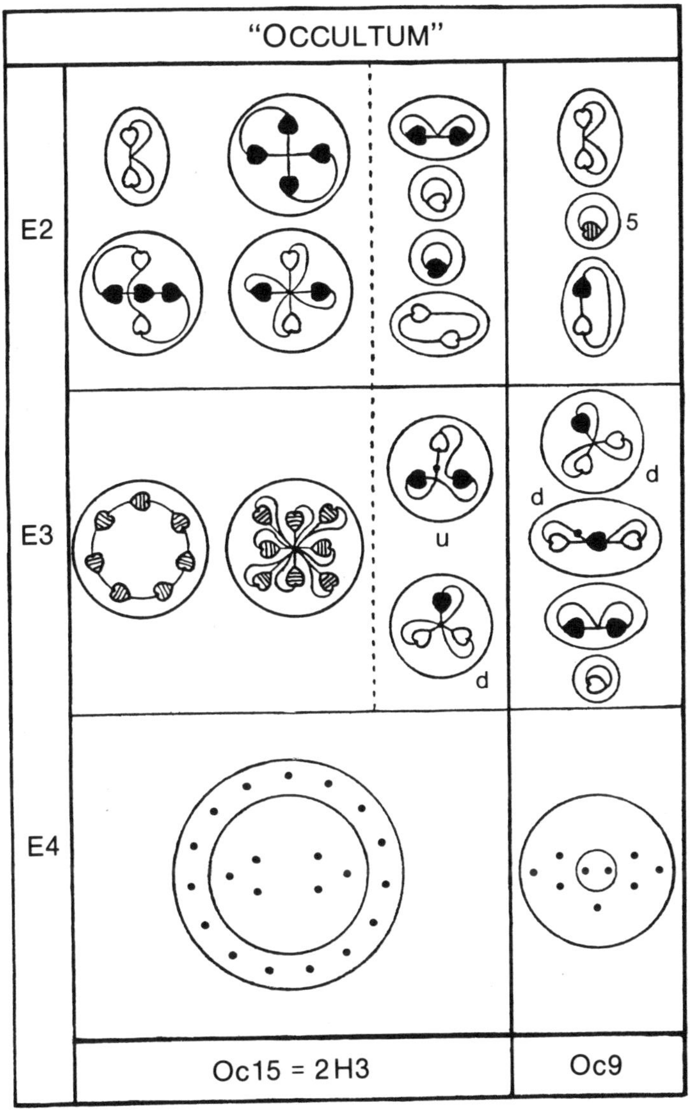

Fig. 7.21

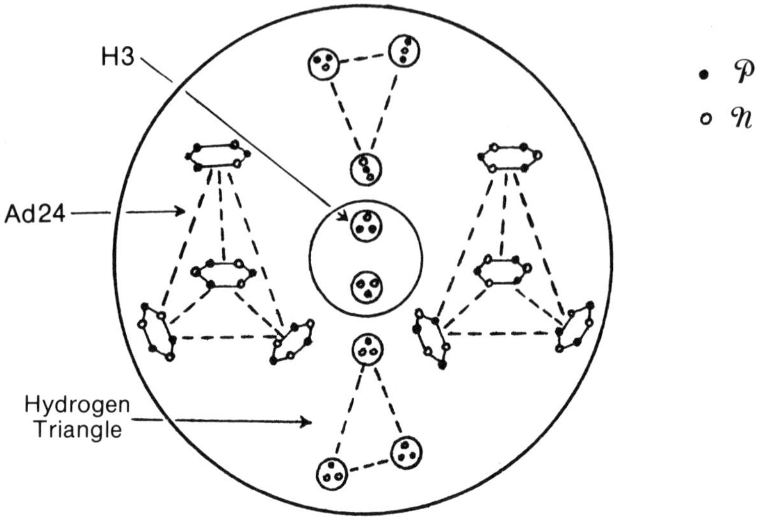

Fig. 7.22. M.P.A. of helium

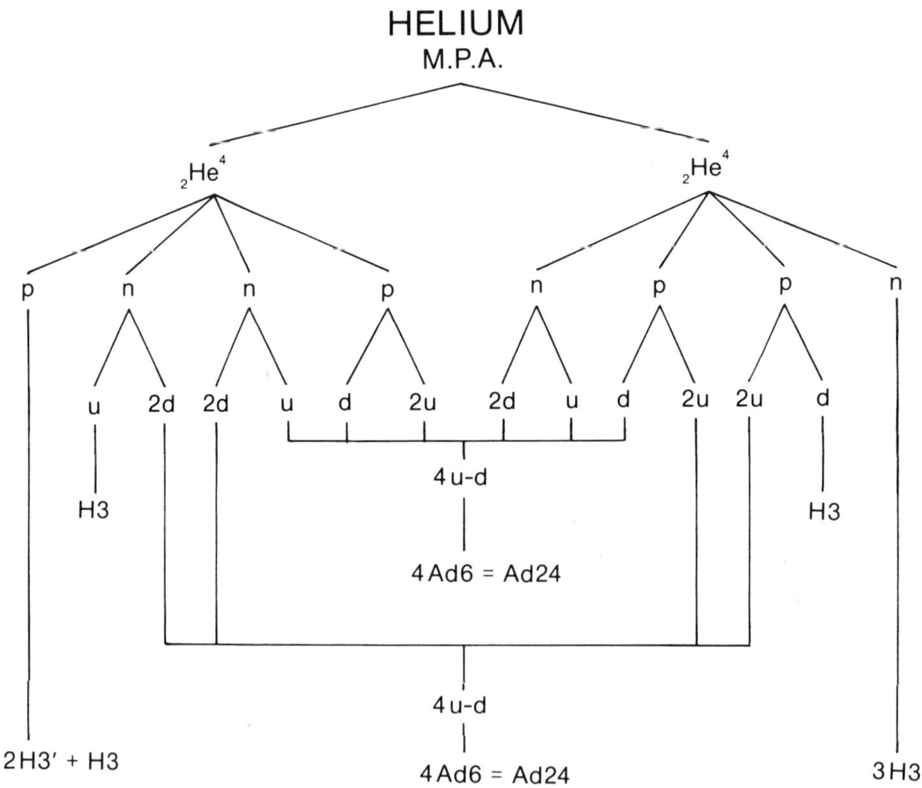

Fig. 7.23

quark), each embedded in an SU(3) Higgs vacuum domain (H3). Since u quarks are "positive" triplets and d quarks are "negative" triplets, the predicted identity of these H3 groups is confirmed by the statement that "in the centre of all the two groups of 3 Anu, being positive and negative, satisfy each other."[18] The disintegration of the various groups proceeds in the manner observed for the corresponding groups in the M.P.A.'s of hydrogen and deuterium.

LITHIUM M.P.A.

Lithium is the first member of the Spike Group of elements. Figure 7.24 shows that its M.P.A. consists of a central, spike-shaped aggregate of sixty-three U.P.A.'s, at the base of which there is a globe containing four spheres, each enclosing a tetrahedron of four U.P.A.'s (Li4). Eight Ad6 groups project outward from the globe. The spike formation revolves rapidly on its axis, carrying these groups with it. The globe is at the centre of a platelike structure that revolves swiftly in the opposite direction.

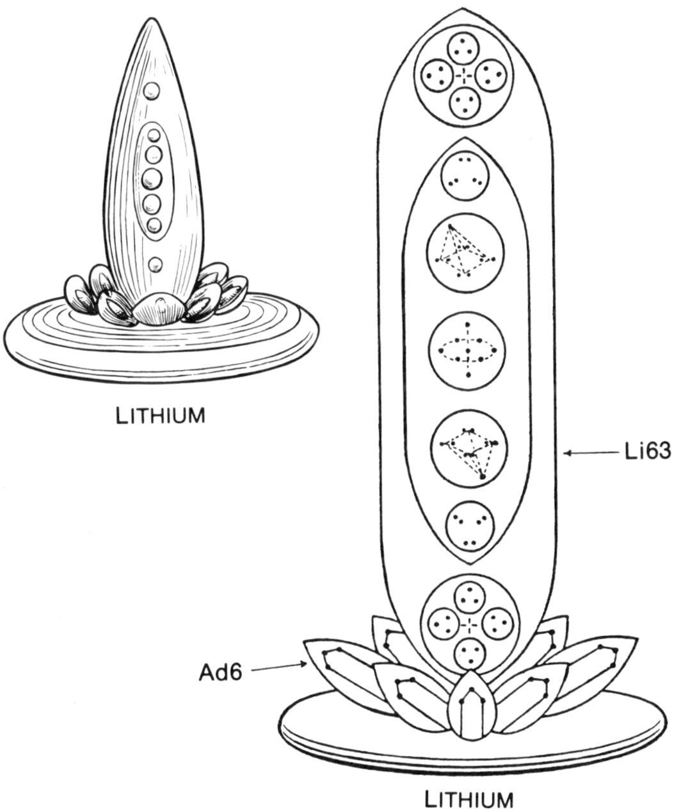

Fig. 7.24. M.P.A. of lithium

The M.P.A. is formed from two $_3Li^7$ nuclei, which provide twenty u and twenty-two d quarks (fig. 7.25). Of these, eight u and eight d quarks form eight u-d diquarks (Ad6). Each carries an electric charge of $+1/3$, so that mutual Coulomb repulsion results in their octagonal arrangement about the globe, to which they are bound by the strong interaction between their strings. The remaining twelve u and fourteen d quarks appear in the M.P.A. as the omegons in the globe and spike. In the globe there are two clusters of three 𝒫 omegons and one 𝒩 omegon (Li4) and two clusters of three 𝒩 and one 𝒫 omegon (Li4), each embedded in an SU(4) Higgs vacuum

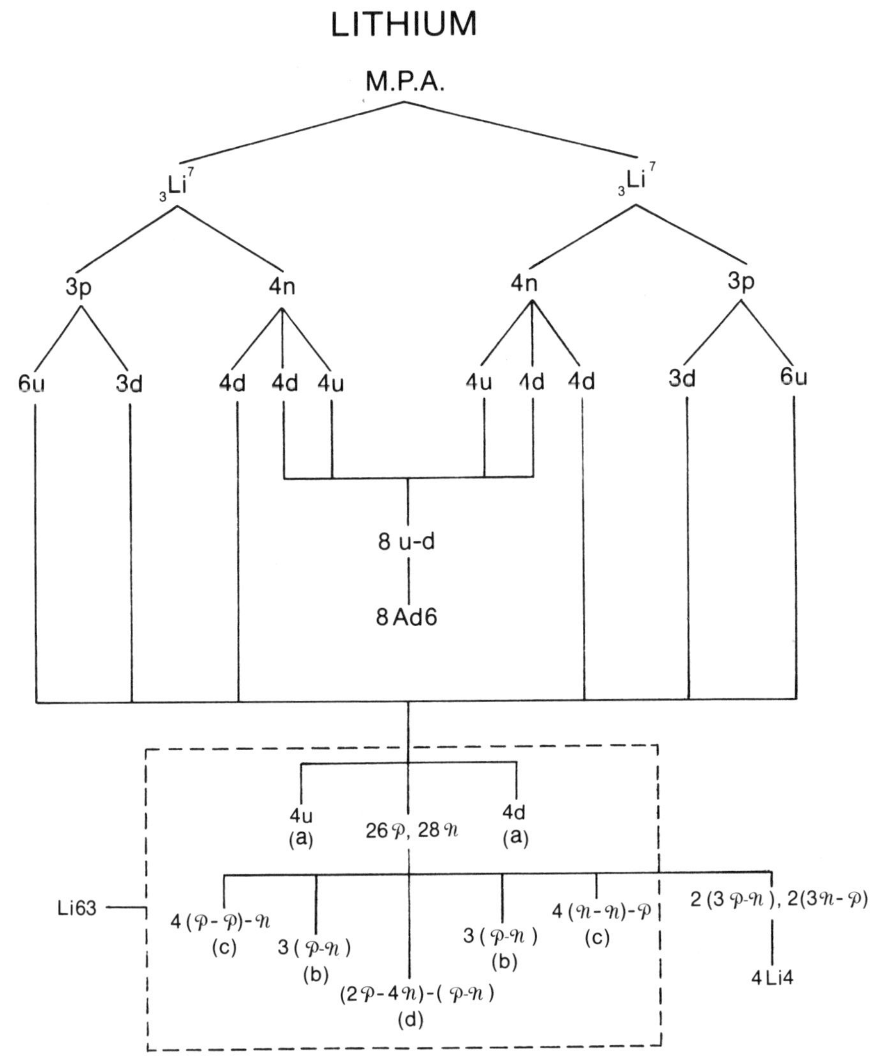

Fig. 7.25

domain. In the Li63 (shown in figure 7.26), there are four u and four d quarks at opposite ends of the spike form, a tetrahedral cluster of 𝒫-𝒫 diomegons with an 𝒩 omegon as the nucleus, a similar cluster of 𝒩-𝒩 with a 𝒫 omegon as nucleus, two groups of three 𝒫-𝒩 diomegons, and, in the centre, two 𝒫 and four 𝒩 omegons, which revolve in a common orbit about a 𝒫-𝒩 diomegon, not a group of three U.P.A.'s, as recorded. It is predicted that the extra U.P.A. is an error of observation; that is, the Li63 should contain sixty-two U.P.A.'s, not sixty-three. It is identified thus:

$$\text{Li63} = 30\mathcal{P} + 32\mathcal{N} .$$

The products of the disintegration of the lithium M.P.A. are shown in figure 7.27. Note:

1. Two (+) and two (−) Li4 make up the globe.
2. A (+) and a (−) triplet make up the Ad6 group.
3. Four (+) and four (−) triplets (*a*) are in the Li63 group.
4. The tetrahedra *c* consist of four (+) and four (−) duads.
5. The group *b* consists of three duads whose form (shown at stage 2) was identified in chapter 5 as indicating a 𝒫-𝒩 diomegon.

All these features agree with the quark/omegon composition given to the groups. The group *d* should have one fewer U.P.A.

BERYLLIUM M.P.A.

Beryllium has the simplest M.P.A. of the Tetrahedral Group A elements. It consists of a tetrahedral array of four funnels, each containing forty U.P.A.'s and radiating from a central globe (Be4) of four U.P.A.'s (fig. 7.28). A funnel contains four ovoids (Be10), each being made up of ten U.P.A.'s present as two triplets and a quartet (fig. 7.29).

The M.P.A. is formed from two $_4\text{Be}^9$ nuclei (fig. 7.30). The Be4 globe is predicted to contain two U.P.A.'s (𝒩 omegons), not four as counted. This is the only discrepancy between theory and micro-psi observations here. A Be10 consists of either two u quarks and a tetrahedral bound state of four

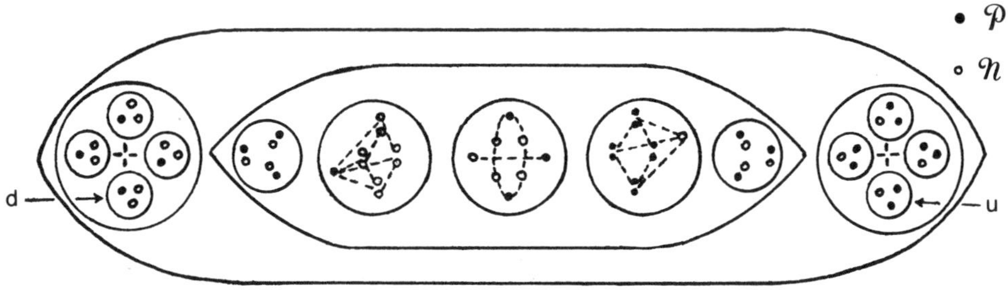

Fig. 7.26. Li63 group

LITHIUM
E2
E3
E4
2−
2+
d
8−
u
8+
a
4+
a
4−
u
6
d
b
4
2
c
4
1
d
u-d
−
+
8
1
a
2
d
1
b
2
c
2
1
8
b
c
d
c
b
a
a
Globe
Petal
Spike Li63

Fig. 7.27

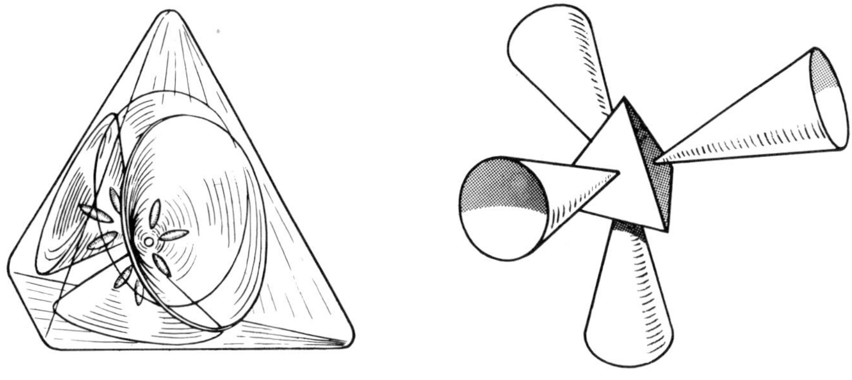

Fig. 7.28. M.P.A. of beryllium

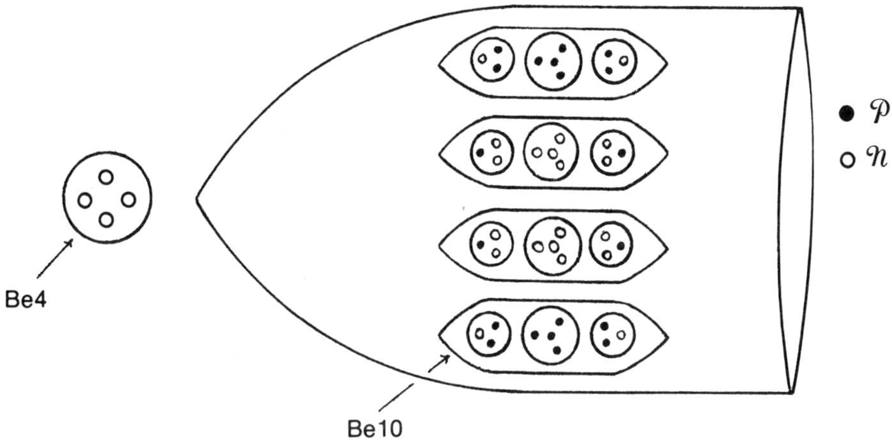

Fig. 7.29. Funnel and central globe

𝒫 omegons or two d quarks and a bound state of four 𝒩 omegons, the four omegons being bound by SU(4) strings. Cohesion between the quarks and the group of four is provided by the interaction between their strings, and the quarks are colourless, being clusters of three SU(3) monopoles. Two types of Be10 are predicted:

$$(+)\ \text{Be10} = \text{u},\ 4\mathcal{P},\ \text{u}\ ,$$

$$(-)\ \text{Be10} = \text{d},\ 4\mathcal{N},\ \text{d}\ ,$$

differing in the sign of the charges of corresponding groups. The disintegration diagram (fig. 7.31) confirms this, for it shows a (+) and (−) Be10. In the former, the U.P.A.'s point outward; in the latter they point inward. This is indicative of positively and negatively charged groups, as predicted above.

BERYLLIUM

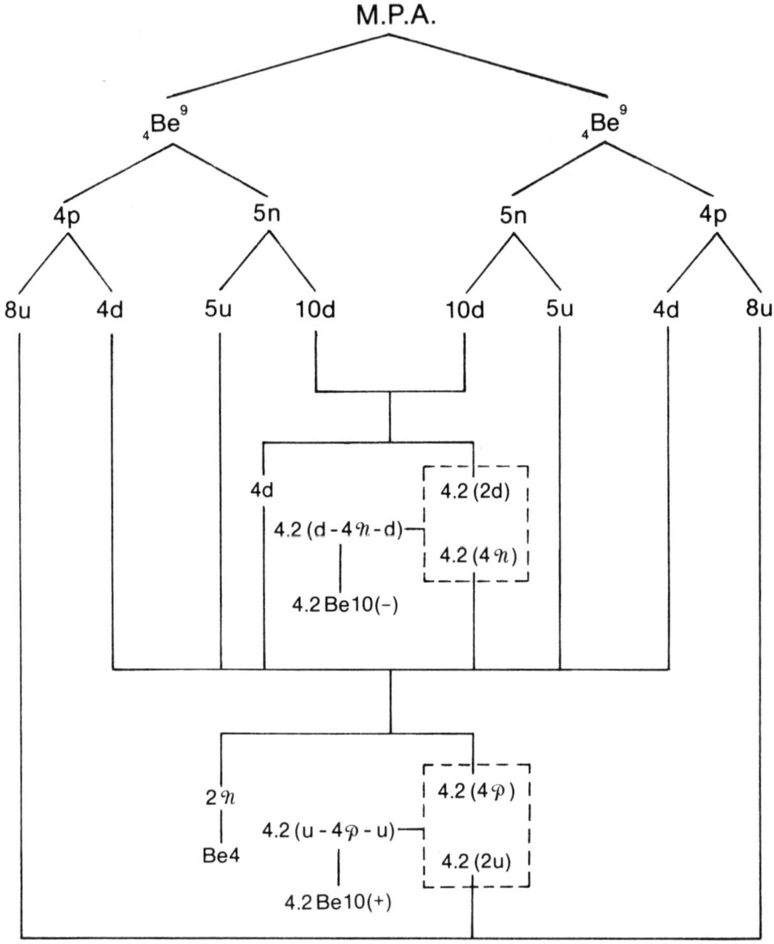

Fig. 7.30

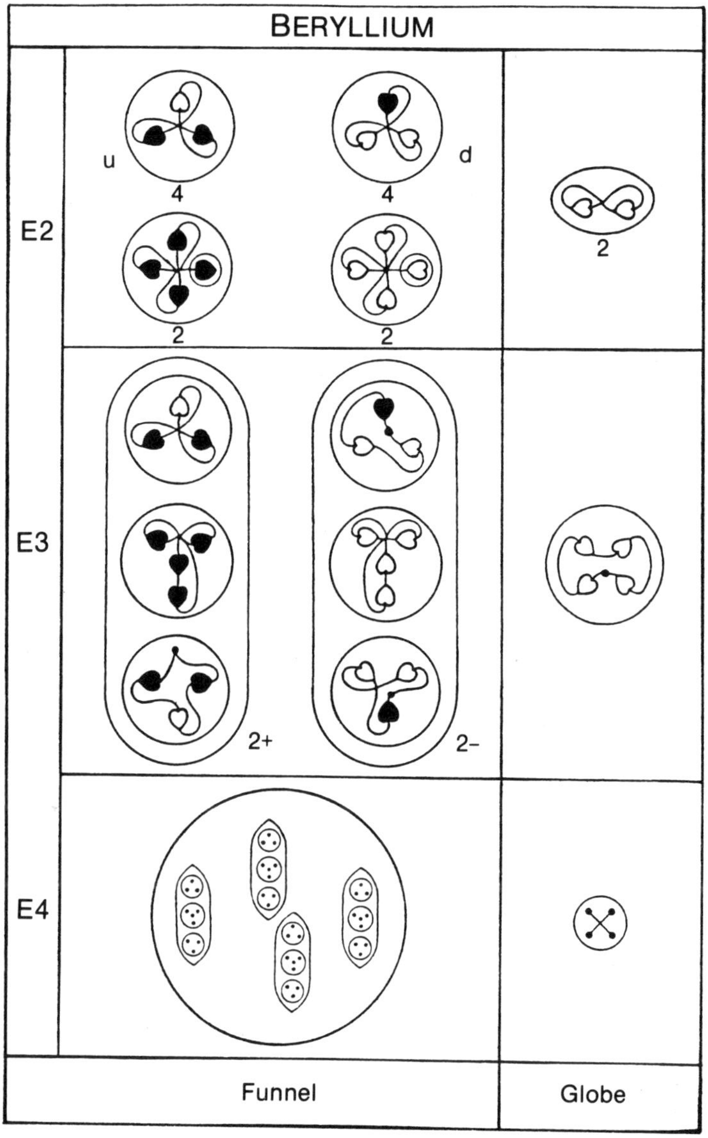

Fig. 7.31

BORON M.P.A.

Boron has the simplest M.P.A. of all Cube Group A elements. It consists of a cubic array of six funnels emanating from a central globe (fig. 7.32). A funnel contains an Ad6 group and four ovoids, each of which contains two Hydrogen Triplets (H3) (fig. 7.33). The globe has four spheres, each containing five U.P.A.'s (B5).

The M.P.A. is formed from two ${}_5B^{11}$ nuclei, which provide thirty-two u and thirty-four d quarks (fig. 7.34). Each funnel contains either a u-u or a d-d diquark (Ad6), two pairs of u quarks (H3), and two pairs of d quarks (H3). These quark pairs are not diquarks because the quarks are not directly bound by strings but are coupled by the interaction between their internal strings that bind the omegons—the analogue of the binding between nucleons in nuclei. The other two u and four d quarks break up, and their omegons form two bound states of three 𝒫 and two 𝒩 omegons and two bound states of one 𝒫 and three 𝒩 omegons. These are the B5 groups. In-

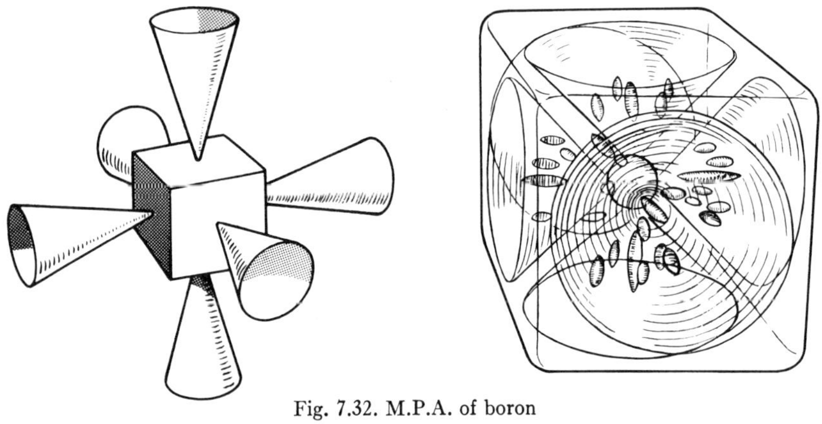

Fig. 7.32. M.P.A. of boron

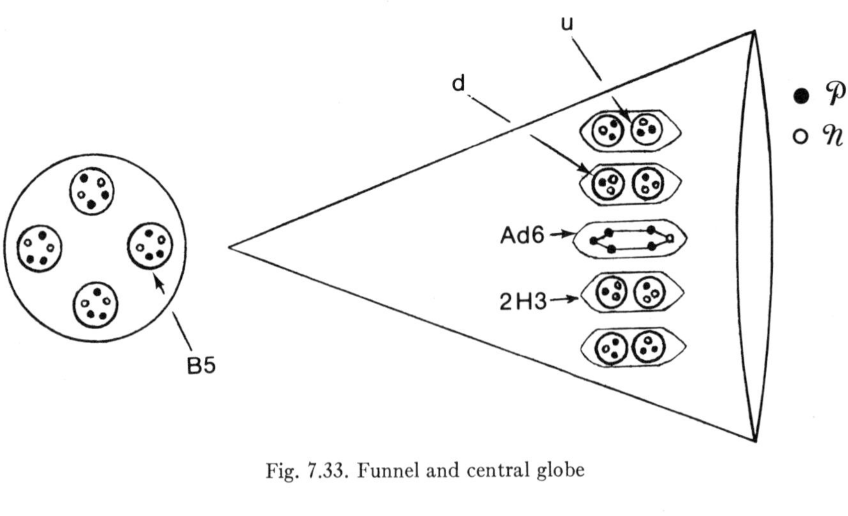

Fig. 7.33. Funnel and central globe

stead of the reported four B5 groups, two B5 and two quartets of U.P.A.'s are predicted to be in the central globe. A B5 group is a bound system of five omegon magnetic monopoles, each with magnetic charges of ± 2 embedded in an SU(5) Higgs vacuum domain (fig. 7.35). It is a u-d diquark without an $\mathcal{N}$ omegon. The disintegration diagram (fig. 7.36) confirms this, for the

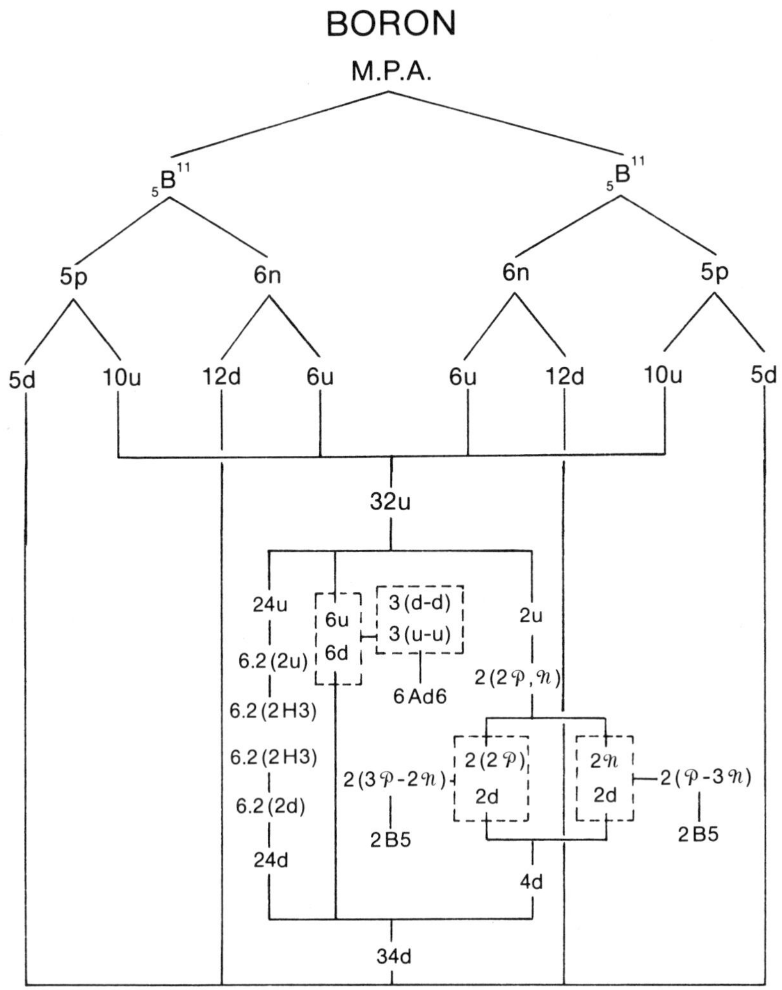

Fig. 7.34

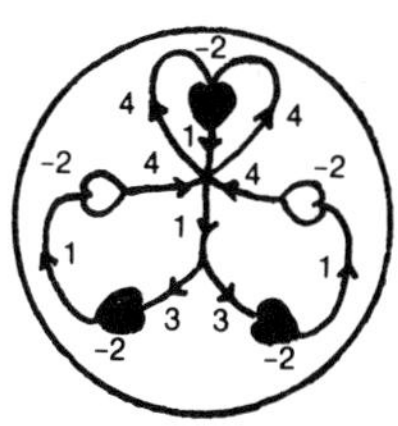

Fig. 7.35. Magnetic charges and flux quanta in the B5 group

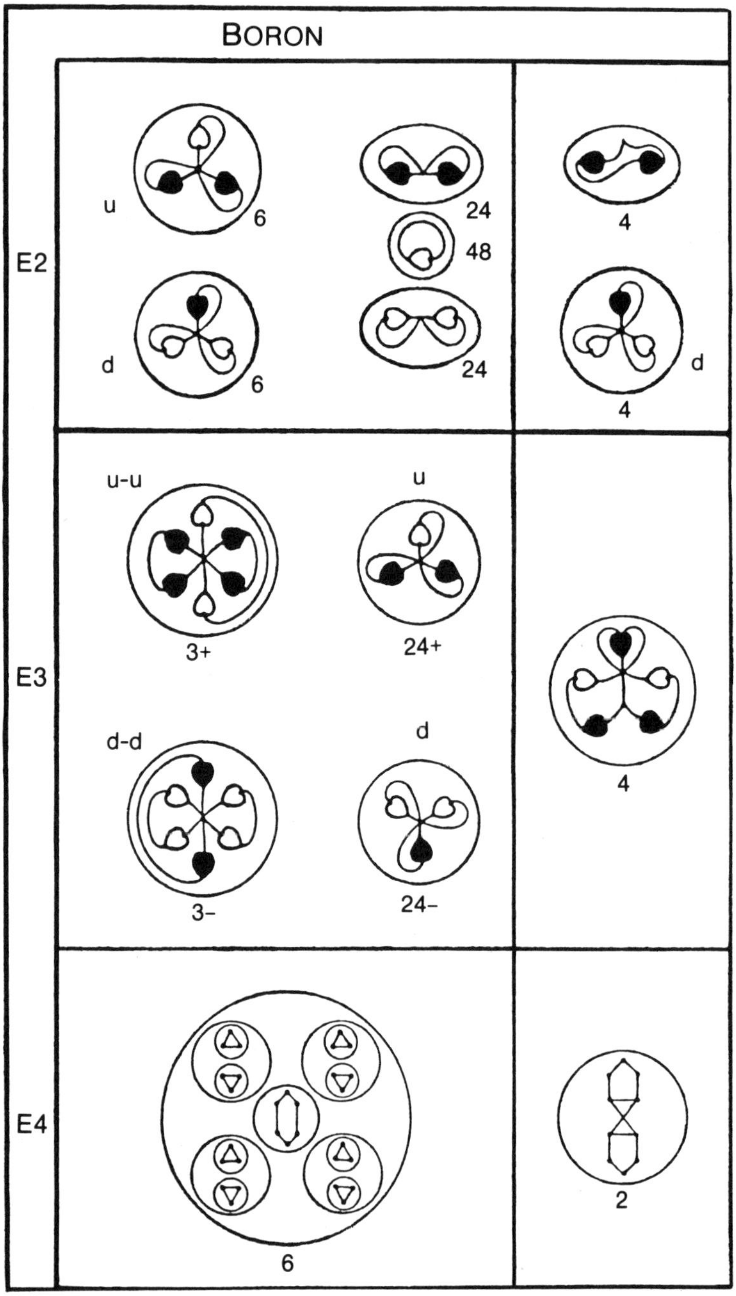

Fig. 7.36

B5 splits up at stage 2 into a triplet and a duad, the string configuration of the latter showing that the monopoles have two units of magnetic charge. It also confirms that the triplet is (−), that is, a d quark, and that the duad is (+), that is, a ℘-℘ diomegon. Finally, it shows that six funnels contain twenty-four (+) and twenty-four (−) triplets, that is, twenty-four u and twenty-four d quarks, in agreement with what is required by the Omegon Model.

CARBON M.P.A.

Of the elements that belong to the Octahedron Group A, carbon has the simplest M.P.A. It consists of an octahedral array of eight funnels, projecting from a central sphere that contains four free U.P.A.'s (fig. 7.37). There are two types of funnels: the C27 funnel contains three Ad6 groups, two H3′ triplets, and one H3 triplet; the C26 funnel contains two Ad6, one B5, two H3, and one H3′ (fig. 7.38).

The M.P.A. is formed from two $_6C^{12}$ nuclei (fig. 7.39). In each of the C27 funnels there are three u-d diquarks (Ad6) and three u quarks (2H3′ +

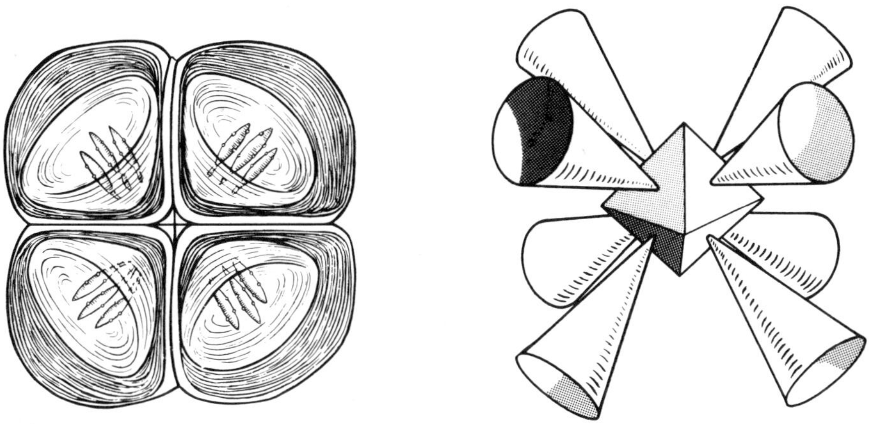

Fig. 7.37. M.P.A. of carbon

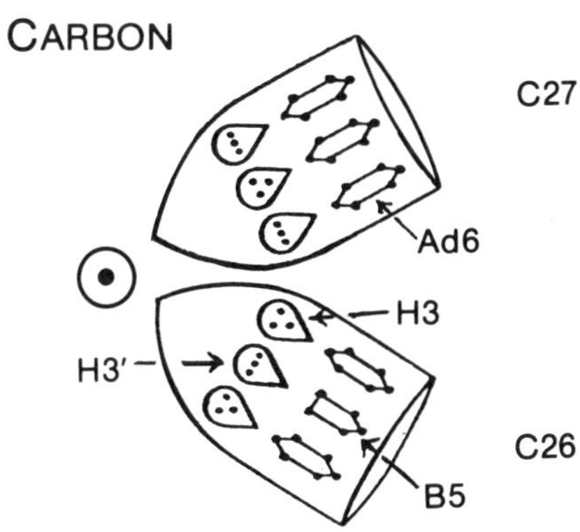

Fig. 7.38. A pair of C26 and C27 funnels

H3). These have a total electric charge of +3. Since the M.P.A. has a charge of +12, each of the remaining identical funnels would contain twenty-seven U.P.A.'s, with a total charge of zero unless the M.P.A. had a central core. But, as explained earlier in this chapter, the groups in a funnel cannot be neutral. Hence, one or more U.P.A.'s of negative charge must be absent from each funnel in order to leave the remaining bodies with a net positive charge. According to the micro-psi account: "In the centre of the octahedron is a globe containing four Anu, each within its own wall; these lie on the dividing lines of the faces and each holds a pair of funnels together.

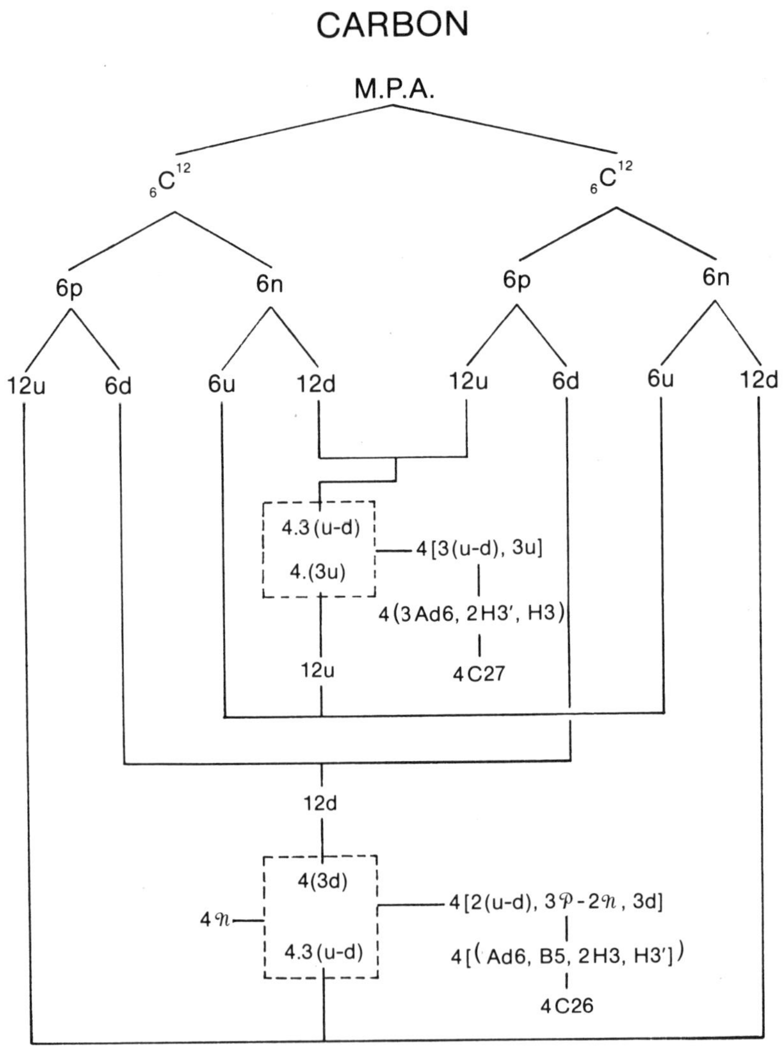

Fig. 7.39

It seems as though this Anu had been economically taken from one Ad6 in the funnels, to form the link."[19] Since it is "within its own wall," it must be a free omegon, that is, one that is not strongly bound to other omegons by strings. It follows that the link must be Coulombic and that the free U.P.A. is an 𝔑 omegon (fig. 7.40). This prediction is confirmed by the disintegration diagram (fig. 7.41) for a pair of funnels. It shows that the B5 breaks up at stage 2 into a (−) triplet (d quark) and a (+) duad (𝒫-𝒫 diomegon). Since the Ad6 group is a u-d diquark, the U.P.A. "economically taken from one Ad6" must be an 𝔑 omegon. Note also that the five Ad6 groups split up at stage 2 into five (+) triplets (u quarks) and five (−) triplets (d quarks), in agreement with their predicted composition.

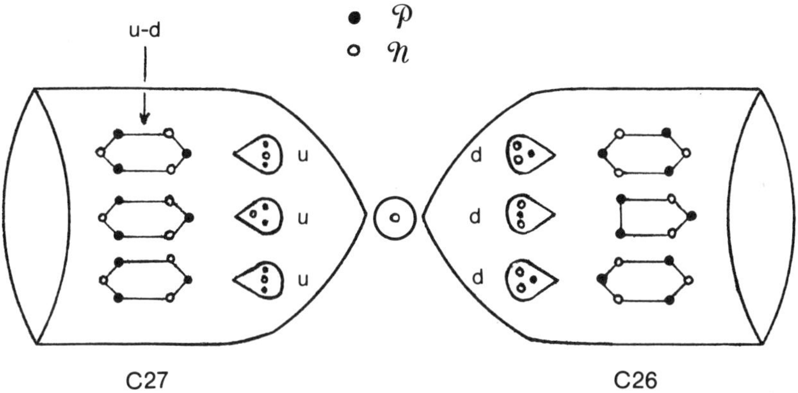

Fig. 7.40

NITROGEN M.P.A.

The M.P.A. is shaped like a sphere and contains six bodies with 261 U.P.A.'s (fig. 7.42). A balloon-shaped body (N110) floats in the middle of the sphere. It contains six smaller spheres arranged in two horizontal rows, each sphere enclosing seven duads (N14). It also contains a long ovoid made up of various groups of U.P.A.'s. Below the N110 is a cluster of seven spheres (N63), each containing three Hydrogen Triplets (N9). Around the N110 are two spheres, each containing five groups of two duads (N20), and two spheres that each contain four globes with six U.P.A.'s in each one (N24).

If the M.P.A. were formed from two $_7N^{14}$ nuclei, the discrepancy between the counted number of U.P.A.'s and the predicted number of omegons would be nine. During the discussion in chapter 4 of the histograms displaying the predicted errors and their frequencies of occurrence, it was pointed out that nitrogen is one of three elements whose M.P.A.'s have

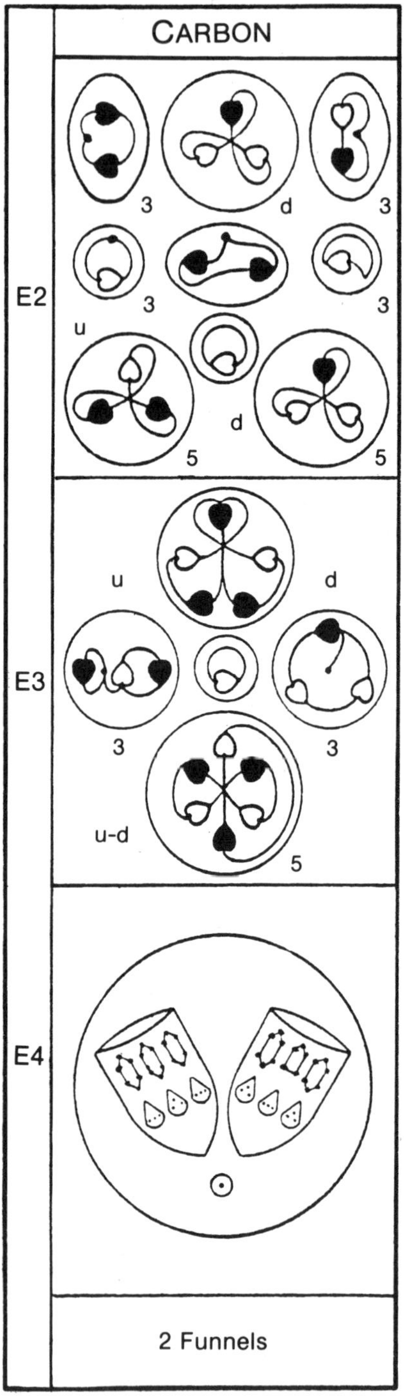

Fig. 7.41

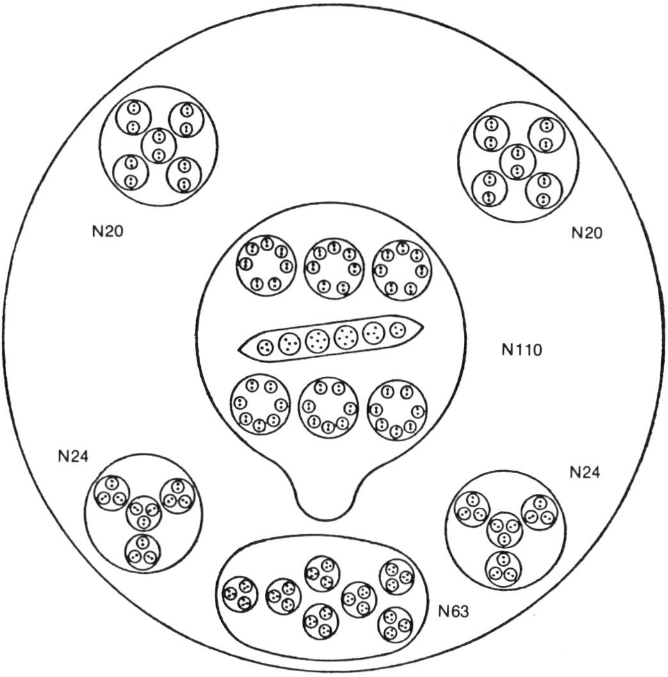

Fig. 7.42. M.P.A. of nitrogen

populations consistent with their formation from two nuclei that differ in mass number by one unit, that is, in a neutron made up of nine omegons. The gas was one of the first elements in the atmosphere that were examined in 1895 by Besant and Leadbeater. Although the relative abundance of the $_7N^{15}$ isotope is only 0.365 percent, compared with 99.635 percent for $_7N^{14}$,[20] the possibility exists that the nitrogen M.P.A., as originally described, was the result of micro-psi selection of a nitrogen molecule made up of both nuclides. The high degree of symmetry of the groups in the M.P.A. makes an observational error of nine difficult to explain in terms of smaller errors for these groups. It is unlikely, therefore, that the extra nine are due to overcounting. With the assumption that both nuclides participated in the formation of the M.P.A., it will now be demonstrated that every reported detail is consistent with the basic theory.

The M.P.A. is formed from a $_7N^{14}$ and a $_7N^{15}$ nucleus (fig. 7.43). Seven neutrons make up the N63 group, an N9 group being a neutron made up of two d quarks and a u quark. This is confirmed by the disintegration diagram (fig. 7.44), which shows that each N9 breaks up at stage 2 into two (−) triplets and one (+) triplet. Four u quarks and four d quarks break up into diomegons and form a cluster of four bound states of three $\mathcal{P}$-$\mathcal{N}$ diomegons:

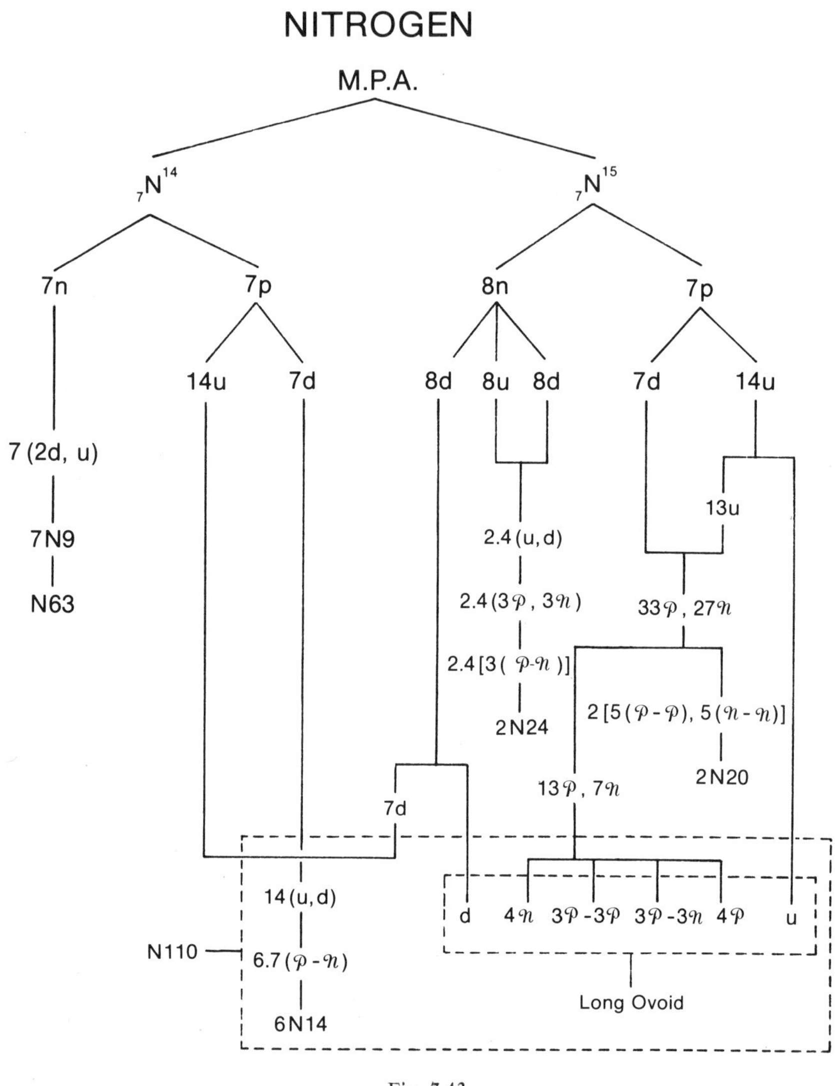
NITROGEN
M.P.A.
$_7N^{14}$
$_7N^{15}$
7n
7p
8n
7p
14u
7d
8d
8u
8d
7d
14u
7 (2d, u)
7N9
N63
13u
2.4 (u, d)
2.4 (3𝒫, 3𝒩)
33𝒫, 27𝒩
2.4 [3(𝒫-𝒩)]
2N24
2 [5(𝒫-𝒫), 5(𝒩-𝒩)]
2N20
13𝒫, 7𝒩
7d
14 (u, d)
N110
6.7 (𝒫-𝒩)
6N14
d
4𝒩
3𝒫-3𝒫
3𝒫-3𝒩
4𝒫
u
Long Ovoid

Fig. 7.43

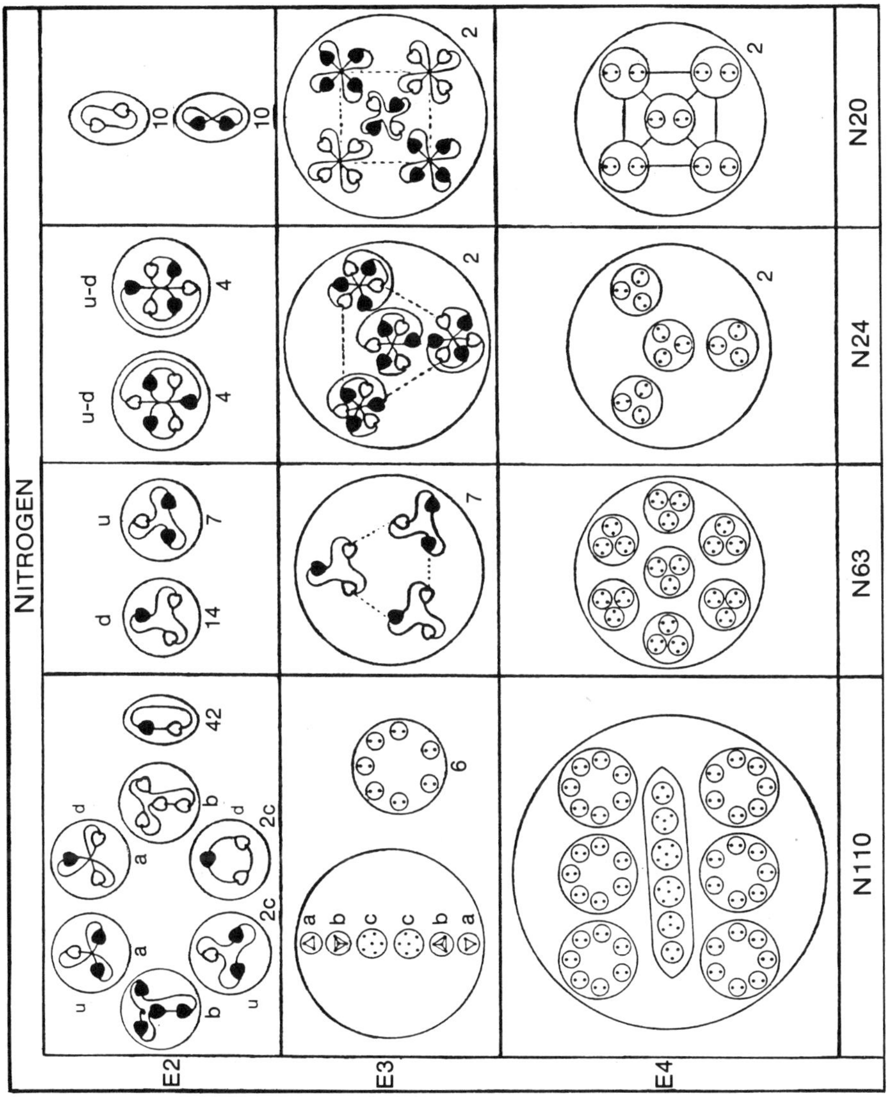

Fig. 7.44

$$
\begin{aligned}
N24 &= 4u + 4d \\
&= 4(2\mathcal{P}, \mathcal{N}) + 4(\mathcal{P}, 2\mathcal{N}) \\
&= 4.3(\mathcal{P}\text{-}\mathcal{N}) \,.
\end{aligned}
$$

Each duad is a pair of SU(2) monopoles bound by Nielsen-Olesen vortices. Mutual interaction of their strings binds them together in groups of three. Each group has an electric charge of $+1/3$, and they are bound in a tetrahedral array by the strong coupling between their strings (fig. 7.45). This prediction is confirmed by the following observation: "On the E_3 level, each assumes a tetrahedral form, with six Anu at each point."[21] The disintegration diagram shows that a group of three duads rearranges itself at stage 3 into an Ad6. This is possible because the group is made up of the omegons in a u and a d quark, and an Ad6 is an u-d diquark. An N24 has an electric charge of $+4/3$. An N20 consists of five $\mathcal{P}$-$\mathcal{P}$ and five $\mathcal{N}$-$\mathcal{N}$. They are bound in a square pyramidal array (fig. 7.46). The isospin states of these diomegons are confirmed by figure 7.44, which shows that the two N20 break up at stage 2 into ten (+) and ten (−) duads, the former being $\mathcal{P}$-$\mathcal{P}$ and the latter being $\mathcal{N}$-$\mathcal{N}$. An N20 has an electric charge of $+10/9$. The N110 is made up of the following quarks and omegons:

$$N110 = 6.7(\mathcal{P}\text{-}\mathcal{N}) + u + 4\mathcal{P} + 3\mathcal{P}\text{-}3\mathcal{N} + 3\mathcal{P}\text{-}3\mathcal{P} + 4\mathcal{N} + d \,.$$

The first term is the six groups of seven duads that surround the long ovoid in the centre of the nitrogen M.P.A. (fig. 7.47). The disintegration diagram confirms the composition given above for them, for it indicates that the N110 releases 42 (6×7) identical duads at stage 2, their type having been previously identified as $\mathcal{P}$-$\mathcal{N}$. The other terms are the groups in the long ovoid. There are u and d quarks, two tetrahedra of four $\mathcal{P}$ and four $\mathcal{N}$

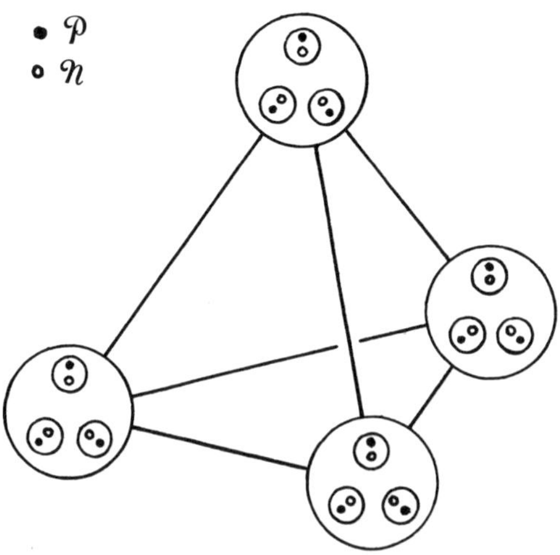

Fig. 7.45. N24 group

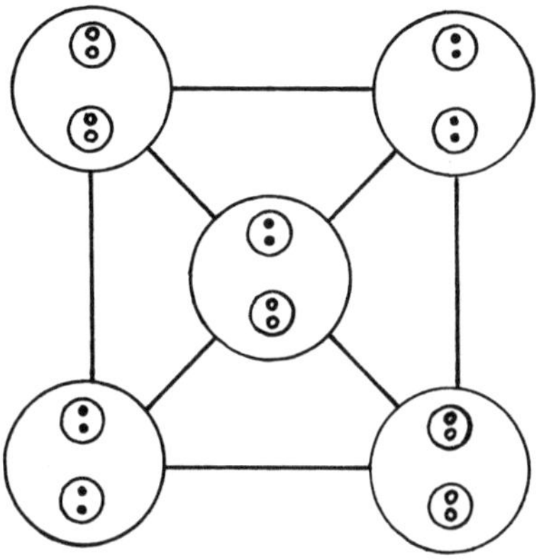

Fig. 7.46. N20 group

omegons, and two sextets of omegons (fig. 7.48). The disintegration diagram confirms that the long ovoid has (+) and (−) triplets (*a*) and (+) and (−) tetrahedra (*b*). The N110 contains fifty-eight 𝒫 and fifty-two 𝒩 omegons. It has an electric charge of +82/9. The total charge of the bodies in the M.P.A. is

$$\begin{aligned} \text{M.P.A.} &= \text{N110} + 2\ \text{N20} + 2\ \text{N24} + \text{N63} \\ +14 &= +82/9 + 2(+10/9) + 2(4/3) + 0 \\ &= +14\,, \end{aligned}$$

that is, twice the charge of a nitrogen nucleus.

In conclusion, every detail of the M.P.A. of nitrogen has been accounted for on the basis that it is assembled from a $_7N^{14}$ and a $_7N^{15}$ nucleus.

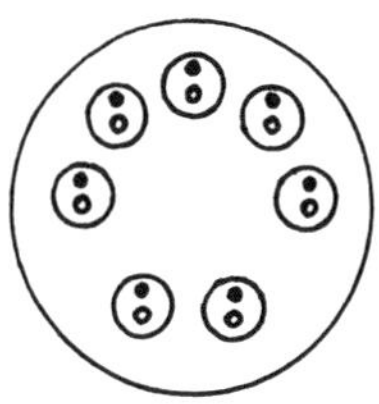

Fig. 7.47. N14 globe

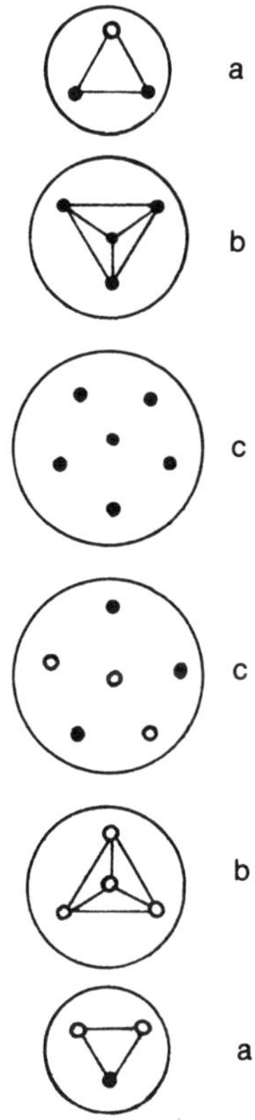

Fig. 7.48

OXYGEN M.P.A.

The form of this M.P.A. is quite unlike that of any other element. Enclosed in an ovoid (fig. 7.49) are two spiral-shaped bodies revolving rapidly in opposite directions about a common axis parallel to their length and passing through their centre. Each spiral body is a chain of fifty-five

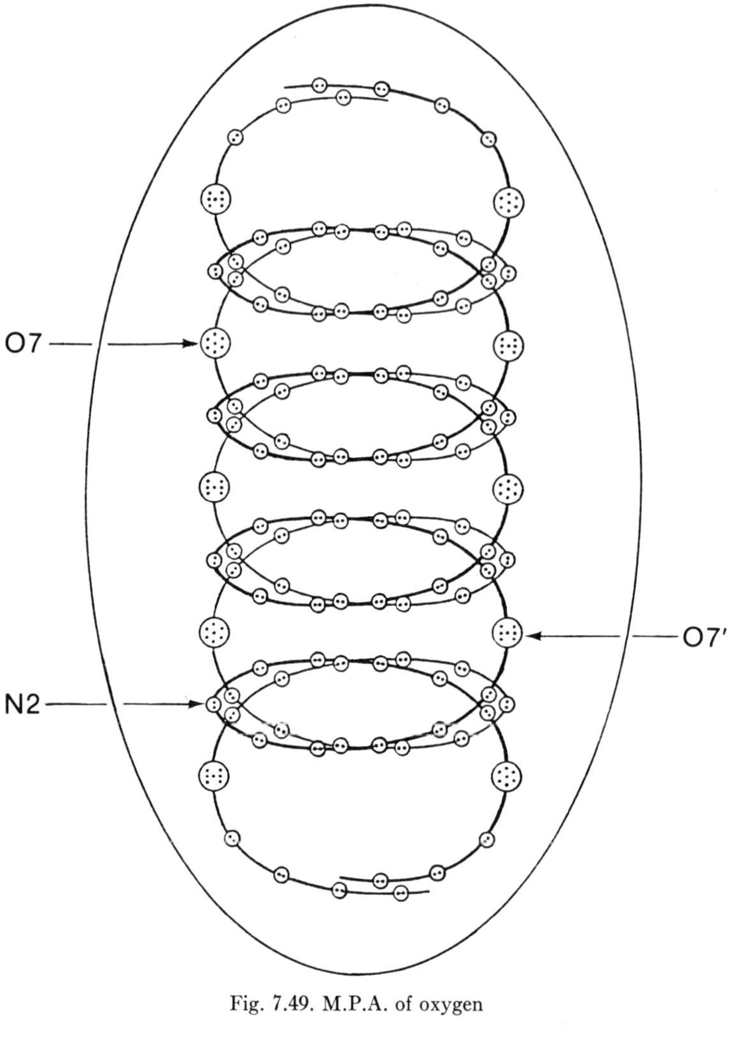

Fig. 7.49. M.P.A. of oxygen

much smaller spheres containing two U.P.A.'s (N2). Each body is shaped like a spring with five coils and has a brilliantly lit point shining on each coil (fig. 7.50). Higher magnification reveals that they are globes containing seven U.P.A.'s that are bound together by continuous lines of force. They divide the spiral body into five sections, each consisting of a globe with six groups of N2 above it and five groups of N2 below it. The principal difference between the two spirals is in their globes. In one spiral, denoted (+), the globes (called "O_7") contain a central (+) U.P.A. surrounded by a face-centred cubic array of six U.P.A.'s, three (+) and three (−) (fig. 7.51). In the other spiral, denoted (−), the globes (called "O_7'") contain a central (−) U.P.A., and three (+) and three (−) U.P.A.'s are arranged in pairs

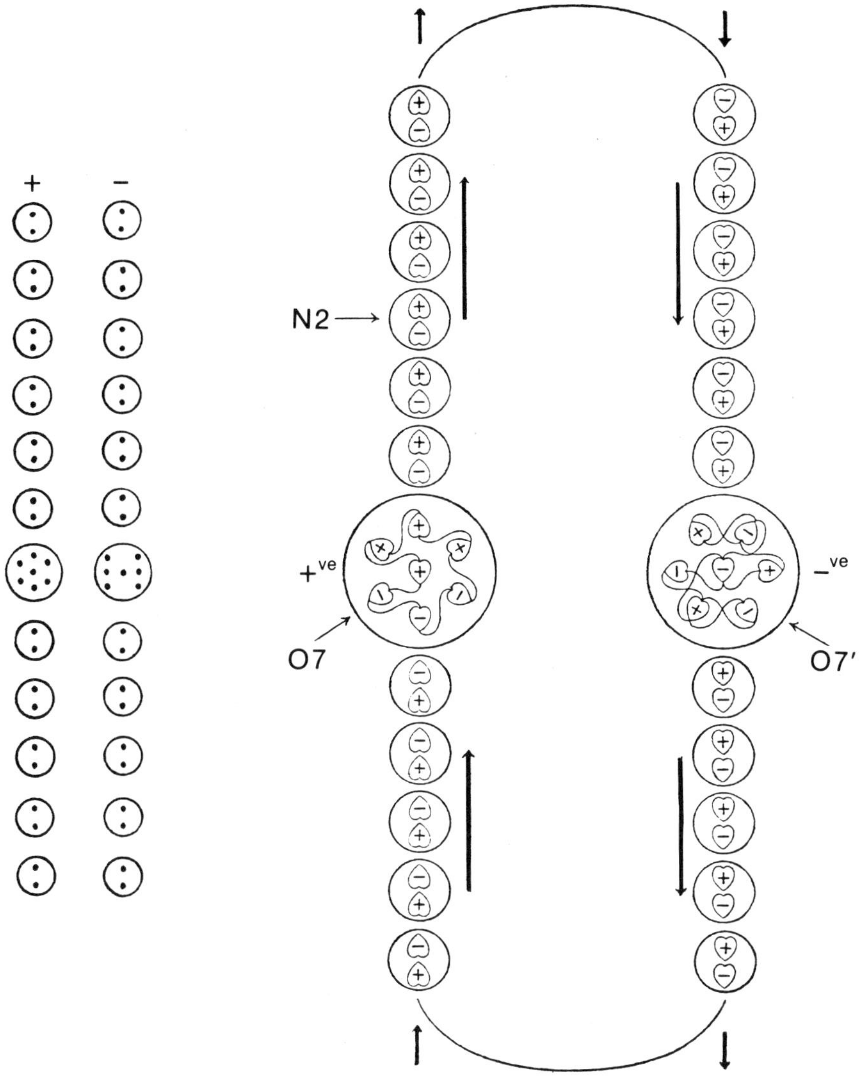

Fig. 7.50. Section of the M.P.A. of oxygen

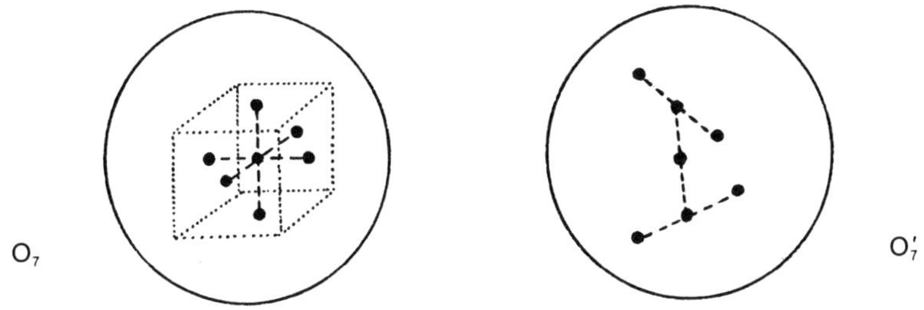

Fig. 7.51

about it, two pairs being perpendicular to each other and inclined to the third pair, which lies on a line with the central U.P.A. A remarkably dynamic interplay of forces is reported to maintain great rigidity in the helical form of the M.P.A. It has the illusory appearance of a solid cylinder, because of the rapid rotation of the (+) and (−) spirals in opposite directions about a common axis.

Like those of hydrogen, helium, and nitrogen, the M.P.A. of oxygen cannot be assigned to any of the seven classes of M.P.A.'s. No reasons are offered to explain why micro-psi examination of an oxygen molecule disintegrates the nuclei into strings of dipoles joined end to end and interspersed by clusters of seven omegons bound by vortex strings (see fig. 7.50). Discussion will be confined to a mathematical demonstration of the compatibility between its composition in terms of (+) and (−) U.P.A.'s and Hypotheses 1 and 2. Oxygen is the only element apart from hydrogen where this specification is complete, and it provides a stringent test of the Omegon Model prediction of monopole charges inside a proton or neutron.

The M.P.A. is formed from two $_8O^{16}$ nuclei, which provide 144 $\mathcal{P}$ and 144 $\mathfrak{N}$ omegons (fig. 7.52). The disintegration diagram (fig. 7.53) indicates that the N2's in the (+) spiral are $\mathcal{P}$-$\mathcal{P}$ diomegons and that the N2's in the (−) spiral are $\mathfrak{N}$-$\mathfrak{N}$ diomegons. The composition of the O_7 and O_7' groups is unknown. A total of 145 U.P.A.'s were counted in each spiral, and it must be presumed that two U.P.A.'s too many were counted. The spirals have the following composition:

	(+) Spiral		(−) Spiral
O_7 (× 5):	4 (+), 3 (−)	O_7' (× 5):	4 (−), 3 (+)
N2 (× 55):	1 (+), 1 (−)	N2 (× 55):	1 (+), 1 (−)

It is unlikely that the error occurred in only one spiral, for this removes the obvious symmetry displayed by the spirals. Therefore, one too many U.P.A.'s were counted in each. It is also unlikely that the error is in the N2 groups, for this would leave one monopole unpaired in each spiral. One too many U.P.A.'s in an O_7 and one too many in an O_7' were counted. Furthermore, it is likely that these were a (+) and a (−) because the M.P.A. would then contain 144 (+) and 144 (−) U.P.A.'s—just the number predicted by Hypothesis 2, since

$$\text{M.P.A.} = 144\,(+) + 144\,(-) = 16[9\,(+) + 9\,(-)]\,,$$

and the M.P.A. is formed from thirty-two nucleons, two of which may contain nine (+) and nine (−) U.P.A.'s, according to the earlier discussion of the hydrogen M.P.A. The corrected O_7 and O_7' groups will be called "O_6"

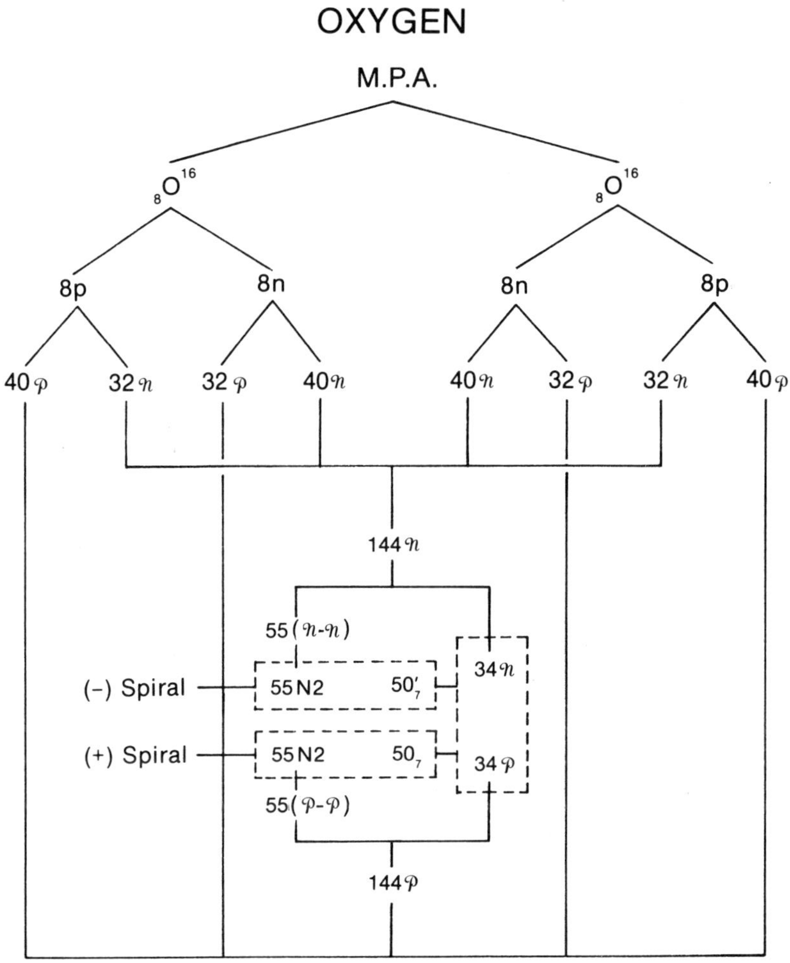
OXYGEN
M.P.A.
$_8O^{16}$
$_8O^{16}$
8p
8n
8n
8p
40 p
32 n
32 p
40 n
40 n
32 p
32 n
40 p
144 n
55 (n-n)
(–) Spiral
55 N2
50′7
34 n
(+) Spiral
55 N2
507
34 p
55 (p-p)
144 p

Fig. 7.52

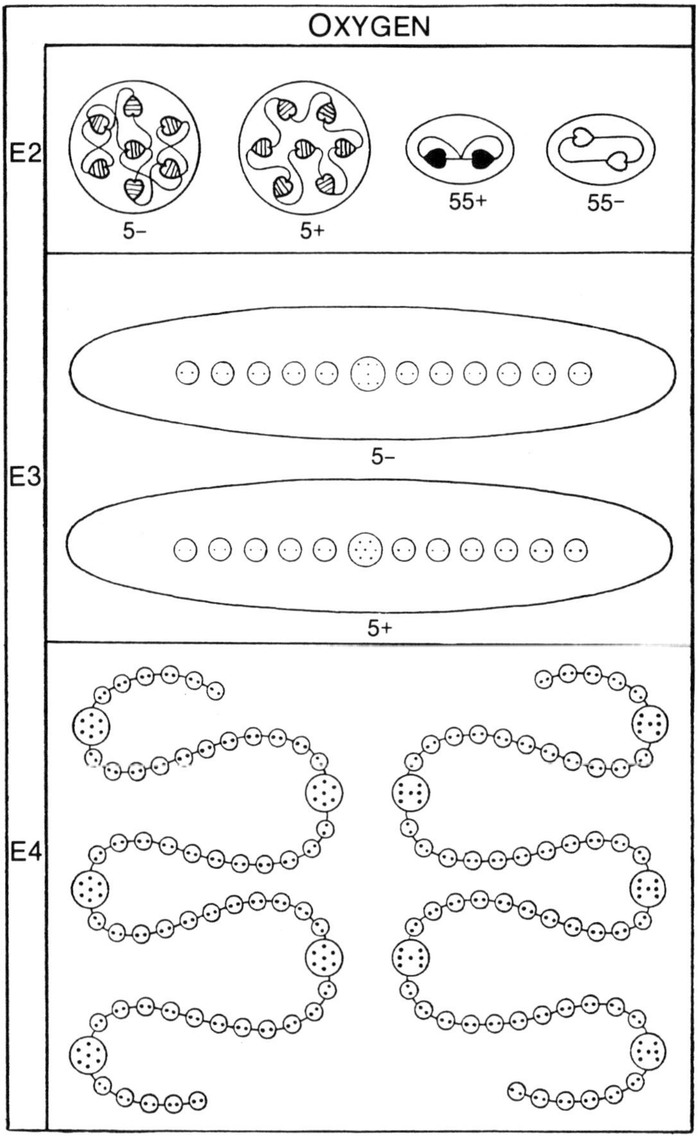

Fig. 7.53

and "O_6'," respectively. The predicted (correct) composition of the spirals is shown below:

(+) Spiral		(−) Spiral	
O_7 (× 4):	4 (+), 3 (−)	O_7' (× 4):	4 (−), 3 (+)
O_6 (× 1):	3 (+), 3 (−)	O_6' (× 1):	3 (−), 3 (+)
N2 (× 55):	1 (+), 1 (−)	N2 (× 55):	1 (+), 1 (−)
Total:	74 (+), 70 (−)	Total:	74 (−), 70 (+)

The O_7 is a stable, magnetically neutral bound state of seven monopoles embedded in an SU(7) Higgs vacuum domain and joined by Nielsen-Olesen vortex strings. Let X, X', Y, and Y' be the number of monopoles of charge 1, 2, −1, and −2, respectively. Then

$$X + X' = 4 ;$$

$$Y + Y' = 3 ;$$

$$X + 2X' - Y - 2Y' = 0 \pmod 7 .$$

This results in

$$X - Y = 2 \pmod 7 .$$

There are three allowed solutions:

(A) $X = X' = 2 ; \quad Y = 0 ; \quad Y' = 3 .$

(B) $X = 3 ; \quad X' = Y = 1 ; \quad Y' = 2 .$

(C) $X = 4 ; \quad X' = 0 ; \quad Y = 2 ; \quad Y' = 1 .$

The O_7' is also a cluster of seven monopoles bound by SU(7) strings. For this we have

$$Y - X = 2 \pmod 7)$$

and the allowed solutions are those above, with X and Y interchanged (and X' and Y' interchanged). These will be called (A'), (B'), and (C'). The O_6 and O_6' groups must be systems of six monopoles bound by SU(6) strings. For either group,

$$X + X' = 3 ;$$

$$Y + Y' = 3 ;$$

$$X + 2X' - 2Y' = 0 \pmod 6 .$$

Then

$$X - Y = 0 \pmod 6 .$$

There are four allowed solutions:

$$\text{(a)}\ X = 3\,; \quad X' = 0\,; \quad Y = 3\,; \quad Y' = 0\,.$$

$$\text{(b)}\ X = 2\,; \quad X' = 1\,; \quad Y = 2\,; \quad Y' = 1\,.$$

$$\text{(c)}\ X = 1\,; \quad X' = 2\,; \quad Y = 1\,; \quad Y' = 2\,.$$

$$\text{(d)}\ X = 0\,; \quad X' = 3\,; \quad Y = 0\,; \quad Y' = 3\,.$$

Are any of the solutions (A)–(C) and (a)–(d) compatible with Hypothesis 2 and the monopole charge composition of a nucleon? According to the Omegon Model, the magnetic charges in a nucleon are 2 (+2), 3 (+1), 3 (−2), and 1 (−1) or 3 (+2), 1 (+1), 2 (−2), and 3 (−1), the number before the parentheses denoting the number of each charge. There should be sixteen of each type forming the M.P.A. The charges in the M.P.A. are 80 (+2), 64 (+1), 80 (−2), and 64 (−1). Figure 7.53 shows that the spirals disintegrate at stage 2 into two types of duads, each having a (+) and a (−) U.P.A. One has the string configuration of a pair of SU(2) monopoles with charges +1 and −1, and the other has charges +2 and −2. The M.P.A. has fifty-five of each. There must be, therefore, 25 (+2), 25 (−2), 9 (+1), and 9 (−1) monopoles in the O_7, O_7', O_6, and O_6' groups. Let A, B, and C be the number of O_7 groups corresponding to the solutions (A), (B), and (C), respectively. Let A', B', and C' be the number of O_7' groups corresponding, respectively, to the solutions (A′), (B′), and (C′). Let a, b, c, and d be the number of monopoles of charge +1, +2, −1, and −2, respectively, present in an O_6 or O_6' group. Then

$$E + F + b = 21\,, \quad b = d\,, \quad a - c\,, \quad \text{and} \quad a + b = 6\,,$$

where

$$E = 2A + B\,, \quad F = 2A' + B'\,.$$

Since

$$A + B + C = 4 = A' + B' + C'\,,$$

it follows that

$$0 \le E \le 8\,, \quad 0 \le F \le 8\,, \quad \text{and} \quad 0 \le E + F \le 16\,.$$

Therefore,

$$5 \le b \le 6\,.$$

There are three possible solutions:

(1) $a = 1, b = 5$: $\quad E = F = 8\,,$

that is, $A = A' = 4$, $B = B' = C = C' = 0$. This corresponds to solutions (c) and (d) for the O_6 and O_6' groups.

(2) $a = 0, b = 6$: $\quad E = 7 \quad \text{and} \quad F = 8\,,$

that is, $A = 3$, $B = 1$, $A' = 4$, $B' = C = C' = 0$. This corresponds to solution (d) for the O_6 and O'_6 groups.

$$(3)\ a = 0, b = 6: \qquad E = 8 \quad \text{and} \quad F = 7\,,$$

that is, $A = 4$, $B = 0$, $A' = 3$, $B' = 1$, $C = C' = 0$. This also corresponds to solution (d) for the O_6 and O'_6 groups. The magnetic charges of the various groups in the M.P.A. implied by solutions (1), (2), and (3) are given below:

		(+) Spiral		(−) Spiral
(1)	O_7 (× 4):	2 (2), 2 (1), 3 (−2)	O'_7 (× 4):	3 (2), 2 (−1), 2 (−2)
	O_6 (× 1):	2 (2), 1 (1), 1 (−1), 2 (−2)	O'_6 (× 1):	3 (2), 3 (−2)
	N2 (× 55):	1 (1), 1 (−1)	N2 (× 55):	1 (2), 1 (−2)
(2)	O_7 (× 3):	2 (2), 2 (1), 3 (−2)	O'_7 (× 4):	3 (2), 2 (−1), 2 (−2)
	O_7 (× 1):	1 (2), 3 (1), 1 (−1), 2 (−2)	O'_6 (× 1):	3 (2), 3 (−2)
	O_6 (× 1):	3 (2), 3 (−2)	N2 (× 55):	1 (2), 1 (−2)
	N2 (× 55):	1 (1), 1 (−1)		
(3)	O_7 (× 4):	2 (2), 2 (1), 3 (−2)	O'_7 (× 3):	3 (2), 2 (−1), 2 (−2)
	O_6 (× 1):	3 (2), 3 (−2)	O'_7 (× 1):	2 (2), 1 (1), 3 (−1), 1 (−2)
	N2 (× 55):	1 (1), 1 (−1)	O'_6 (× 1):	3 (2), 3 (−2)
			N2 (× 55):	1 (2), 1 (−2)

It has been demonstrated that the (±) U.P.A. structure of the oxygen M.P.A. is compatible with both Hypothesis 2 and the Omegon Model prediction of the magnetic charge composition of a nucleon.

Finally, the electric charges carried by the (+) and (−) spirals are calculated. The (+) spiral is made up of 55 $\mathcal{P}$-$\mathcal{P}$, X $\mathcal{P}$, and $(34 - X)$ $\mathcal{N}$; the (−) spiral is made up of 55 $\mathcal{N}$-$\mathcal{N}$, $(34 - X)$ $\mathcal{P}$, and X $\mathcal{N}$. The charges of the $\mathcal{P}$ and $\mathcal{N}$ omegons are $+5/9$ and $-4/9$, respectively. The charges Q_+ and Q_- of the (+) and (−) spirals are

$$Q_+ = 46 + X\,, \qquad Q_- = -30 - X\,.$$

Since $X \geq 0$, the (+) and (−) spirals are positively and negatively charged, respectively. The value of X is unknown.

FLUORINE M.P.A.

The M.P.A. has a cylindrical body containing two N110 spheres and a pointed end formed by eight funnels that converge to a point (fig. 7.54). Each funnel contains four small groups made up of three quartets of U.P.A.'s (two Be4 and an Li4) and a hydrogen triplet (H3′).

The omegons in two ${}_9F^{19}$ nuclei form the M.P.A. (fig. 7.55). Four of the funnels each contain a bound state of three $\mathcal{N}$ omegons and a $\mathcal{P}$ omegon (Li4), a d quark (H3′), and two clusters of four omegons (four $\mathcal{P}$ and four

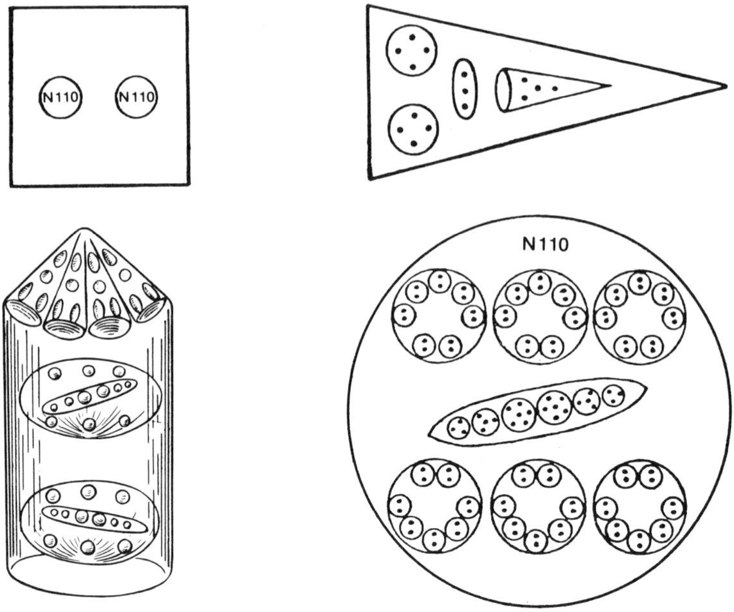

Fig. 7.54. M.P.A. of fluorine

𝒩) (Be4). The other four funnels each contain a bound state of three 𝒫 omegons and an 𝒩 omcgon (Li4), a u quark (H3′), and two clusters of four omegons (four 𝒫 and four 𝒩) (Be4). Figure 7.56 shows the identified U.P.A.'s in these two types of funnels. The composition of the latter type is also indicated by the disintegration diagram (fig. 7.57), which shows that a funnel has a (+) triplet (u quark), (+) and (−) Be4 groups that split up at stage 2 into (+) and (−) duads (𝒫-𝒫 and 𝒩-𝒩), and an Li4 that breaks up at stage 2 into a (+) duad (𝒫-𝒫) and a duad that has been identified in earlier analyses of M.P.A.'s as a 𝒫-𝒩 diomegon. One N110 group contains fifty-eight 𝒫 and fifty-two 𝒩 (the same composition as that found for the N110 in the nitrogen M.P.A.), and the other contains fifty-two 𝒫 and fifty-eight 𝒩, that is, it is the former group with 𝒫 and 𝒩 omegons interchanged. The investigators counted 340 U.P.A.'s in the M.P.A., but 342 omegons should be present. The discrepancy of 2 is accountable if an extra 𝒩 is present in each N110. The possibility that the same type of group in different M.P.A.'s could differ slightly in population was anticipated by the investigators, who remarked: "In the heavier elements, such as gold, with 3,546 Anu, it would have been impossible to count each Anu without quite unnecessary waste of time, when making a preliminary investigation. Later, it may be worth while to count each division separately, as in some we noticed that two groups, at first sight alike, differed by 1 or 2 Anu."[22]

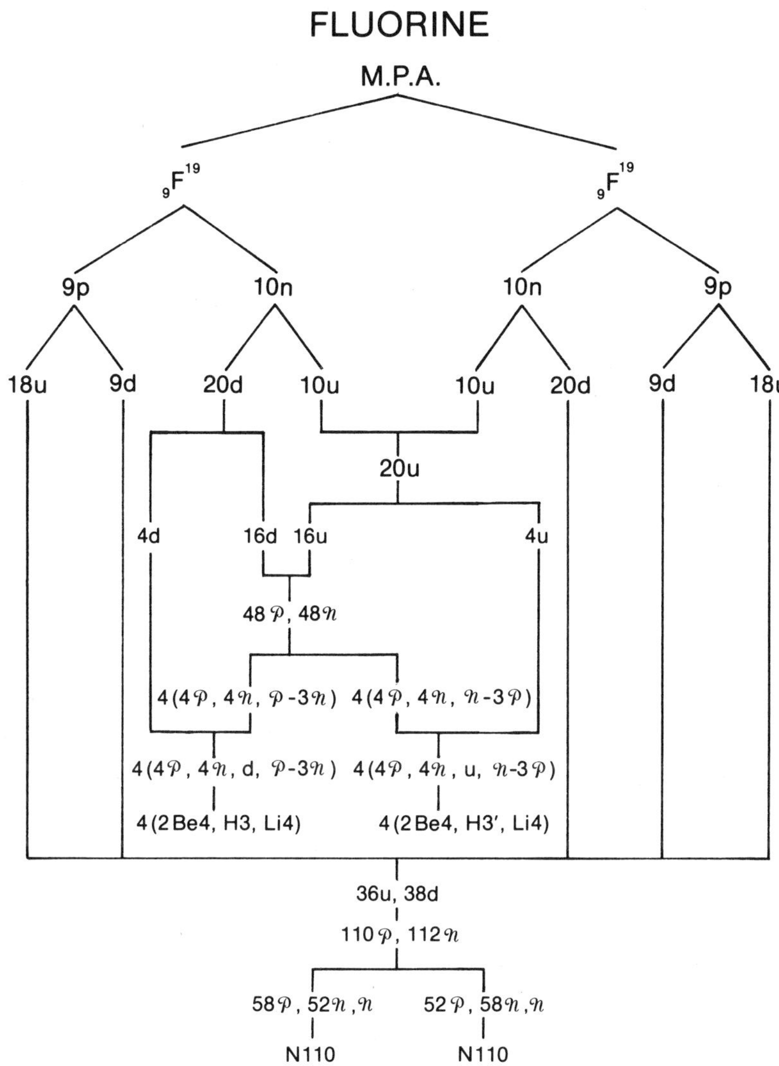
FLUORINE
M.P.A.
$_9F^{19}$
$_9F^{19}$
9p
10n
10n
9p
18u
9d
20d
10u
10u
20d
9d
18u
20u
4d
16d
16u
4u
48𝒫, 48𝒩
4(4𝒫, 4𝒩, 𝒫-3𝒩)
4(4𝒫, 4𝒩, 𝒩-3𝒫)
4(4𝒫, 4𝒩, d, 𝒫-3𝒩)
4(4𝒫, 4𝒩, u, 𝒩-3𝒫)
4(2Be4, H3, Li4)
4(2Be4, H3′, Li4)
36u, 38d
110𝒫, 112𝒩
58𝒫, 52𝒩, 𝒩
52𝒫, 58𝒩, 𝒩
N110
N110
Fig. 7.55

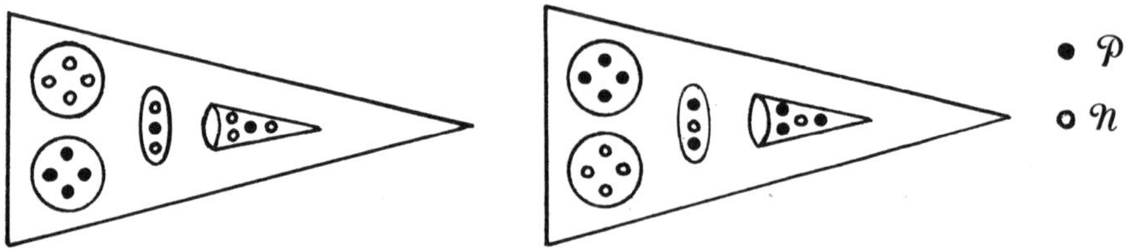

Fig. 7.56. Two types of funnel

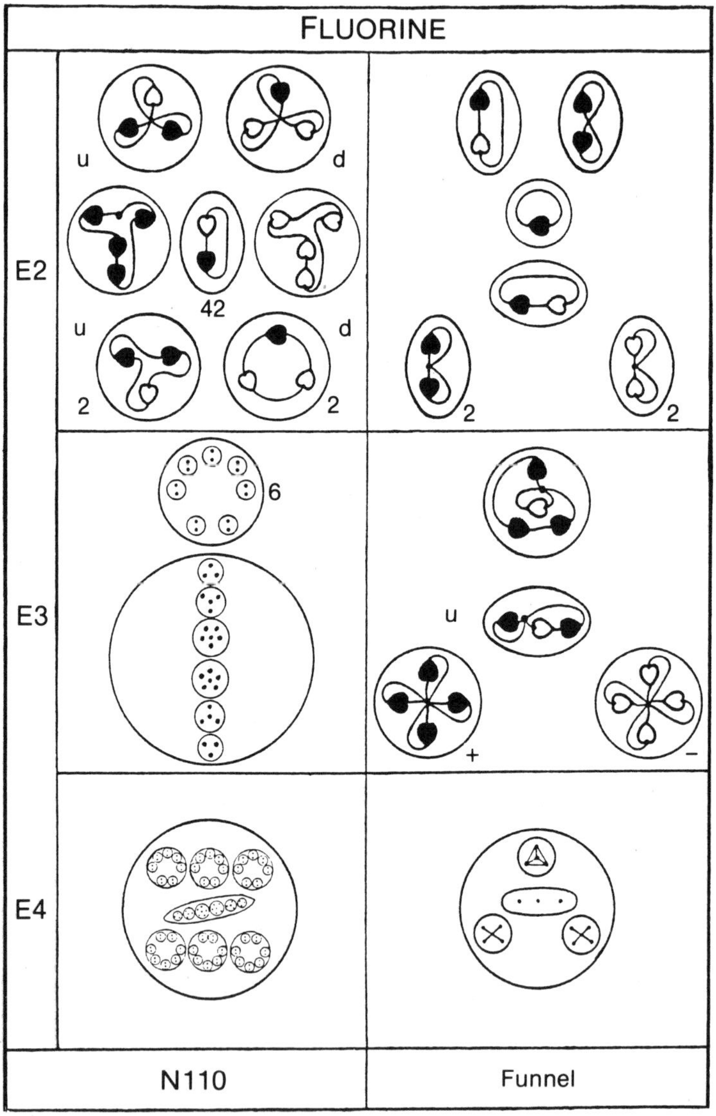

Fig. 7.57

NEON M.P.A.

As a member of the Star Group, neon has the simplest M.P.A. It has the appearance of a flat, six-armed star (fig. 7.58). All the arms are identical and radiate from a central sphere (Ne120) enclosing five interpenetrating tetrahedral clusters of Ad6 groups. Each tetrahedron is an Ad24 group similar to those present in the M.P.A.'s of deuterium and helium. The Ad6 groups are fixed at the twenty corners of a regular dodecahedron (fig. 7.59) formed by the apices of the five tetrahedra. Each arm contains three bodies: a group of two Hydrogen Triplets (H3), a cluster of three Li4 tetrahedra, and a group of five spheres (Ne22), four of which enclose quartets of U.P.A.'s and one of which contains six U.P.A.'s. Figure 7.60 shows the central Ne120 and the particles in one arm.

Two $_{10}Ne^{20}$ nuclei, providing sixty u and sixty d quarks, form the M.P.A. (fig. 7.61). Of these, twenty u and twenty d quarks combine in pairs to form twenty u-d diquarks (Ad6). The Ne120 consists of twenty u-d di-

Fig. 7.58

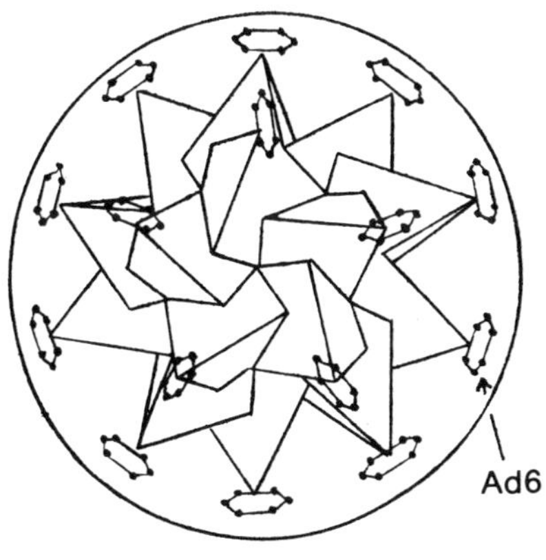

Fig. 7.59. Ne120

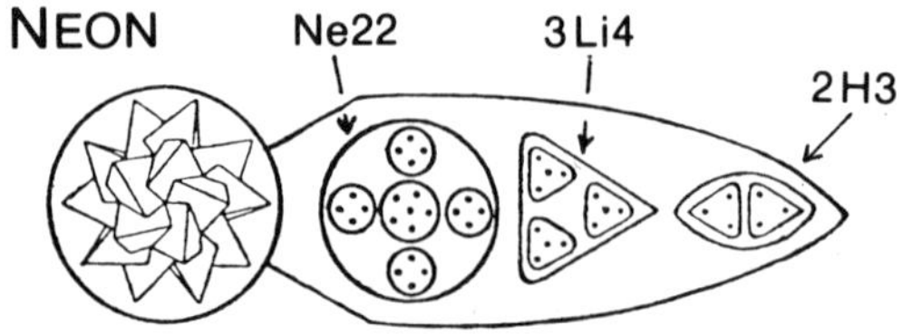

Fig. 7.60. Central globe and one arm of M.P.A. of neon

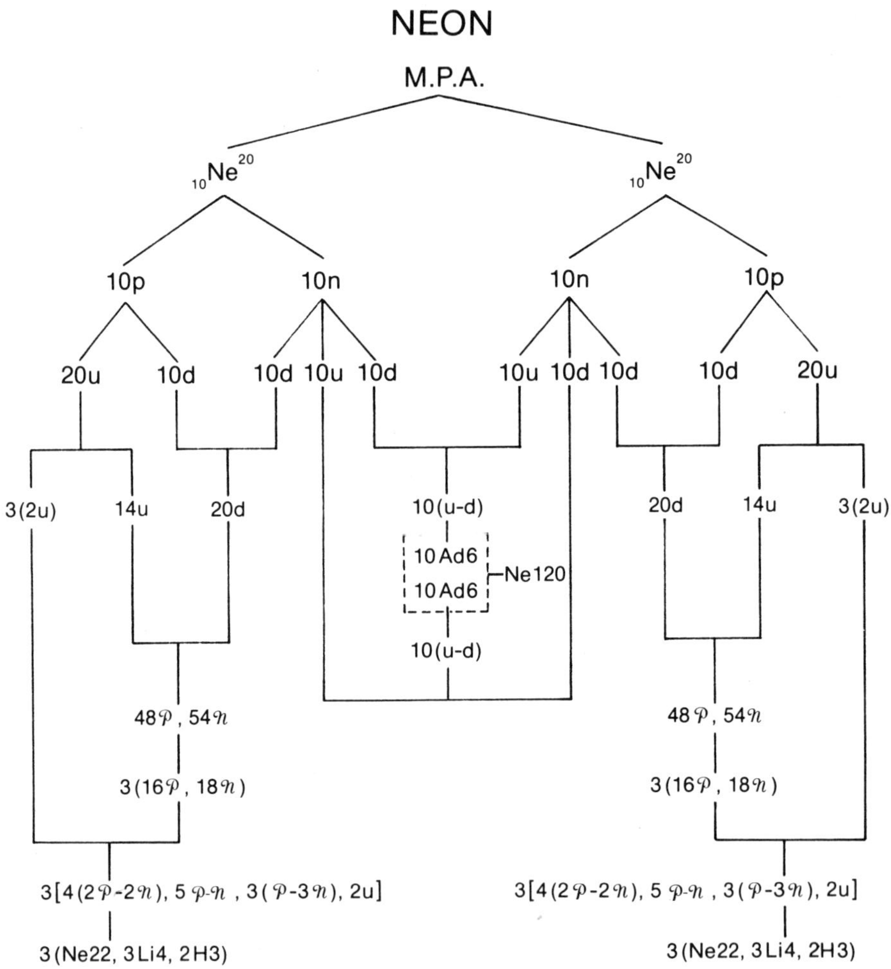

Fig. 7.61

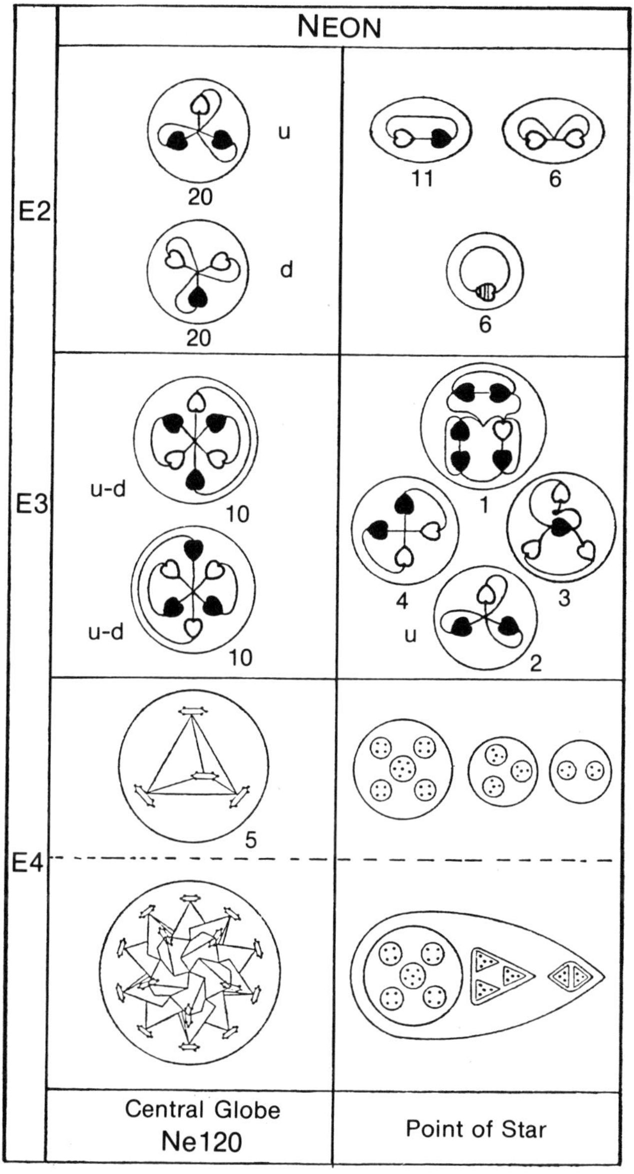
NEON
E2
u
20
d
20
11
6
6
E3
u-d
10
u-d
10
1
4
3
u
2
5
E4
Central Globe
Ne120
Point of Star

Fig. 7.62

quarks, bound as five clusters of four spinless bosons in an SU(6) Higgs vacuum domain as a result of their mutual, short-range strong interaction. Five tetrahedral arrays of identical diquarks with electric charges of $+1/3$ minimize their Coulomb interaction energy by maximizing the distances between diquarks. The regular solid figure allowing this is a dodecahedron with twenty corners equally spaced apart. The symmetry of the arrangement of Ad6 groups in the Ne120 is due to Coulomb repulsion of their identical charges. The Ne120 is the central core of all other inert gas M.P.A.'s, and it is the quark analogue of the especially stable, closed-shell structure of the doubly magic $_{20}Ca^{40}$ nucleus. That there are twenty u and twenty d quarks in the Ne120 is shown by the disintegration diagram (fig. 7.62). The group breaks up at stage 2 into twenty (+) and twenty (−) triplets. The two H3 triplets in the arm are predicted to be u quarks, and figure 7.62 indicates that they are (+) triplets. The Li4 groups consist of three bound systems of one $\mathcal{P}$ and three $\mathcal{N}$ omegons, and the Ne22 consists of four identical $2\mathcal{P}$-$2\mathcal{N}$ bound states with a bound system of five $\mathcal{P}$ omegons and one $\mathcal{N}$ omegon at its centre (fig. 7.63).

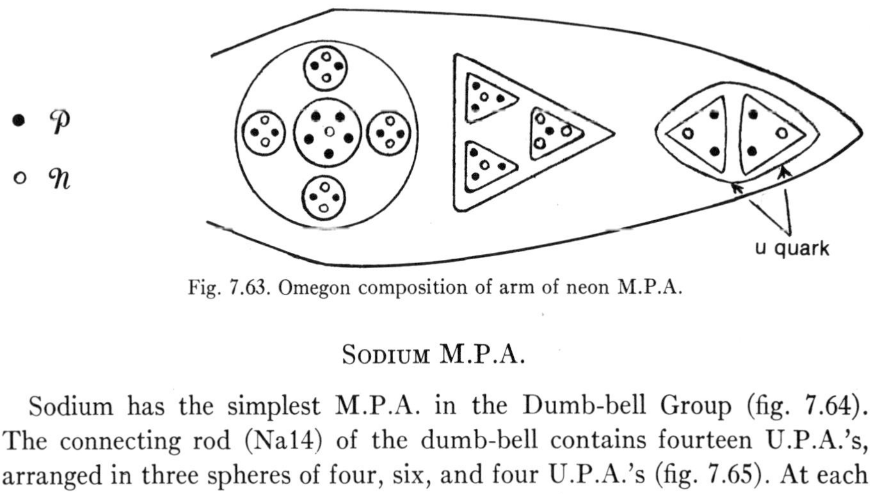

Fig. 7.63. Omegon composition of arm of neon M.P.A.

SODIUM M.P.A.

Sodium has the simplest M.P.A. in the Dumb-bell Group (fig. 7.64). The connecting rod (Na14) of the dumb-bell contains fourteen U.P.A.'s, arranged in three spheres of four, six, and four U.P.A.'s (fig. 7.65). At each end of the rod are two concentric spheres (Na10), the inner one containing four U.P.A.'s, the outer one enclosing six U.P.A.'s. Twelve funnels radiate symmetrically from each Na10. Each encloses four groups of U.P.A.'s.

The M.P.A. is formed from two $_{11}Na^{20}$ nuclei (fig. 7.66). The 384 omegons in sixty-four u and sixty-four d quarks are distributed equally among the twenty-four funnels. Six of the twelve funnels at one end of the dumb-bell contain twelve $\mathcal{P}$ and four $\mathcal{N}$ omegons per funnel. Each of the other six contains twelve $\mathcal{N}$ and four $\mathcal{P}$ omegons. Figure 7.67 shows the composition of one of the former funnels. The remaining four u and six d quarks disinte-

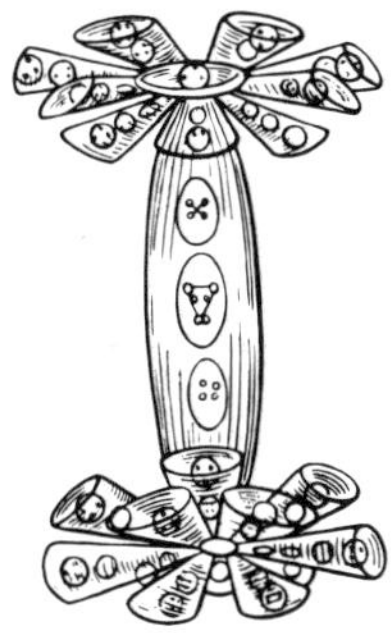

Fig. 7.64. M.P.A. of sodium

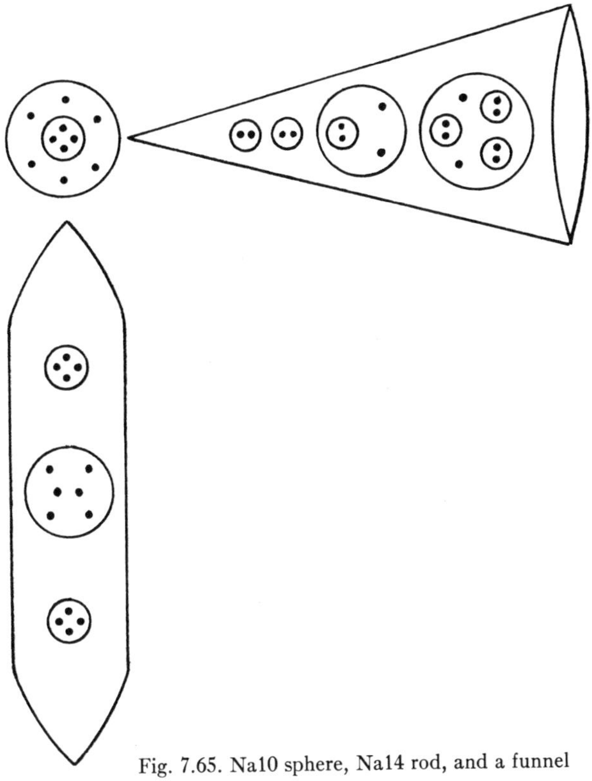

Fig. 7.65. Na10 sphere, Na14 rod, and a funnel

grate into fourteen $\mathcal{P}$ and sixteen $\mathcal{N}$ during the formation of the M.P.A., and these fill the rod and Na10 globes. The inner sphere should contain a $\mathcal{P}$-$\mathcal{N}$ diomegon and not a group of four U.P.A.'s, as reported. The outer sphere encloses a hexagonal array of three $\mathcal{P}$ and three $\mathcal{N}$ omegons (fig. 7.68). The Na14 rod contains a bound system of two $\mathcal{P}$ and four $\mathcal{N}$ omegons and two bound states of two $\mathcal{P}$ and two $\mathcal{N}$, arranged in the form of a cross (fig. 7.69). The disintegration diagram (fig. 7.70) shows successive stages in the breakup of a funnel, globe, and rod. The two crosses in the rod break up at stage 2 into four identical duads that are $\mathcal{P}$-$\mathcal{N}$ diomegons. The group of six U.P.A.'s splits up into a (−) duad ($\mathcal{N}$-$\mathcal{N}$ diomegon) and two $\mathcal{P}$-$\mathcal{N}$ diomegons, confirming the predicted composition of the Na14 rod. The outer sphere of six U.P.A.'s splits up into (+) and (−) triplets, that is, a u and a d quark comprising three $\mathcal{P}$ and three $\mathcal{N}$ omegons, in agreement with the prediction above. The groups released by a funnel also confirm the assignment of isospin states shown in figure 7.67. The discrepancy of four between the 418 U.P.A.'s counted in the M.P.A. and the 414 omegons predicted is due to an overcounting by 2 of the particles in each Na10 globe.

SODIUM

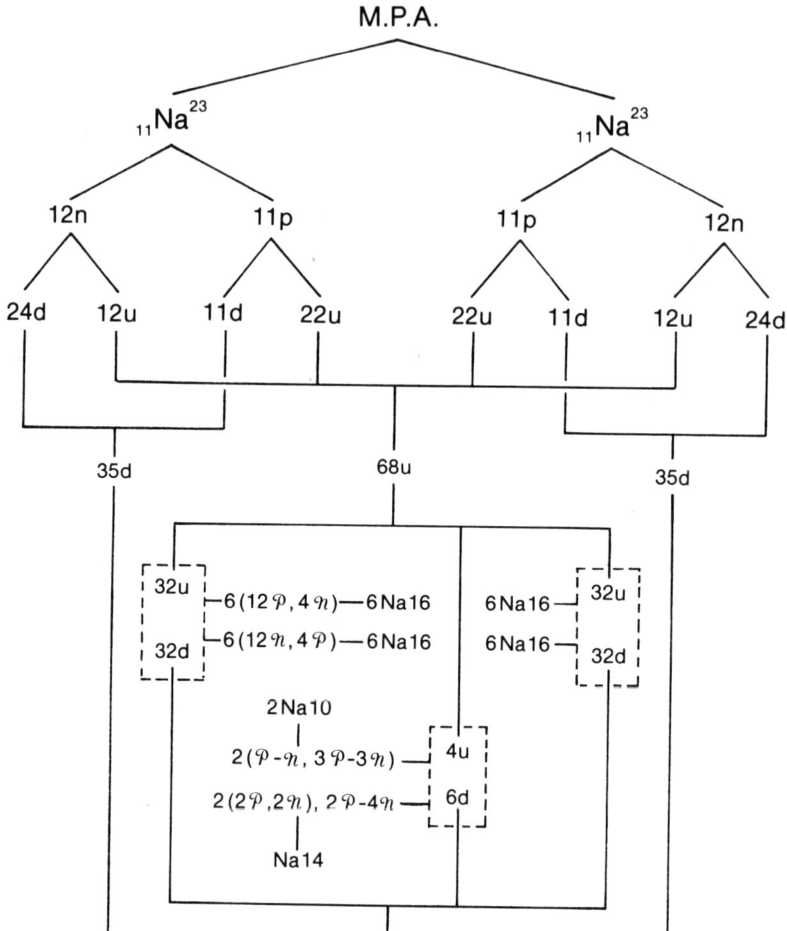

Fig. 7.66

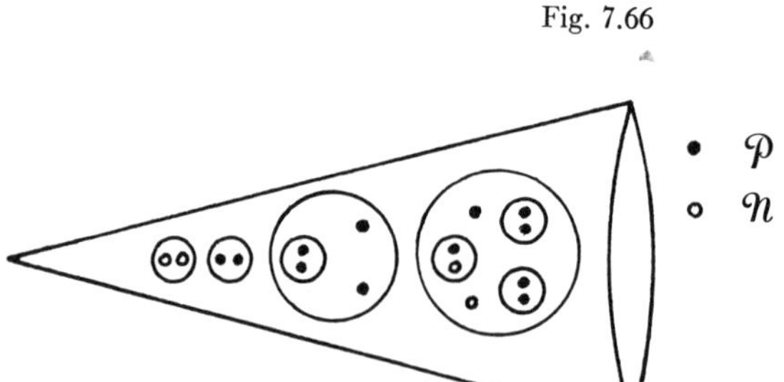

Fig. 7.67

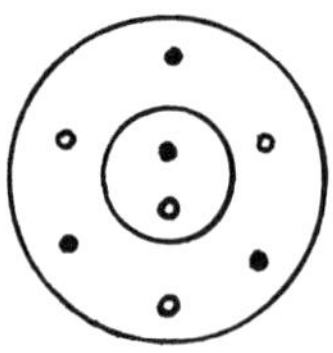

Fig. 7.68. Corrected Na10

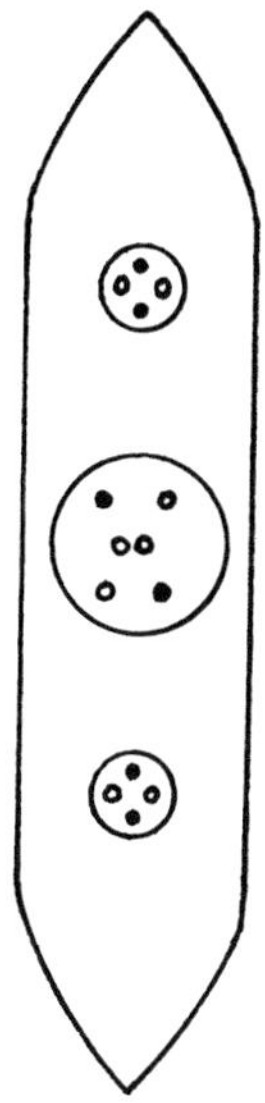

Fig. 7.69. Na14 rod

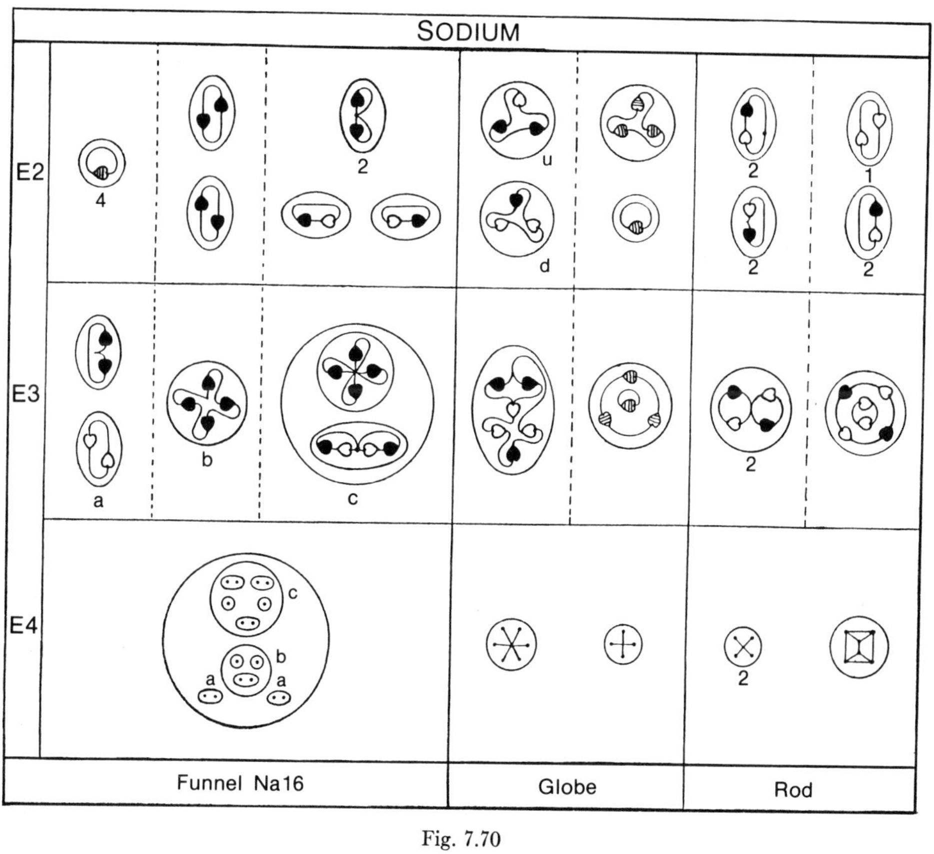

Fig. 7.70

MAGNESIUM M.P.A.

The M.P.A. is a tetrahedral array of four similar funnels (fig. 7.71). A funnel contains three identical bodies, each of which has three similar ovoids enclosing twelve U.P.A.'s (Mg12). They are made up of three spheres of two U.P.A.'s (N2), seven U.P.A.'s (I.7), and three U.P.A.'s (H3) (fig. 7.72).

Two $_{12}Mg^{24}$ nuclei form the M.P.A. (fig. 7.73). The omegons in eighteen u and eighteen d quarks fill each funnel, each Mg12 containing the omegons in two u and two d quarks. The N2 is a $\mathcal{P}$-$\mathcal{N}$ diomegon, the I.7 is a bound state of three $\mathcal{P}$ and four $\mathcal{N}$ omegons, and the H3 is a u quark. Thus

$$\begin{aligned} Mg12 &= N2 + I.7 + H3 \\ &= \mathcal{P}\text{-}\mathcal{N} + 3\mathcal{P}\text{-}4\mathcal{N} + u \end{aligned}$$

(see fig. 7.74). The disintegration diagram (fig. 7.75) confirms that the H3 in the Mg12 group is (+), that is, a u quark, and that the N2 is a $\mathcal{P}$-$\mathcal{N}$ diomegon. The I.7 group dissociates at stage 2 into a (+) quartet of U.P.A.'s and a (−) triplet. The former is a bound state of two $\mathcal{P}$ and two $\mathcal{N}$ omegons, and the latter is a d quark. The I.7 therefore consists of three $\mathcal{P}$ and four $\mathcal{N}$ omegons, in agreement with the analysis above.

FIG. 7.71. M.P.A. of magnesium

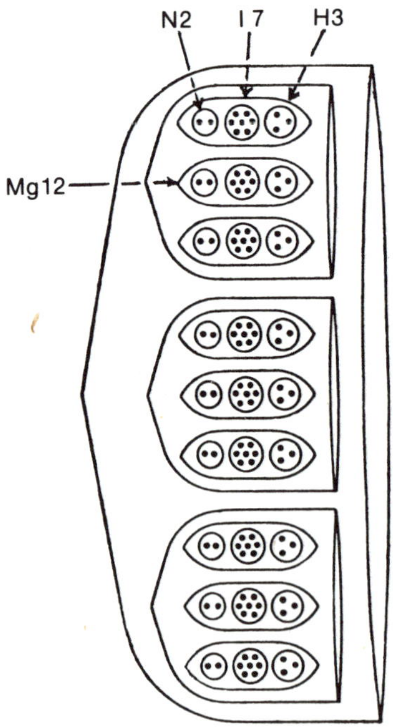

Fig. 7.72. A funnel

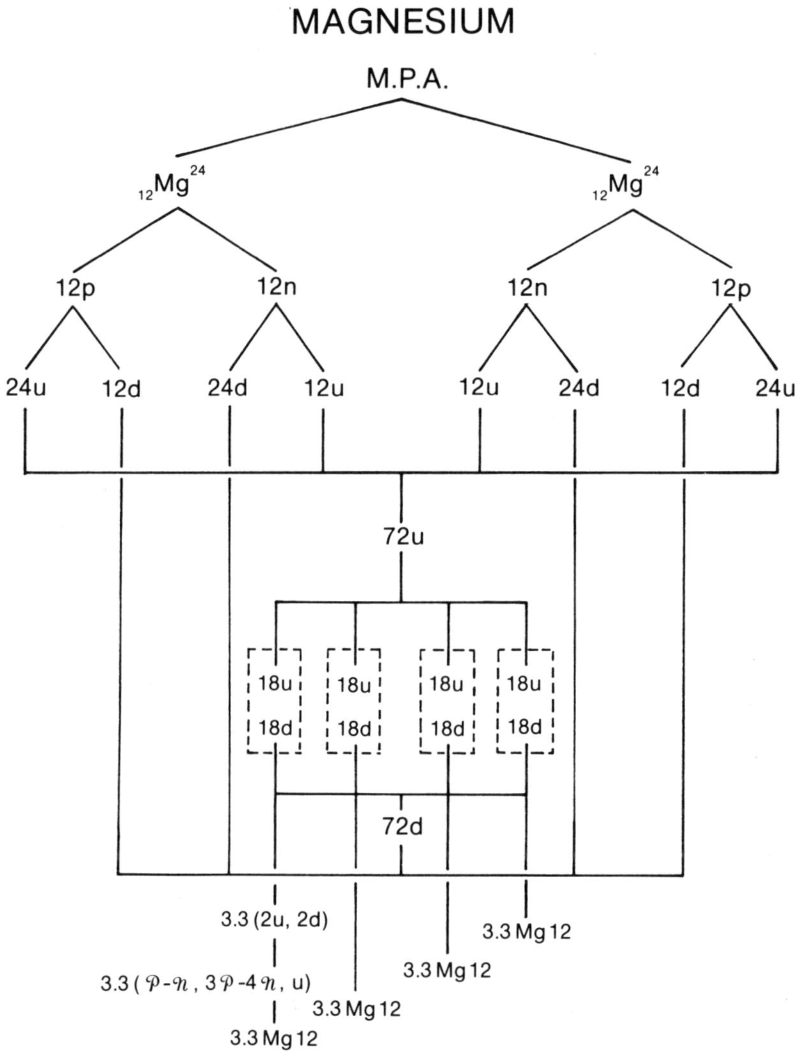
MAGNESIUM
M.P.A.
$_{12}Mg^{24}$
$_{12}Mg^{24}$
12p
12n
12n
12p
24u
12d
24d
12u
12u
24d
12d
24u
72u
18u
18d
18u
18d
18u
18d
18u
18d
72d
3.3 (2u, 2d)
3.3 (𝒫-𝒩, 3𝒫-4𝒩, u)
3.3 Mg 12
3.3 Mg 12
3.3 Mg 12
3.3 Mg 12

Fig. 7.73

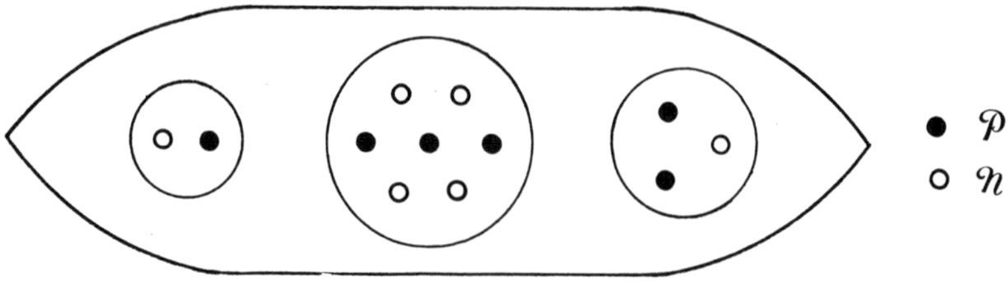

Fig. 7.74. Mg12 group

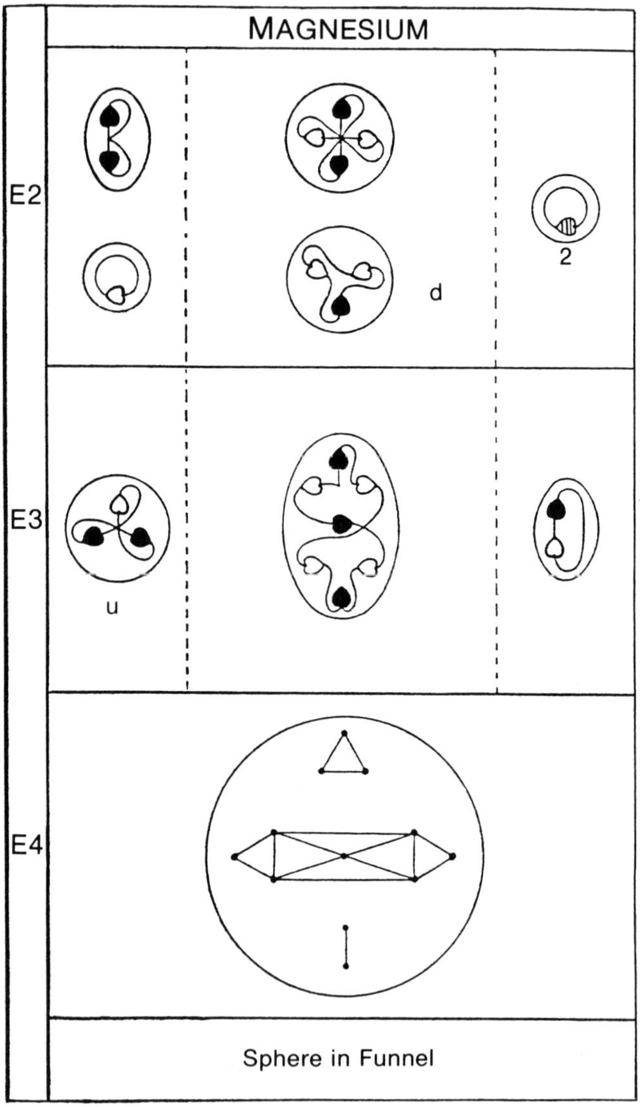

Fig. 7.75

ALUMINIUM M.P.A.

The M.P.A. is the simplest member of Cube Group B. It has the form of a cubic array of six funnels (fig. 7.76). Each funnel contains eight similar ovoids (Al.9) enclosing three spheres of triplets of U.P.A.'s. Below these is an ovoid (Al.9′) containing three spheres. Two enclose duads of U.P.A.'s, and the third encloses a group of five U.P.A.'s (fig. 7.77).

The M.P.A. is formed from two $_{13}Al^{27}$ nuclei, which contain 486 omegons, the same number as the counted number of U.P.A.'s (fig. 7.78). Seventy-two u quarks are equally distributed among the six funnels, four groups of three per funnel, and seventy-two d quarks are distributed in a similar fashion. An Al.9 is a bound state of either three u or three d quarks (fig. 7.79). It is not, however, the N^{*-} or N^{*++} baryon, since the quarks are bound indirectly through the coupling between their internal SU(3) strings and not by the SU(9) strings that act as valence bonds between quarks in a baryon (the quarks in the Al.9 are colourless). The remaining eight u and ten d quarks dissociate into twenty-six $\mathcal{P}$ and twenty-eight $\mathcal{N}$ omegons that are distributed in the funnels as follows: in each of four funnels, the Al.9′ consists of a $\mathcal{P}$-$\mathcal{P}$ and $\mathcal{N}$-$\mathcal{N}$ diomegon and a bound state of two $\mathcal{P}$ and three $\mathcal{N}$ omegons; in the other two, the Al.9′ contains a $\mathcal{P}$-$\mathcal{P}$ and $\mathcal{N}$-$\mathcal{N}$ diomegon and a bound state of two $\mathcal{N}$ and three $\mathcal{P}$ omegons (fig. 7.80). The prediction that an Al.9 has two types and that there are four of each in a funnel is confirmed by the disintegration diagram (fig. 7.81), which shows that the eight Al.9

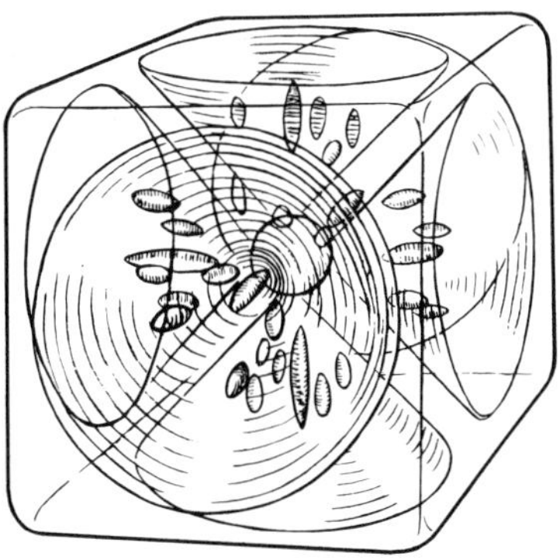

Fig. 7.76. M.P.A. of aluminium

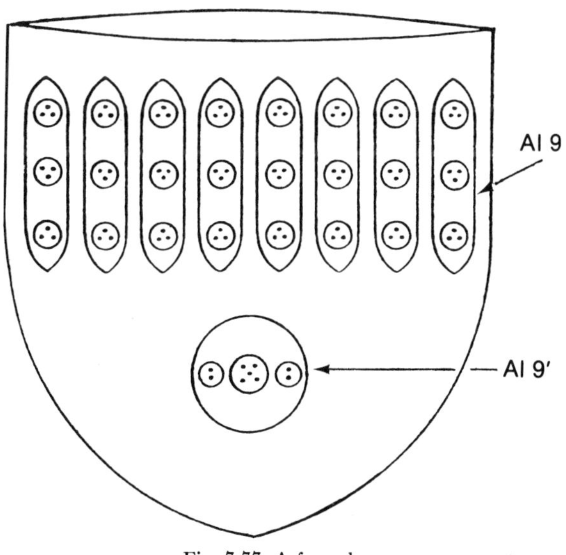

Fig. 7.77. A funnel

ALUMINIUM

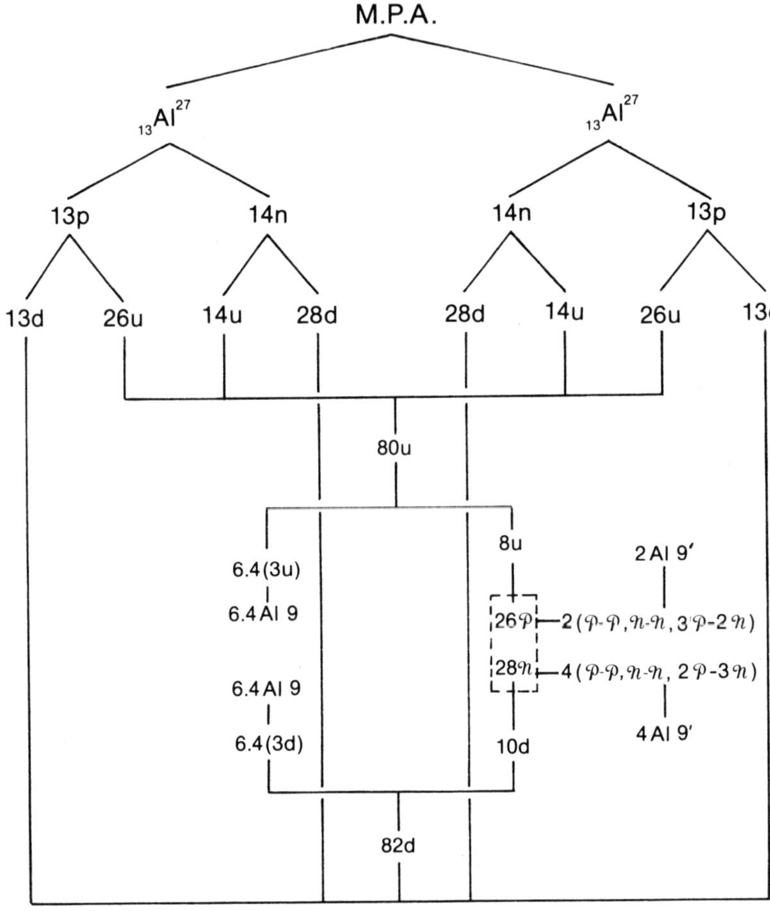

Fig. 7.78

groups separate into two sets of four at stage 3. One type is a (+) group that breaks up at stage 2 into three (+) triplets (u quarks); the other is a (−) group that breaks up into three (−) triplets (d quarks). The Al.9′ is shown as breaking up at stage 2 into a free U.P.A. and four (+) duads. However, this implies a composition of at least eight $\mathcal{P}$ omegons, which is inconsistent with the above predictions. But in the calcium M.P.A. (see later), the disintegration diagram indicates that the Al.9′ groups in the M.P.A. break up into (+) and (−) duads, that is, that the duads are $\mathcal{P}$-$\mathcal{P}$ and $\mathcal{N}$-$\mathcal{N}$ omegons. This composition agrees with the prediction for both the calcium and aluminium M.P.A.'s:

$$\begin{aligned} \text{Al.9}' &= \mathcal{P}\text{-}\mathcal{P} + \mathcal{N}\text{-}\mathcal{N} + 2\mathcal{P}\text{-}3\mathcal{N} \\ &\rightarrow 2(\mathcal{P}\text{-}\mathcal{P}) + 2(\mathcal{N}\text{-}\mathcal{N}) + \mathcal{N}\,. \end{aligned}$$

An error of observation is predicted for the groups released from the Al.9′ (see fig. 7.82).

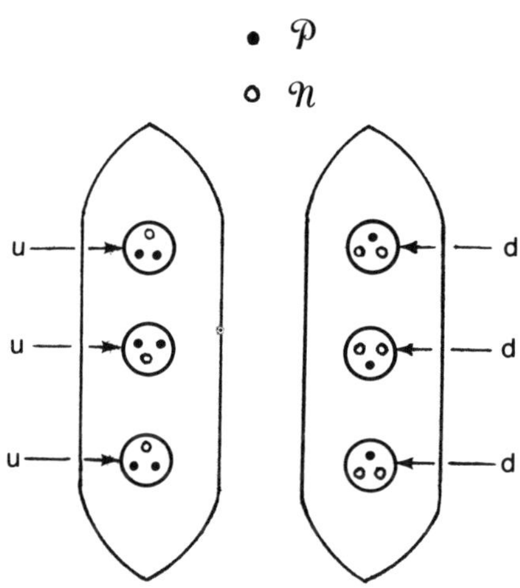

Fig. 7.79. The Al.9 group

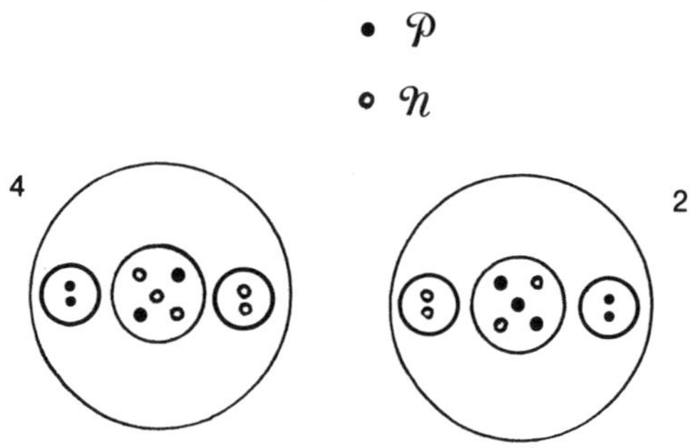

Fig. 7.80. The Al.9′ group

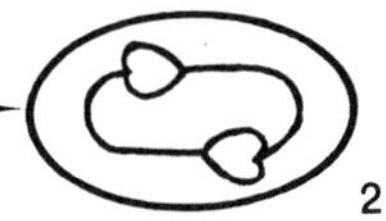

Fig. 7.82

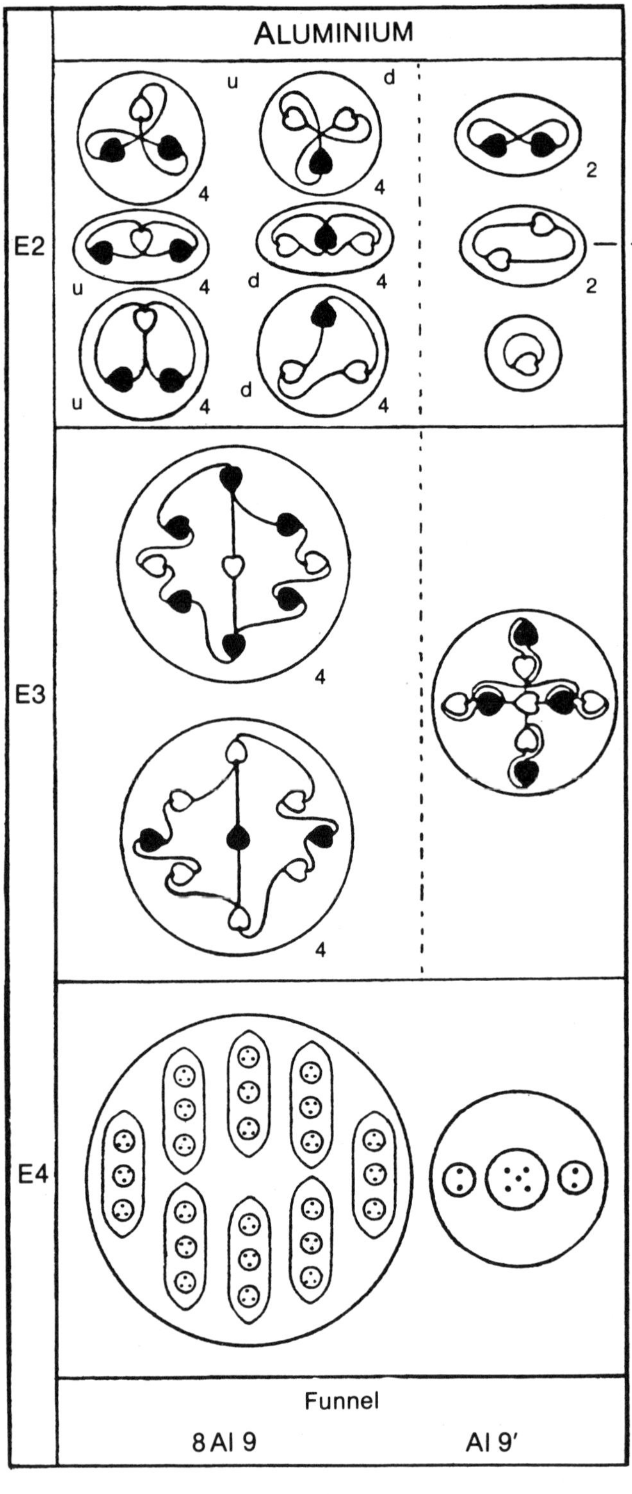

Fig. 7.81

SILICON M.P.A.

Of the Octahedron Group B elements, silicon has the simplest M.P.A. It consists of eight funnels directed towards the faces of an octahedron (fig. 7.83). Each funnel contains a group of five U.P.A.'s (B5) and four ovoids (Si15), each enclosing an Ad6 group, a group of four U.P.A.'s, and a group of five U.P.A.'s similar to that in the middle of an Al.9′ group (fig. 7.84).

The M.P.A. is formed from two ${}_{14}Si^{18}$ nuclei, that is, from eighty-four u and eighty-four d quarks (fig. 7.85). Thirty-two u and thirty-two d quarks are present as thirty-two u-d diquarks (Ad6), four to a funnel (fig. 7.86). The B5 groups in four funnels are bound states of three $\mathcal{P}$ and two $\mathcal{N}$ omegons (identical with the B5 groups in the boron M.P.A.); in the other four, they are bound states of two $\mathcal{P}$ and three $\mathcal{N}$ omegons. The group of four U.P.A.'s is a cross-shaped cluster of two $\mathcal{P}$ and two $\mathcal{N}$ omegons. In two Si15 ovoids, the groups of five U.P.A.'s are made up of three $\mathcal{P}$ and two $\mathcal{N}$ and two $\mathcal{P}$ and three $\mathcal{N}$ omegons; in the other two Si15, there should be quartets of two $\mathcal{P}$ and two $\mathcal{N}$ instead of groups of five. The discrepancy of 16 between the counted number (520) of U.P.A.'s and the predicted population of 504 omegons is due to an overcounting of two U.P.A.'s in each funnel, there being two groups of five and two groups of four instead of four groups of five. The disintegration diagram (fig. 7.87) shows that a B5 breaks up at stage 2 into a (+) duad ($\mathcal{P}$-$\mathcal{P}$) and a (−) triplet (d quark), that is, it is made up of three $\mathcal{P}$ and two $\mathcal{N}$ omegons, in agreement with prediction for four of the funnels. The Ad6 dissociates into (+) and (−) triplets, that is, u and d quarks.

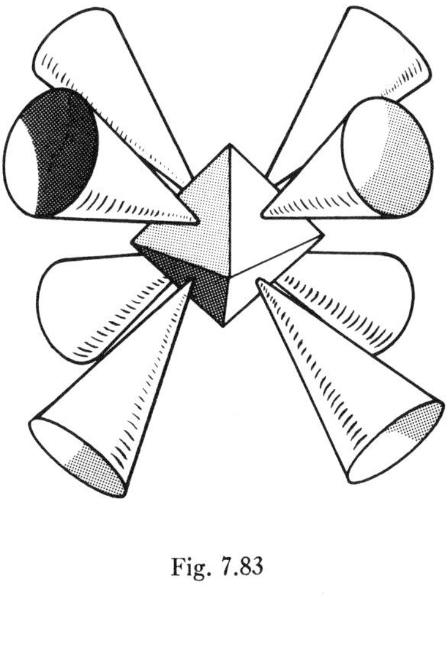

Fig. 7.83

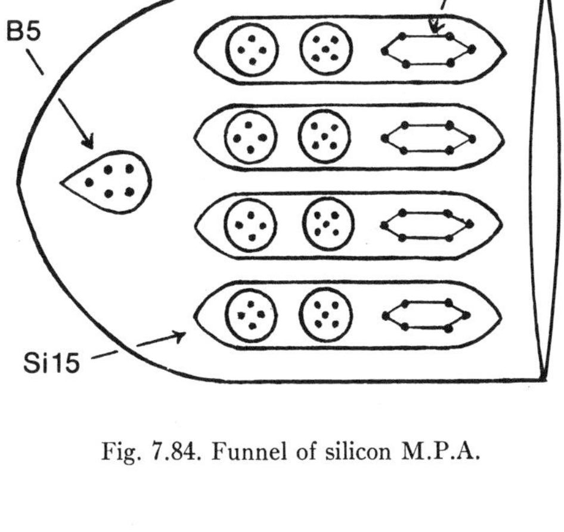

Fig. 7.84. Funnel of silicon M.P.A.

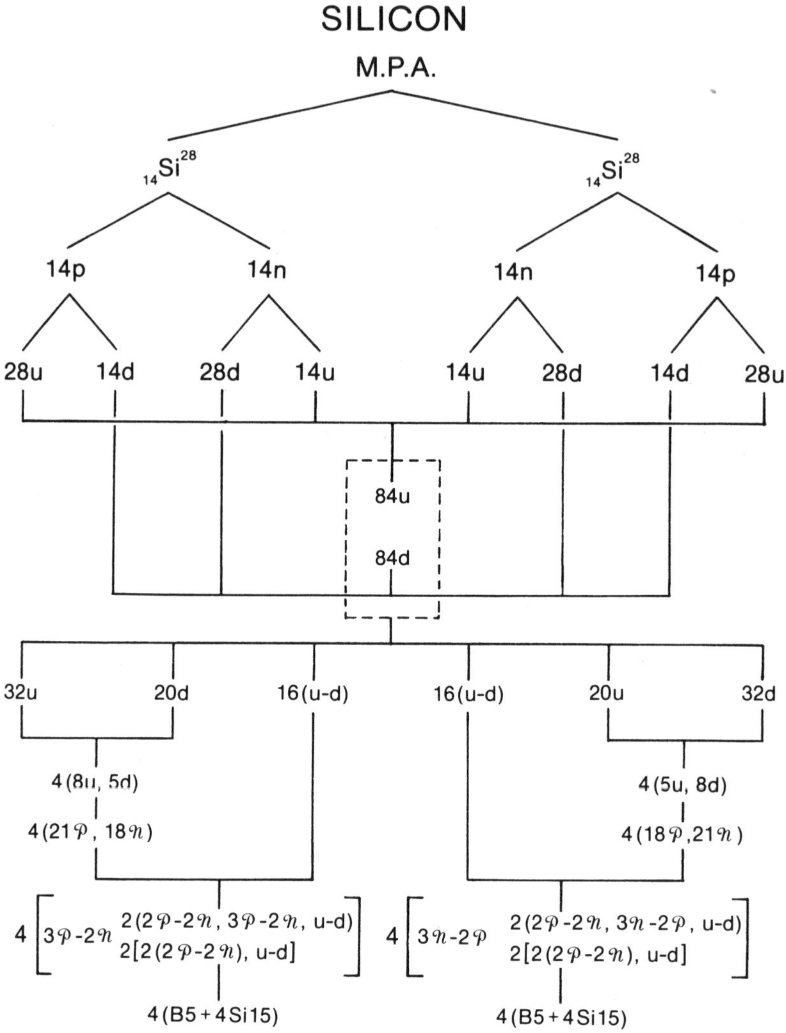

Fig. 7.85

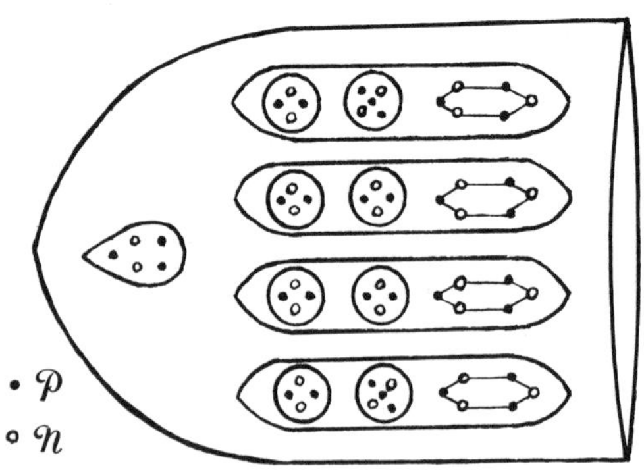

Fig. 7.86. "Corrected" funnel

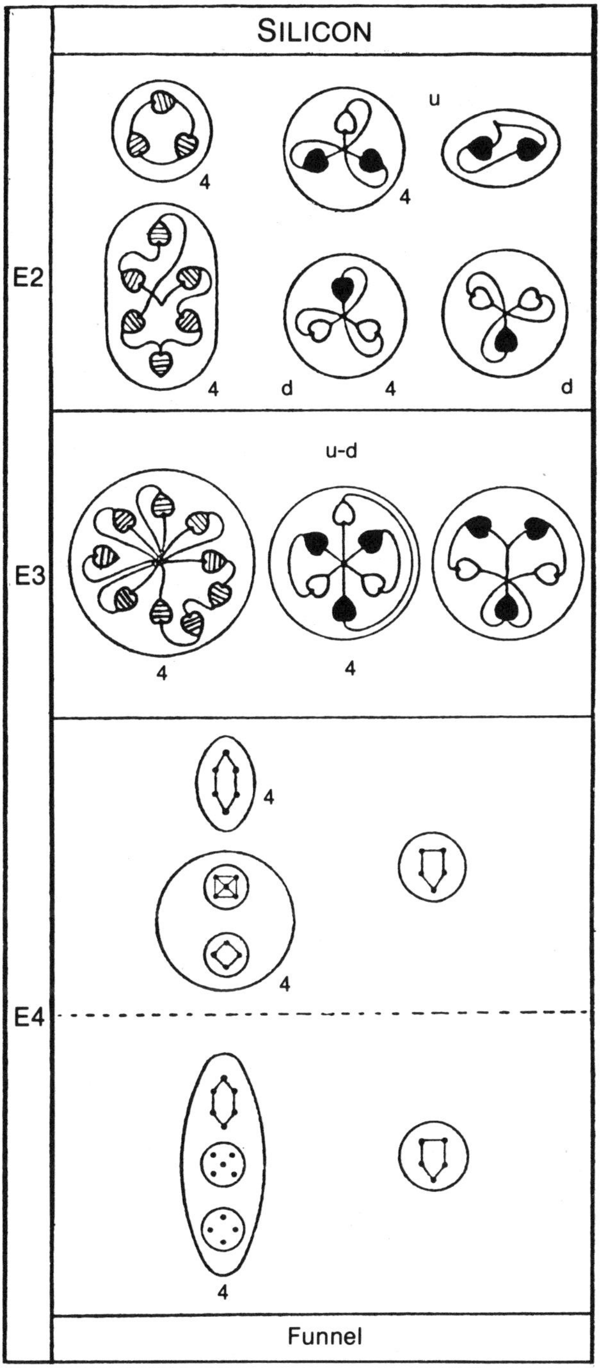

Fig. 7.87

PHOSPHORUS M.P.A.

The M.P.A. is a cubic array of six funnels (fig. 7.88). Each funnel contains two segments. One is made up of a B5 group, three groups of six U.P.A.'s (N6), and three spheres of nine U.P.A.'s (P9), a total of fifty U.P.A.'s. The other segment contains an Li4 group, three Be4 groups, and three P9 groups, making a total of forty-three U.P.A.'s (fig. 7.89).

The M.P.A. contains 558 omegons originally present in two $_{15}P^{31}$ nuclei, the same number as the population of U.P.A.'s (fig. 7.90). The composition of the funnels is given below:

4 funnels: each contains 46 $\mathcal{P}$ and 47 $\mathcal{N}$ provided by 15 u and 16 d quarks:

P9 (× 3):	5$\mathcal{P}$-4$\mathcal{N}$	P9 (× 3):	5$\mathcal{N}$-4$\mathcal{P}$
N6 (× 3):	3$\mathcal{P}$-3$\mathcal{N}$	Be4 (× 3):	2$\mathcal{P}$-2$\mathcal{N}$
B5 (× 1):	3$\mathcal{P}$-2$\mathcal{N}$	Li4 (× 1):	$\mathcal{P}$-3$\mathcal{N}$

Fig. 7.88

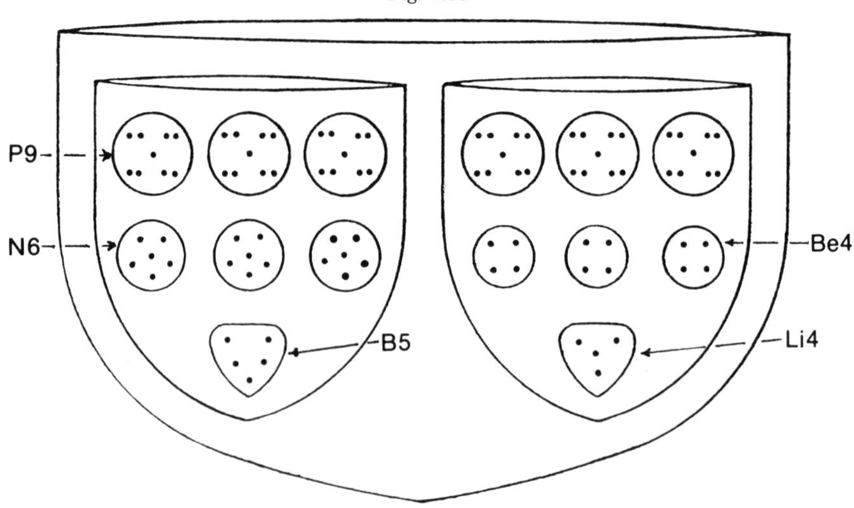

Fig. 7.89

PHOSPHORUS

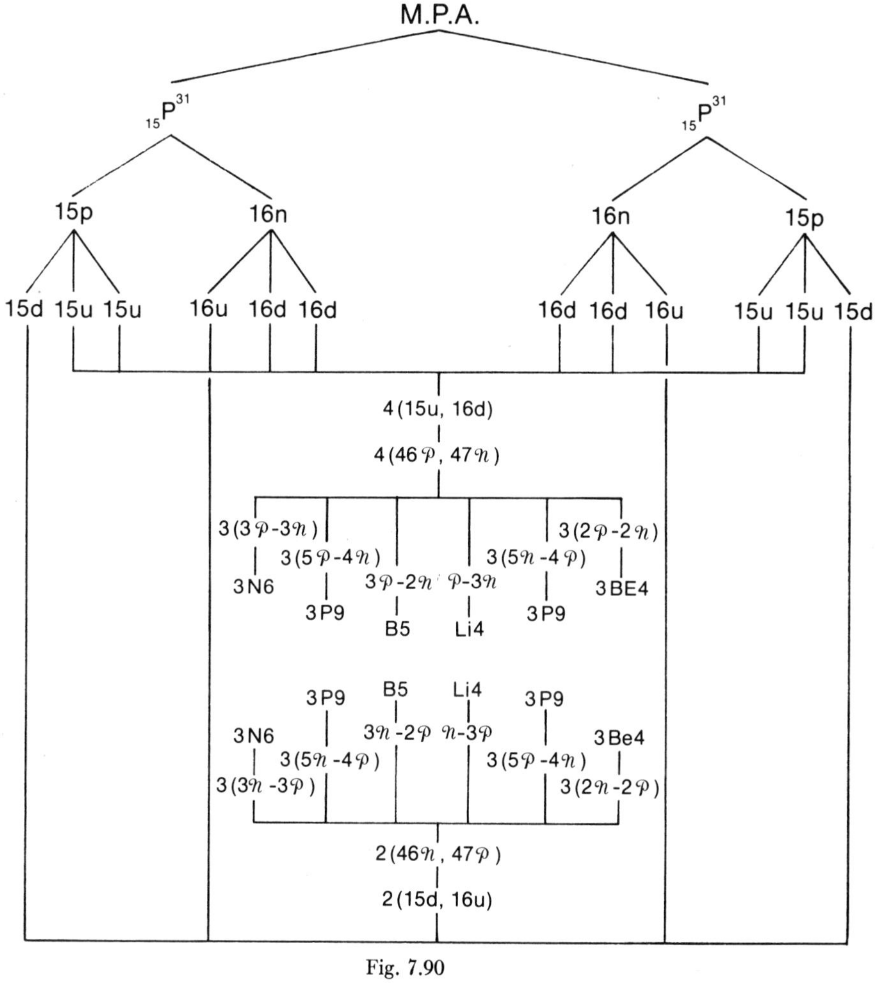

Fig. 7.90

2 funnels: each contains 46 $\mathcal{N}$ and 47 $\mathcal{P}$ provided by 15 d and 16 u quarks:

P9 (× 3):	5$\mathcal{N}$-4$\mathcal{P}$	P9 (× 3):	5$\mathcal{P}$-4$\mathcal{N}$
N6 (× 3):	3$\mathcal{N}$-3$\mathcal{P}$	Be4 (× 3):	2$\mathcal{N}$-2$\mathcal{P}$
B5 (× 1):	3$\mathcal{N}$-2$\mathcal{P}$	Li4 (× 1):	$\mathcal{N}$-3$\mathcal{P}$

Figure 7.91 shows the composition of the various groups. A P9 is a body-centred cubic array of nine omegons, consisting of intersecting tetrahedra of four $\mathcal{P}$ and four $\mathcal{N}$ that are bound by strong, attractive forces to a $\mathcal{P}$ or an $\mathcal{N}$ omegon acting as their nucleus. Coulomb forces between their electric charges determine their rigid body-centred cubic arrangement in space. A P9 is predicted to consist either of five $\mathcal{P}$ and four $\mathcal{N}$ or four $\mathcal{P}$ and five $\mathcal{N}$. The former is consistent with the composition implied by its disintegration diagram (fig. 7.92): it breaks up at stage 2 into three duads that are $\mathcal{P}$-$\mathcal{N}$ diomegons and a (+) duad that is a $\mathcal{P}$-$\mathcal{P}$ diomegon. The P9 groups are not all identical, as implied by the diagram. The U.P.A.'s in an N6 regroup at stage 3 to form an Ad6, which dissociates into a (+) triplet (u quark) and a (−) triplet (d quark) at stage 2. It consists, therefore, of three $\mathcal{P}$ and three $\mathcal{N}$ omegons, in agreement with the prediction above. According to figure 7.92, the three Be4 groups and the Li4 group break up at stage 2 into seven (+) duads ($\mathcal{P}$-$\mathcal{P}$) and a $\mathcal{P}$-$\mathcal{N}$. But the composition given to them above implies that these groups should have broken up into seven $\mathcal{P}$-$\mathcal{N}$ and a (−) duad ($\mathcal{N}$-$\mathcal{N}$). The number seven is assigned to the wrong duad, an error presumably made during the drafting of the original drawings.

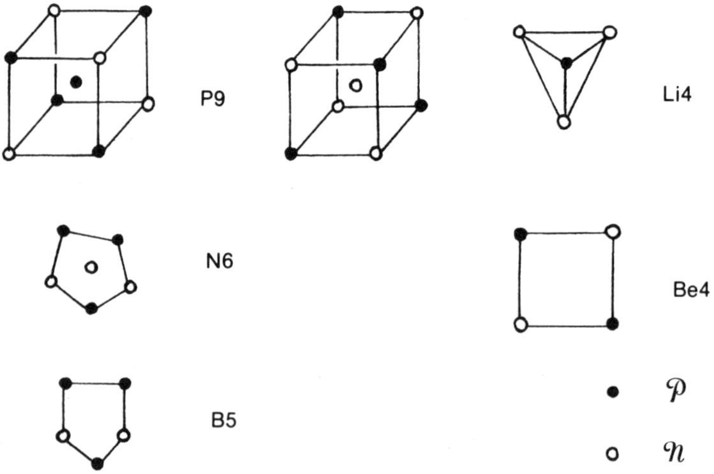

Fig. 7.91

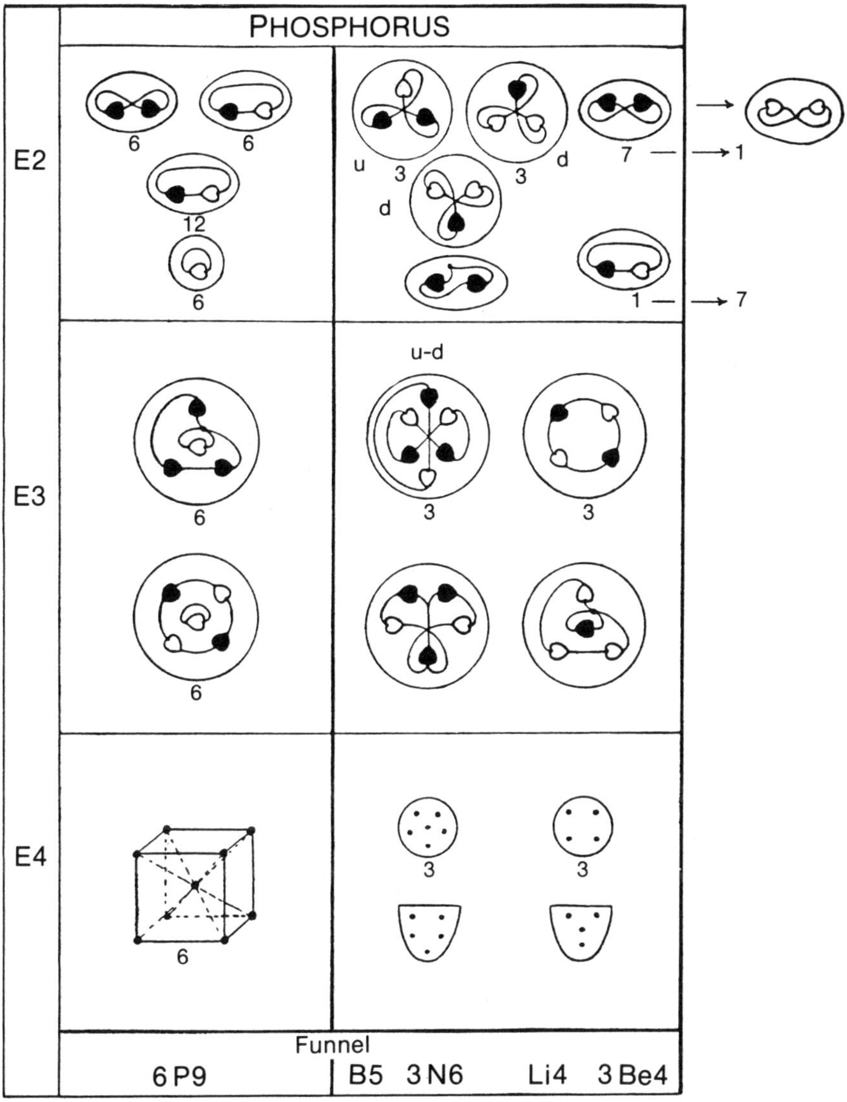
PHOSPHORUS
E2
6
6
12
6
u
3
3
d
d
7
1
1
7
E3
u-d
6
3
3
6
E4
6
3
3
Funnel
6 P9
B5 3 N6
Li4 3 Be4

Fig. 7.92

SULPHUR M.P.A.

The M.P.A. is a tetrahedral array of four funnels, similar to that of magnesium (fig. 7.93). Each funnel has three identical sections, each containing three similar ovoids (S16). An ovoid encloses three small spheres, one containing a duad of U.P.A.'s (N2) and two containing identical groups of seven U.P.A.'s (I.7) like that present in the M.P.A. of magnesium (fig. 7.94).

The M.P.A. is formed from two ${}_{16}S^{32}$ nuclei (fig. 7.95). Ninety-six u and ninety-six d quarks are equally distributed among the funnels. Each section, therefore, contains the omegons in eight u and eight d quarks. An S16 group contains a 𝒫-𝒩 diomegon (N2), a bound state of four 𝒫 and three 𝒩 omegons (I.7), and a bound state of four 𝒩 and three 𝒫 omegons (I.7) (fig. 7.96). The M.P.A. of sulphur differs from that of magnesium solely in the replacement of an H3 triplet in the Mg12 by an I.7 to form the S16. It is instructive to see how this difference is directly related to the difference in the nuclei of the two elements. A sulphur nucleus contains four more protons and four more neutrons than a magnesium nucleus. Therefore, the sulphur M.P.A. has seventy-two more 𝒫 and seventy-two more 𝒩 omegons than the magnesium M.P.A., that is, two more 𝒫 and two more 𝒩 per ovoid. These additional omegons combine with the omegons in the u quark (H3) of an ovoid in the magnesium M.P.A. to form another I.7:

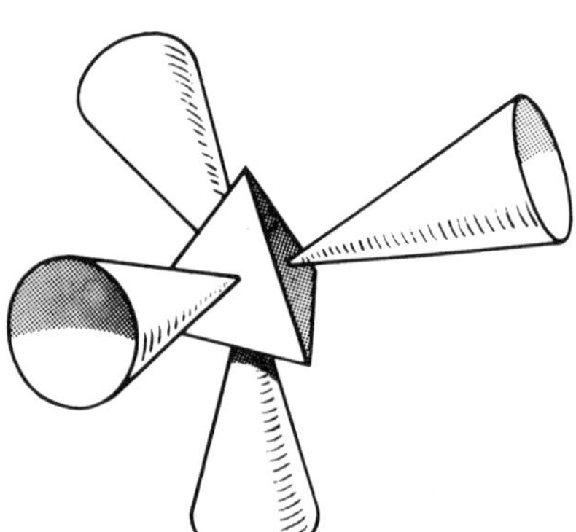

Fig. 7.93

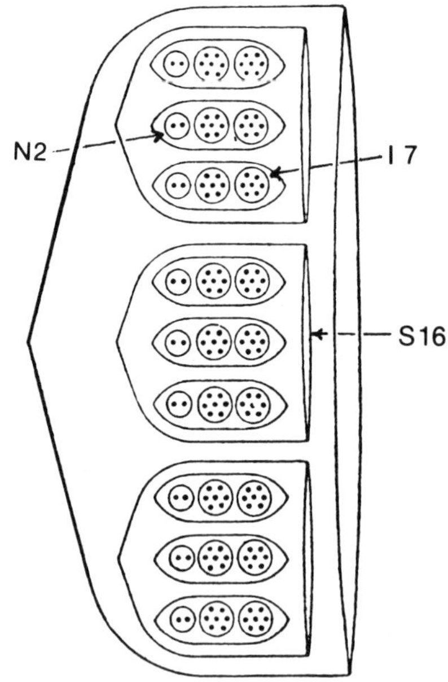

Fig. 7.94

SULPHUR

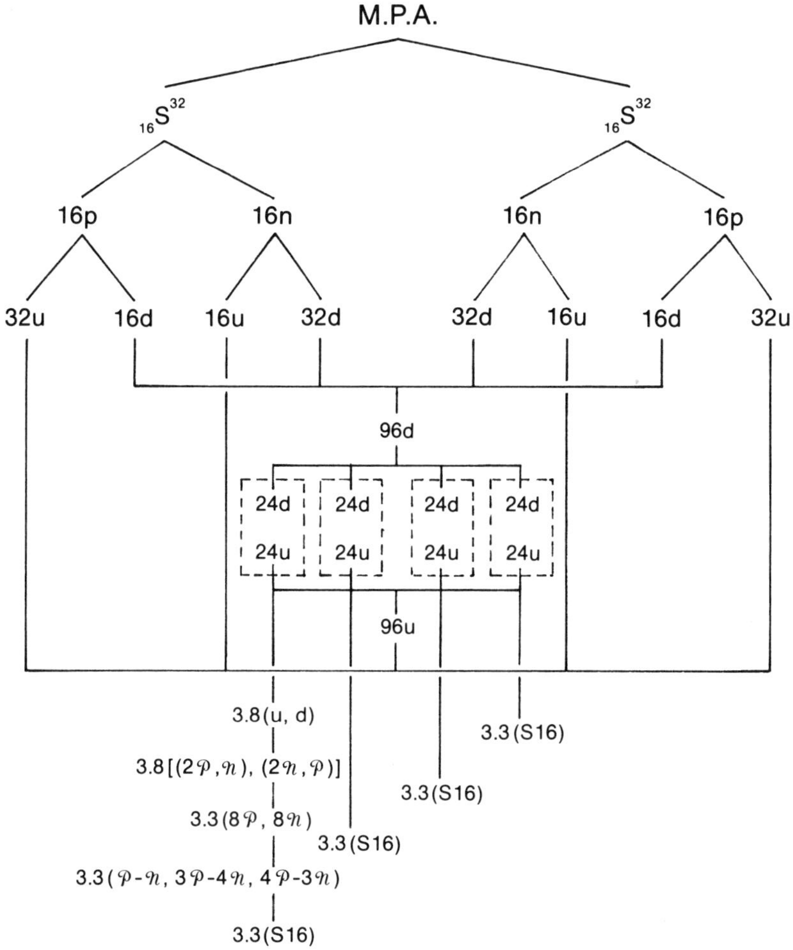

Fig. 7.95

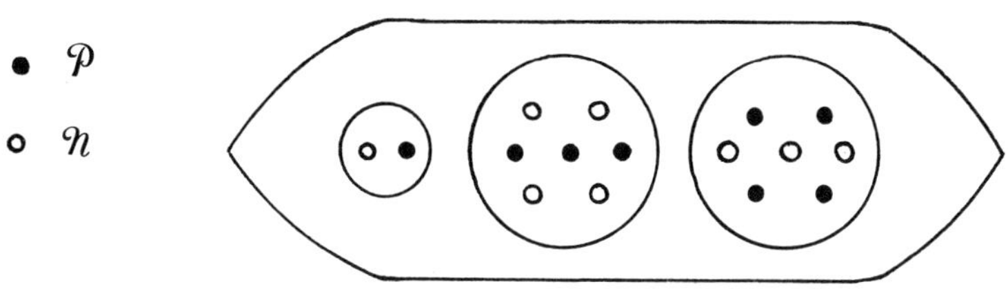

Fig. 7.96. S16 ovoid

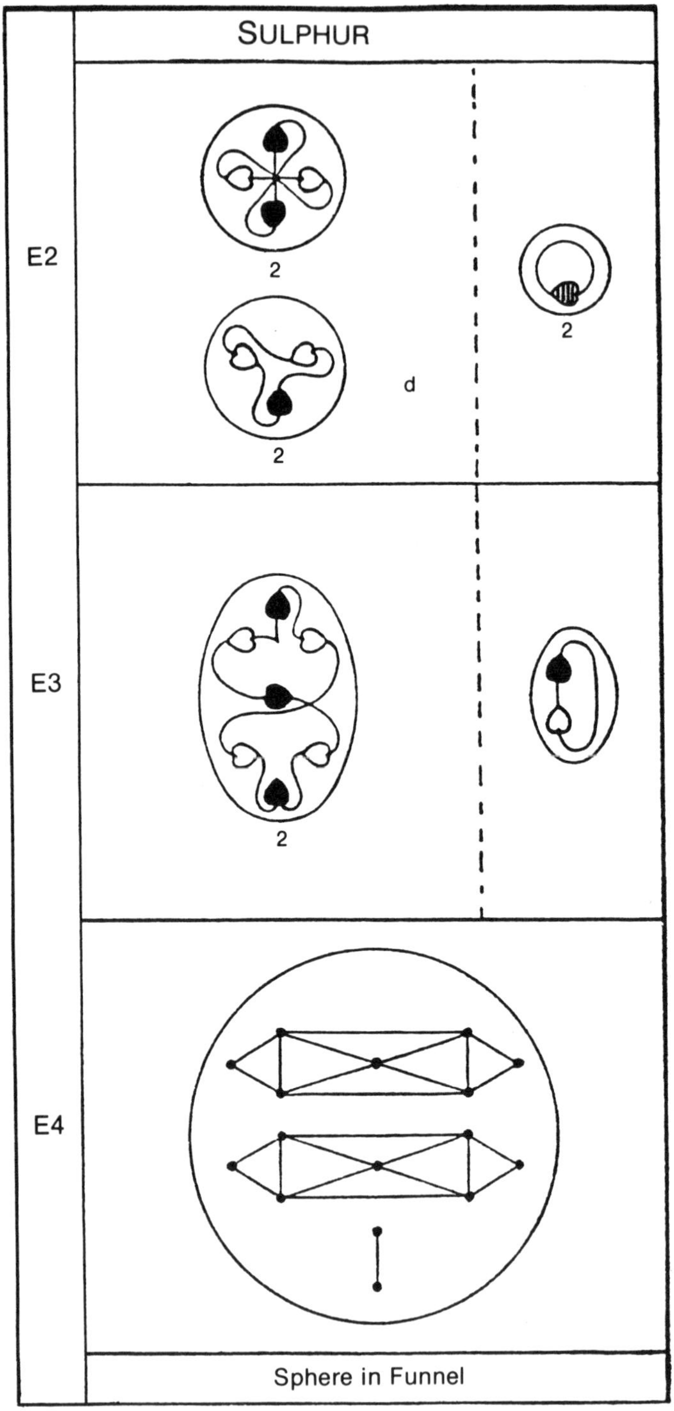

Fig. 7.97

$$\begin{aligned} \text{Mg12} &= \text{N2} + \text{I.7} + \text{H3} \\ &= \mathcal{P}\text{-}\mathcal{N} + 3\mathcal{P}\text{-}4\mathcal{N} + \text{u}\,; \\ \text{Mg12} + 2\mathcal{P} + 2\mathcal{N} &= \mathcal{P}\text{-}\mathcal{N} + 3\mathcal{P}\text{-}4\mathcal{N} + \text{u} + 2\mathcal{P} + 2\mathcal{N} \\ &= \mathcal{P}\text{-}\mathcal{N} + 3\mathcal{P}\text{-}4\mathcal{N} + 3\mathcal{N}\text{-}4\mathcal{P} \\ &= \text{N2} + \text{I.7} + \text{I.7} \\ &= \text{S16}\,. \end{aligned}$$

Figure 7.97, showing the disintegration of a single S16 group, indicates that the N2 duad is a $\mathcal{P}$-$\mathcal{N}$ diomegon (having been identified as such in previous analyses). The I.7 disintegrates into a (+) quartet of U.P.A.'s and a (−) triplet, that is, a bound state of two $\mathcal{P}$ and two $\mathcal{N}$ omegons and a d quark. It therefore consists of four $\mathcal{N}$ and three $\mathcal{P}$ omegons, in agreement with the predicted composition of one of the I.7 groups. The two I.7 groups are not identical.

CHLORINE M.P.A.

Chlorine is the second member of the Dumb-bell Group, after sodium. Its M.P.A. has the form shown in figure 7.98. In the central connecting rod (Cl.19) are five small spheres containing three, four, five, four, and three U.P.A.'s (fig. 7.99). At each end of the rod are Na10 globes like those in the sodium M.P.A. A funnel contains a group of three Hydrogen Triplets (N9) in addition to the contents (Na16) of a sodium funnel. There are twenty-five U.P.A.'s in a funnel, the aggregate being called "Cl.25." There are 639 U.P.A.'s present in the M.P.A.

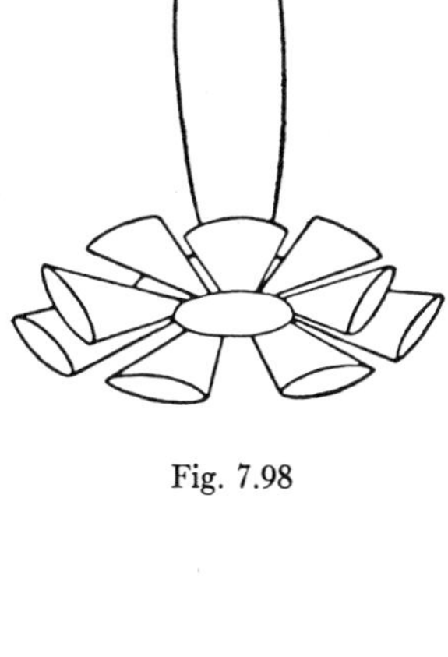

Fig. 7.98

The M.P.A. is formed from two $_{17}Cl^{35}$ nuclei (fig. 7.100). Three hundred omegons in fifty u and fifty d quarks are equally distributed among a set of twelve funnels, twenty-five to a funnel. In the case of the sodium M.P.A., the omegons in thirty-two u and thirty-two d quarks were found to be distributed among a set of twelve Na16 funnels. But here there are eighteen more u and eighteen more d quarks in the set of twelve Cl.25 funnels. Six funnels each contain an extra two u quarks and a d quark (fig. 7.101). The other six funnels each contain two d quarks and a u quark in addition to an Na16 group. These u-u-d and d-d-u bound states are protons and neutrons, respectively (in the analysis of the nitrogen M.P.A., an N9 group was identified as a neutron). The disintegration diagram in figure 7.102 confirms that, for the funnel examined, the N9 breaks up at stage 2 into two (−) and one (+) triplet, that is, two d quarks and a u quark. Two types of Cl.25 groups are predicted:

$$\text{Cl.25}(+) = \text{Na16} + \text{u-u-d}\,,$$

which has an electric charge of $+53/9$, and

$$\text{Cl.25}(-) = \text{Na16} + \text{d-d-u}\,,$$

which has a charge of $-28/9$. Six of each type are present in a set of twelve funnels at either end of the dumb-bell form, the two types alternating as

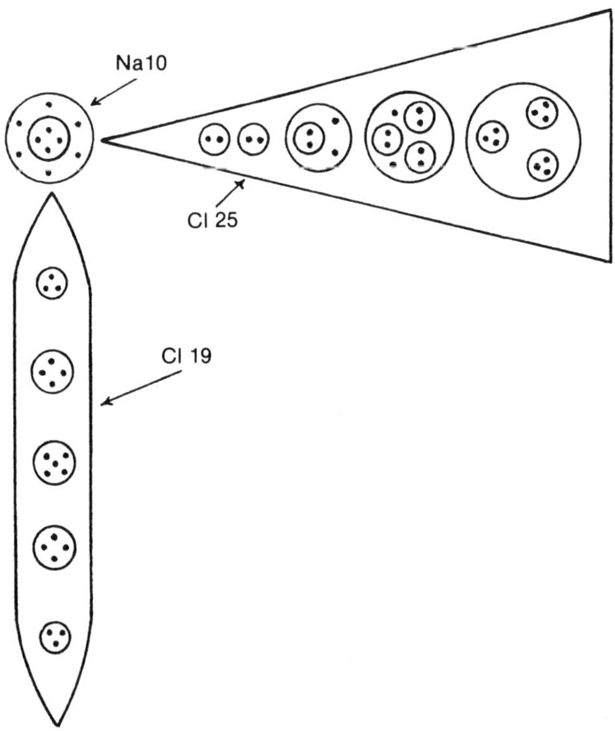

Fig. 7.99. Central rod (Cl.19), Na10 globe, and one funnel of the M.P.A. of chlorine

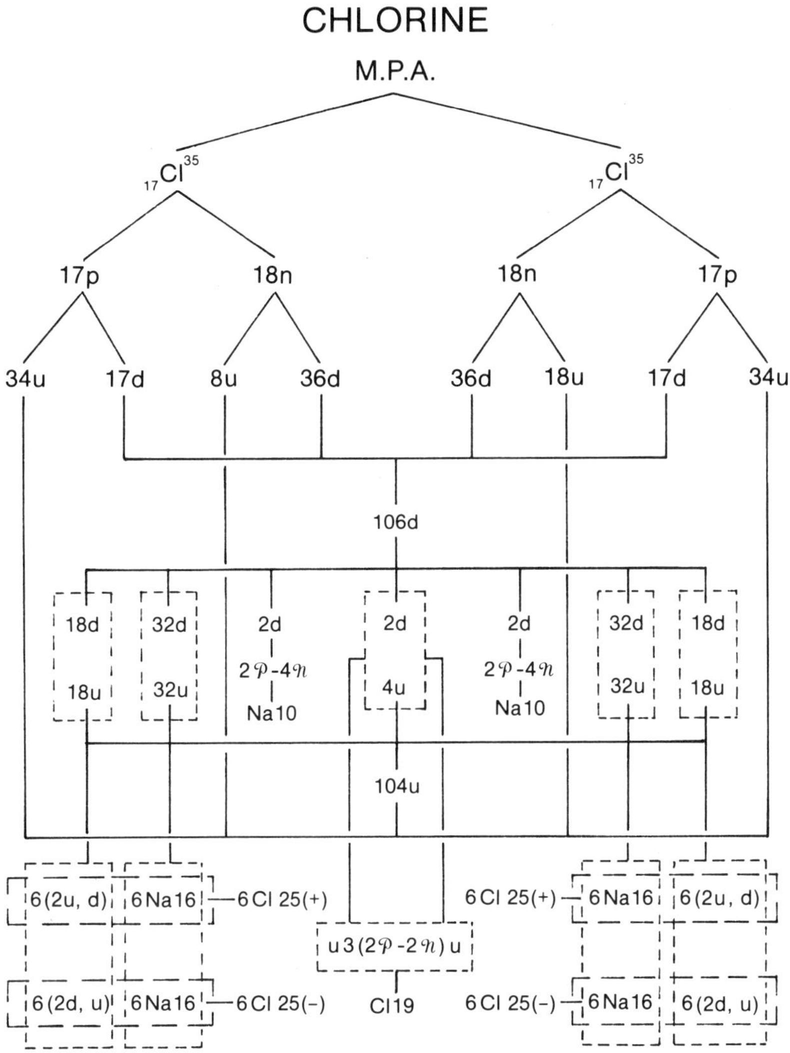

Fig. 7.100

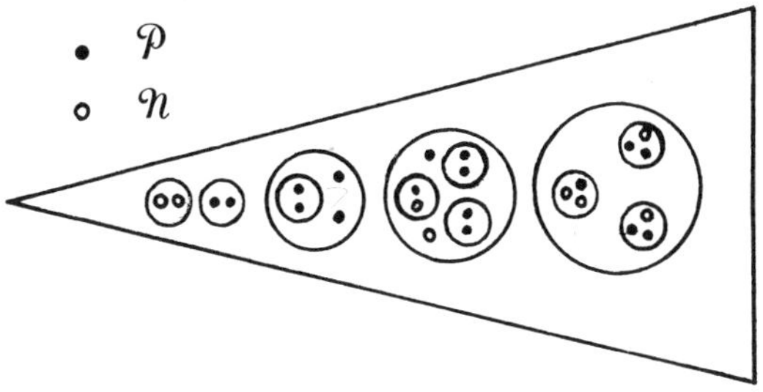

Fig. 7.101. Cl.25(+) funnel

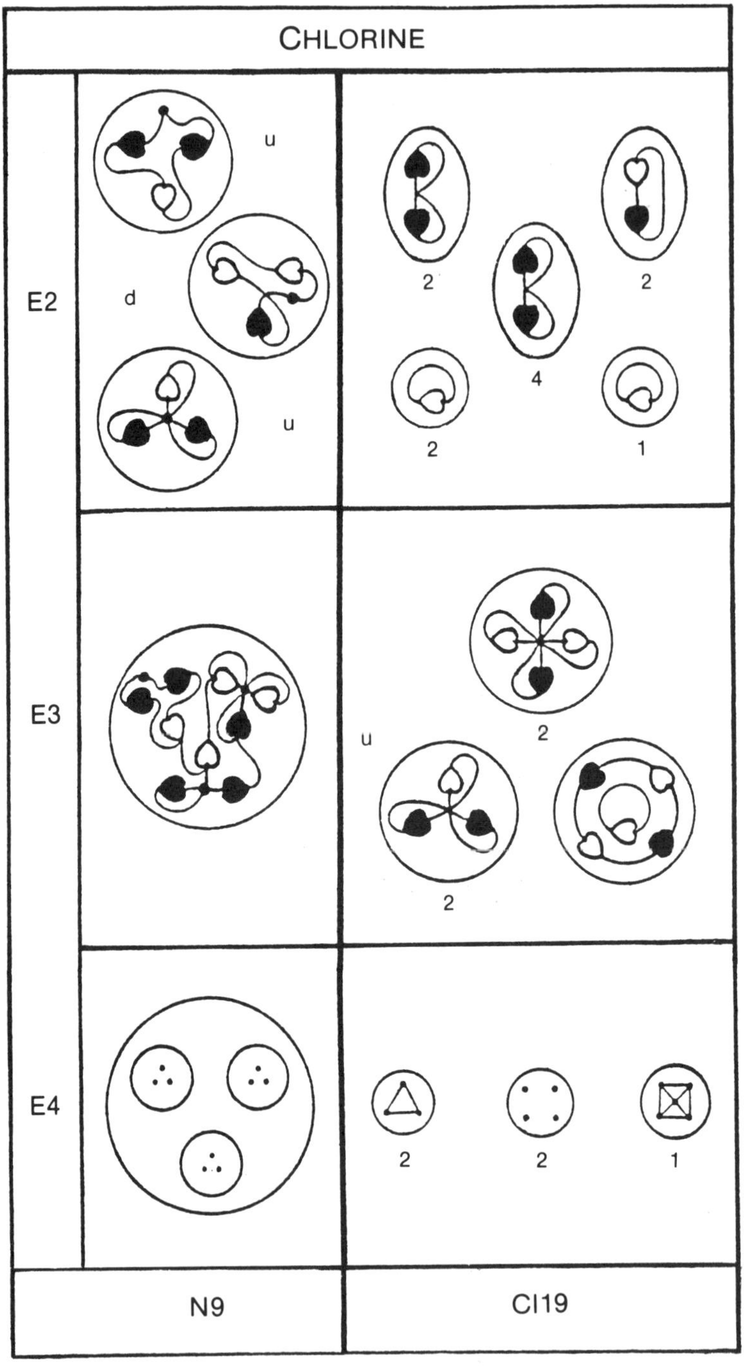

Fig. 7.102

shown in figure 7.103. The Cl.19 rod is made up of two u quarks, two $2\mathcal{P}$-$2\mathcal{N}$ bound states, and a central group of two $\mathcal{P}$ and two $\mathcal{N}$ (fig. 7.104). Figure 7.102 shows two (+) triplets at stage 3 of its disintegration. It also indicates that the two groups of four U.P.A.'s break up into four (+) duads. This must be an error of observation, for they should disintegrate into two (+) and two (−) duads. The reported group of five U.P.A.'s at the centre of the rod should be a group of four, that is, the Cl.19 should contain eighteen U.P.A.'s. The omegons in two d quarks make up an Na10 globe, which should contain six, not ten, U.P.A.'s (fig. 7.105). The discrepancy of 9 between the 630 omegons predicted to be in the M.P.A. and the counted population of 639 arises from an overcounting by 1 for the Cl.19 rod and by 4 for the Na10 globes. An argument that supports this conclusion is as follows: a slightly more complex form of the chlorine M.P.A. was described by the investigators.[23] Each funnel had an extra U.P.A., and each globe contained two octahedral arrays of U.P.A.'s, one inside the other (fig. 7.106). The M.P.A. contained 667 U.P.A.'s. This is the M.P.A. of the $_{17}Cl^{37}$

Fig. 7.103

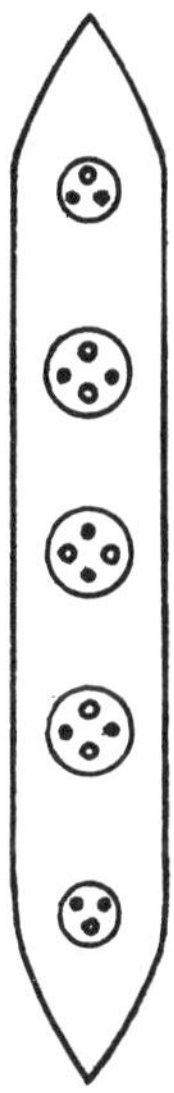

Fig. 7.105. Corrected Na10

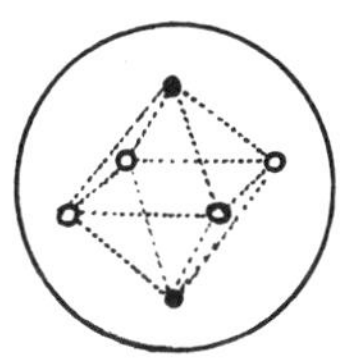

Fig. 7.104. Corrected Cl.19

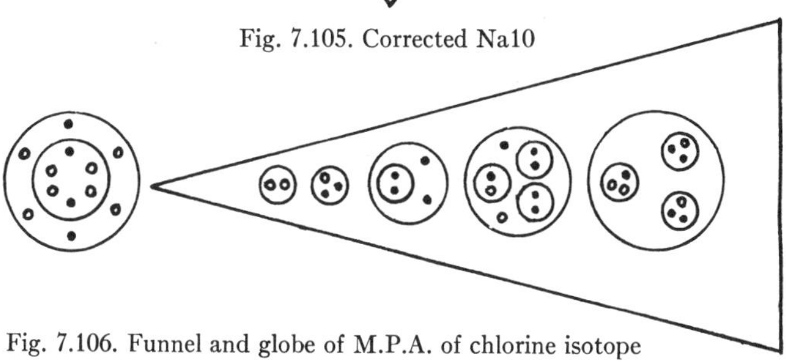

Fig. 7.106. Funnel and globe of M.P.A. of chlorine isotope

isotope, for which there should be 666 U.P.A.'s. If the Cl.19 rod actually contained eighteen U.P.A.'s, the counted and predicted populations would be equal. Furthermore, the difference between the two M.P.A.'s should amount to sixteen $\mathcal{P}$ and twenty $\mathcal{N}$ omegons, contributed by four extra neutrons. Assuming that twelve $\mathcal{P}$ and twelve $\mathcal{N}$ are added to the twenty-four funnels, with an innermost duad ($\mathcal{P}$-$\mathcal{P}$) becoming a triplet (u quark) by the addition of an $\mathcal{N}$, and a duad ($\mathcal{N}$-$\mathcal{N}$) becoming a triplet (d quark) by the addition of a $\mathcal{P}$, then four $\mathcal{N}$ and two $\mathcal{P}$ should be added to each Na10 globe (which already contains two $\mathcal{P}$ and four $\mathcal{N}$, according to the earlier analysis). Each globe in the more complex M.P.A. should contain twelve U.P.A.'s. This agrees with the observations (see fig. 7.106). The investigators also reported[24] that the twelve particles were arranged in two octahedral clusters (fig. 7.107), one inside the other. This is consistent with the predicted composition of each cluster: four $\mathcal{N}$ and two $\mathcal{P}$ omegons.

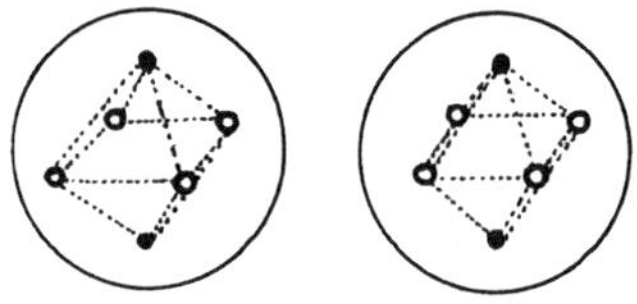

Fig. 7.107

ARGON M.P.A.

Argon is a member of the Star Group. Figure 7.108 shows the general form of its M.P.A. The investigators noted three varieties: "proto-argon," containing 672 U.P.A.'s, the ordinary form with 720 U.P.A.'s (the most frequently observed), and "meta-argon" with 756 U.P.A.'s. In the ordinary

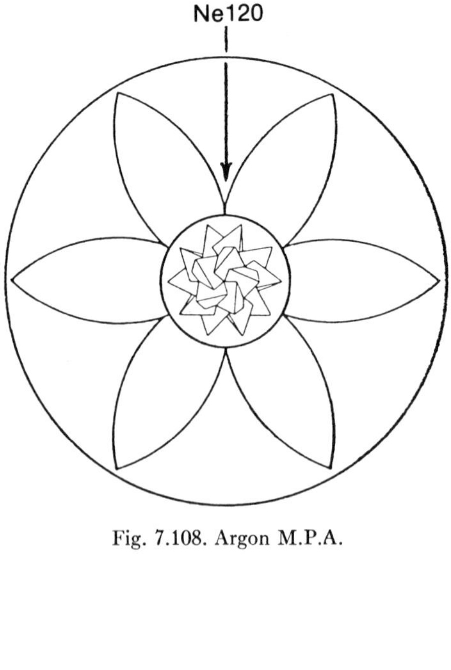

Fig. 7.108. Argon M.P.A.

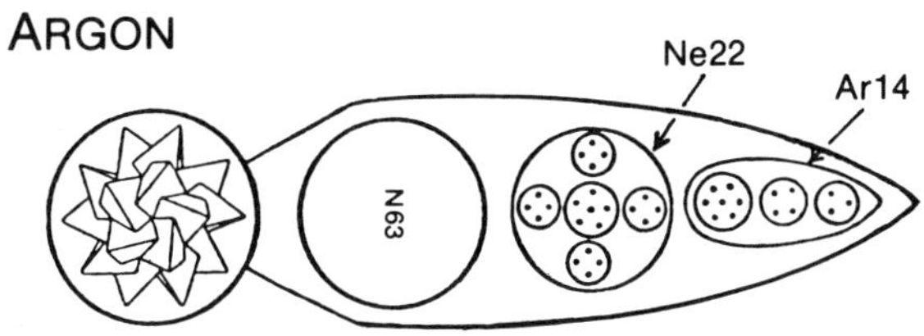

Fig. 7.109. Central globe and one arm

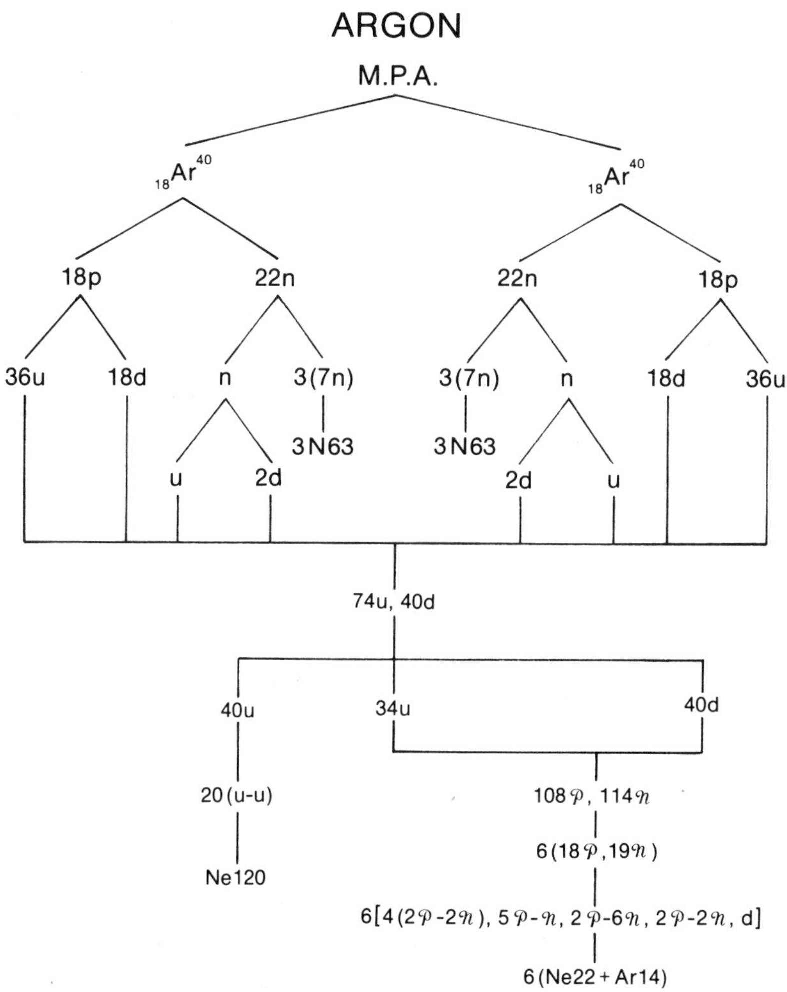

Fig. 7.110

type, an Ne120 acts as the nucleus or core (as it does for all the inert gas elements), and each of the six arms projecting from this centre is made up of an N63 group, an Ne22 group and, at the outer end of the arm, a group of fourteen U.P.A.'s (Ar14). The latter comprises a cluster of seven, a quartet, and a triplet of U.P.A.'s (fig. 7.109).

The ordinary M.P.A. of argon is formed from two ${}_{18}Ar^{40}$ nuclei, this isotope being the most abundant of the isotopes of argon (fig. 7.110). Of the original forty-four neutrons contained in these nuclei, forty-two are present still in the M.P.A., appearing as clusters of seven groups of three Hydrogen Triplets in each arm. This cluster is the N63 group. The Ne120 group consists of twenty u-u diquarks. Thirty-four u and forty d quarks present in nucleons belonging to the original nuclei break up into omegons that then interact with one another to form new bound states. As was found in the case of the neon M.P.A., the Ne22 group is a cluster of four $2\mathcal{P}$-$2\mathfrak{N}$ bound states, surrounding a core made up of five $\mathcal{P}$ omegons and an $\mathfrak{N}$ omegon. The predicted composition of the Ar14 is shown in figure 7.111. An extra

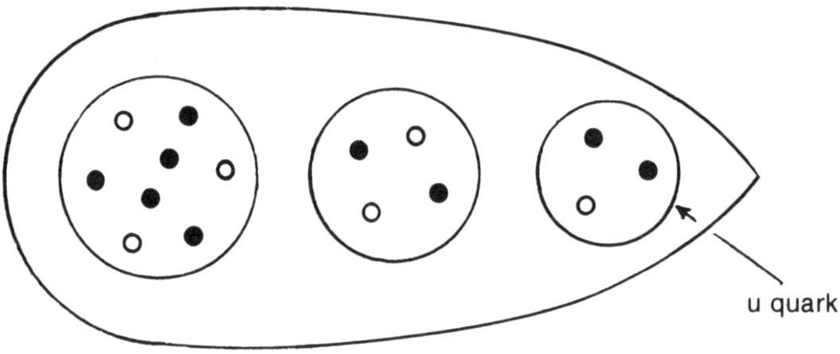

Fig. 7.111. Omegon composition of Ar14

U.P.A. is predicted to belong to the group of seven U.P.A.'s. Since the investigators did not provide a disintegration diagram for argon, it is not possible to confirm the prediction that the Ne120 group, being made up of twenty u-u diquarks, would have been recorded as having broken up into forty (+) triplets (u quarks), instead of twenty (+) triplets and twenty (−) triplets, as noted in the case of the neon M.P.A., where the Ne120 was identified as a bound system of twenty u-d diquarks.

POTASSIUM M.P.A.

The M.P.A. belongs to the Spike Group. It has a central globe made up of an N110 group, surrounded by six spheres containing Li4 groups. Nine Li63 groups are bound to the central globe and project outwards like spikes. There are 701 U.P.A.'s in the M.P.A. (fig. 7.112).

The M.P.A. is formed from two ${}_{19}K^{39}$ nuclei, which provide 116 u and 118 d quarks (fig. 7.113). Of these, sixteen u and twenty-one d quarks

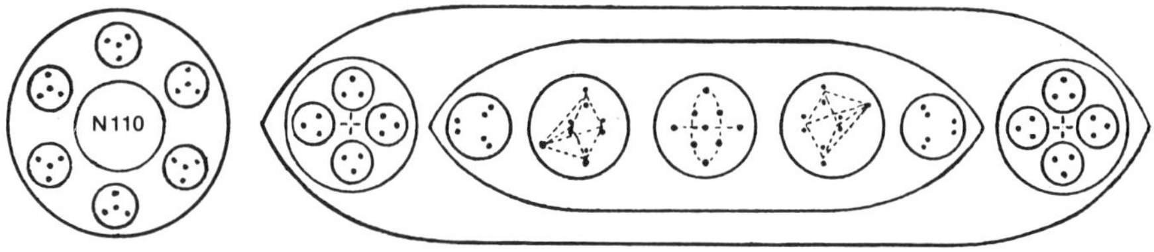

Fig. 7.112. Central globe and one spike of the M.P.A. of potassium

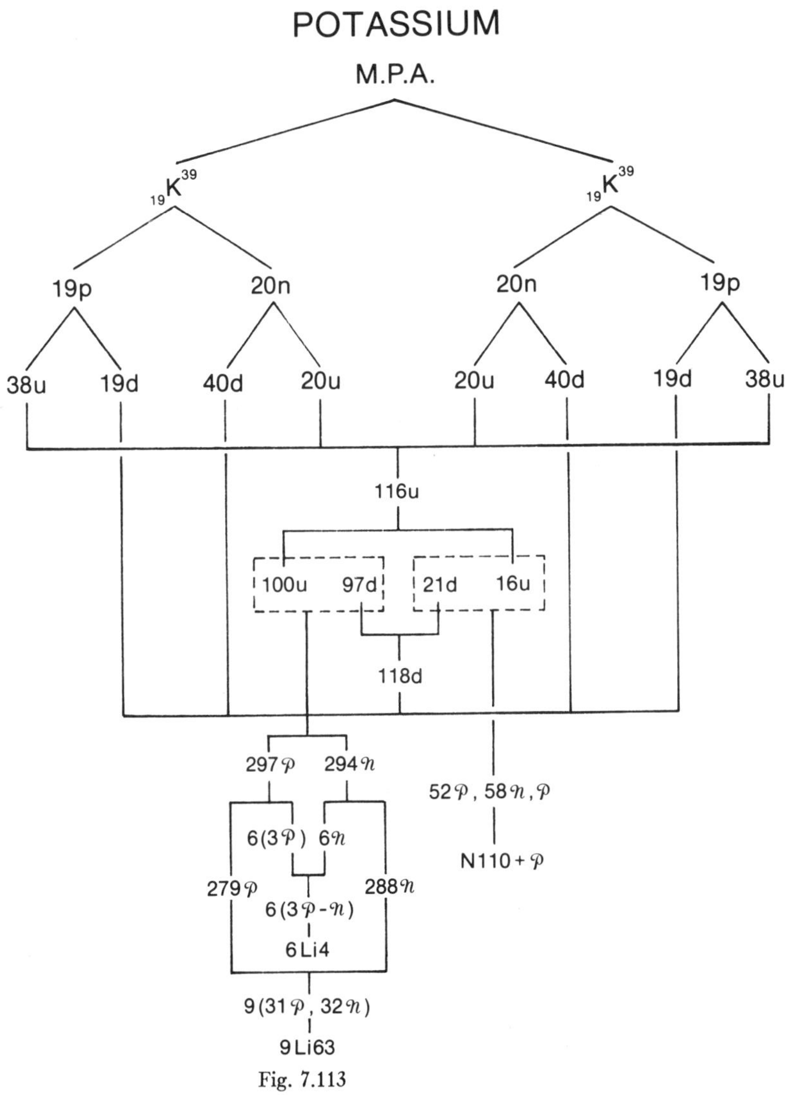

Fig. 7.113

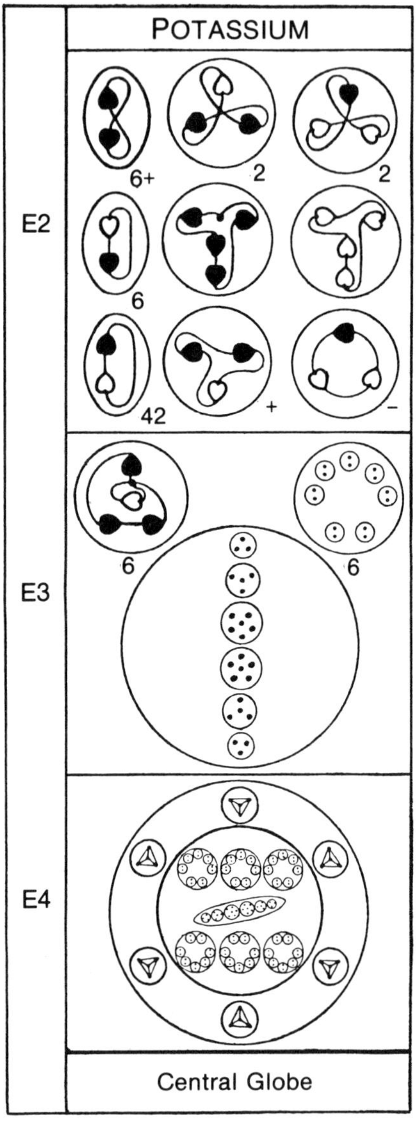

Fig. 7.114

break up into fifty-three $\mathcal{P}$ and fifty-eight $\mathcal{N}$ omegons that regroup to form the N110 group. According to the analysis of the fluorine M.P.A., the N110 can consist of either fifty-eight $\mathcal{P}$ and fifty-two $\mathcal{N}$ or fifty-two $\mathcal{P}$ and fifty-eight $\mathcal{N}$ omegons. Therefore, an extra $\mathcal{P}$ is predicted to be in the N110, having been overlooked by the investigators. This accounts for the discrepancy of 1 between the counted number of U.P.A.'s and the predicted 702 omegons in the M.P.A. Each Li4 group is a bound state of three $\mathcal{P}$ omegons and an $\mathcal{N}$ omegon. Figure 7.114 shows that an Li4 group breaks up at stage 2 into a (+) duad ($\mathcal{P}$-$\mathcal{P}$) and a $\mathcal{P}$-$\mathcal{N}$ duad, that is, it consists of three $\mathcal{P}$ omegons and one $\mathcal{N}$ omegon, in agreement with the analysis. Each Li63 group contains thirty-one $\mathcal{P}$ and thirty-two $\mathcal{N}$ omegons. In the analysis of the lithium M.P.A., the Li63 group was found to have thirty $\mathcal{P}$ and thirty-two $\mathcal{N}$ omegons. Here, an extra $\mathcal{P}$ is added to the central sphere in the Li63, giving it the structure actually observed (fig. 7.115).

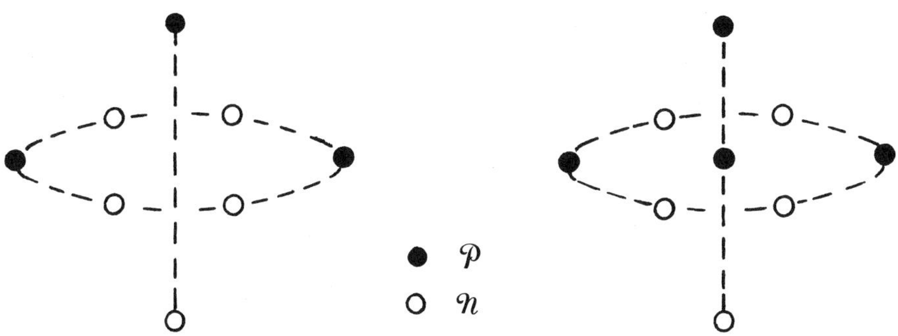

Fig. 7.115. *Left:* Central group in the Li63 groups of the lithium M.P.A. *Right:* Central group in the Li63 groups of the potassium M.P.A.

CALCIUM M.P.A.

The M.P.A. consists of a tetrahedral array of four funnels projecting from a central globe (fig. 7.116). The globe (Ca80) is a pair of concentric spheres that are divided into eight equal segments (fig. 7.117). Eight Li4 groups are symmetrically bound in the inner sphere; eight Ad6 groups are distributed around them in the outer sphere, one to each segment. A funnel contains three spheres, the central one (Ca70) having seven Be10 ovoids and the spheres (Ca45) on either side each containing five Al.9′ ovoids. A funnel contains 160 U.P.A.'s and is denoted "Ca160."

The M.P.A. is formed from two $_{20}Ca^{40}$ nuclei (fig. 7.118). In one Ca45 group, the Al.9′ groups are each composed of the omegons in a neutron that have been set free and have recombined as $\mathcal{P}$-$\mathcal{P}$ and $\mathcal{N}$-$\mathcal{N}$ diomegons and a square pyramidal array of two $\mathcal{P}$ and three $\mathcal{N}$ omegons. In the other Ca45, the omegons in a proton make up an Al.9′ as $\mathcal{P}$-$\mathcal{P}$ and $\mathcal{N}$-$\mathcal{N}$ diomegons and

Fig. 7.116

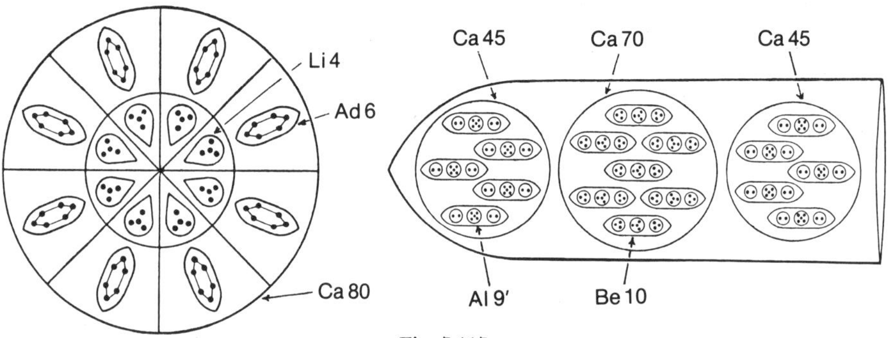

Fig. 7.117

a square pyramidal array of three 𝒫 and two 𝒩 omegons (fig. 7.119). Hence, in the two Ca45 groups there are two types of duads of U.P.A.'s (ten 𝒫-𝒫 and ten 𝒩-𝒩). The disintegration diagram (fig. 7.120) confirms this prediction: in the products of disintegration at stage 3 there are ten (+) duads (𝒫-𝒫) and ten (−) duads (𝒩-𝒩). The Ca80 consists of four u-u and four d-d diquarks (Ad6), surrounding an inner core of four tetrahedral clusters of four 𝒫 and four tetrahedral clusters of four 𝒩 omegons (fig. 7.121). The disintegration diagram confirms this as well, for it shows that the globe breaks up at stage 3 into four (+) Ad6 groups and four (−) Ad6 groups and (+) and (−) Li4 groups. The latter break up at stage 2 into two (+) and two (−) duads; that is, an Li4 group consists of either four 𝒫 or four 𝒩 omegons, in agreement with the analysis. The composition of the Ca70 group is shown below:

2 funnels:	Be10 (× 4):	u-4𝒫-u
	Be10 (× 3):	d-4𝒩-d
2 funnels:	Be10 (× 4):	d-4𝒩-d
	Be10 (× 3):	u-4𝒫-u

CALCIUM

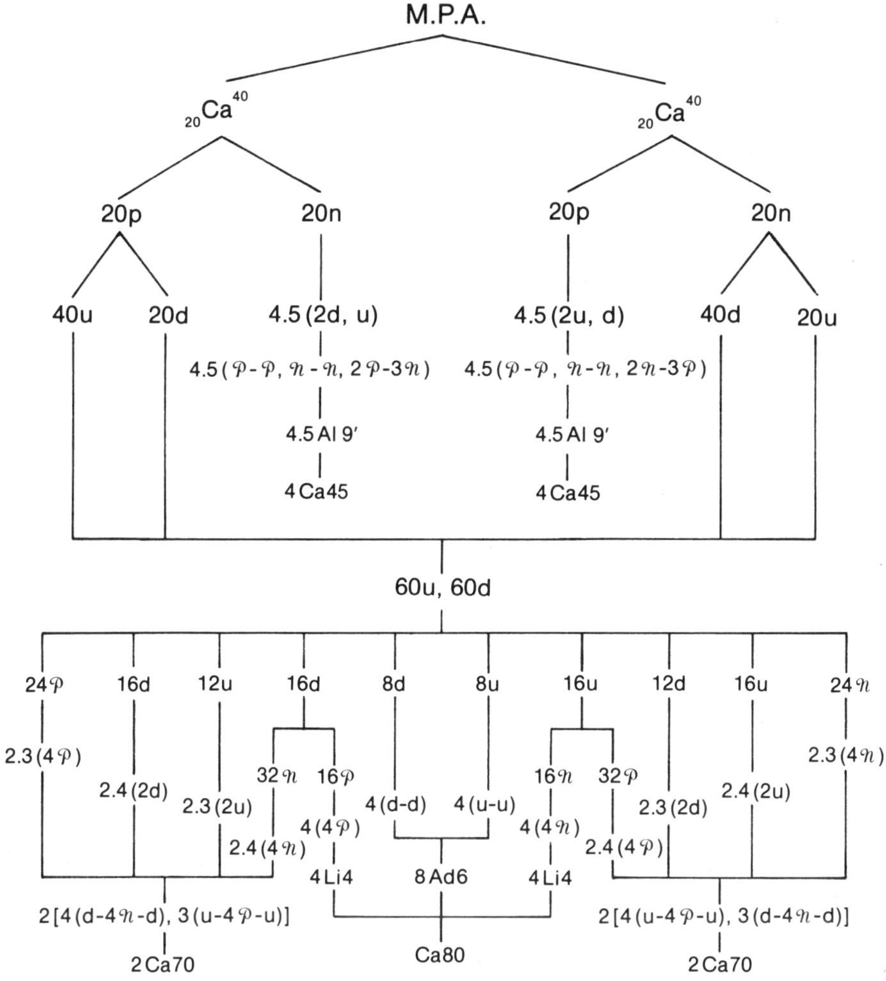

Fig. 7.118

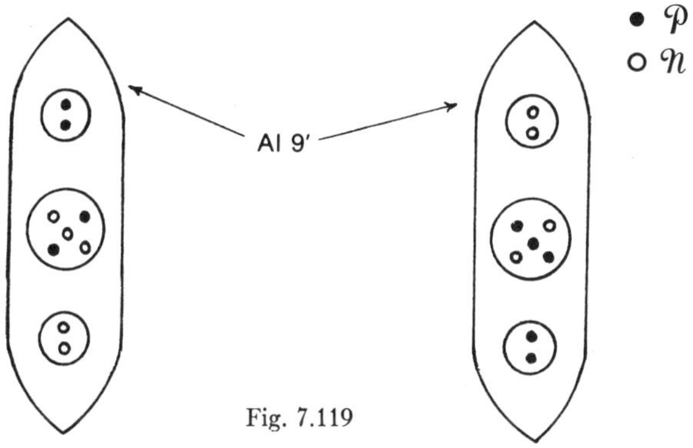

Fig. 7.119

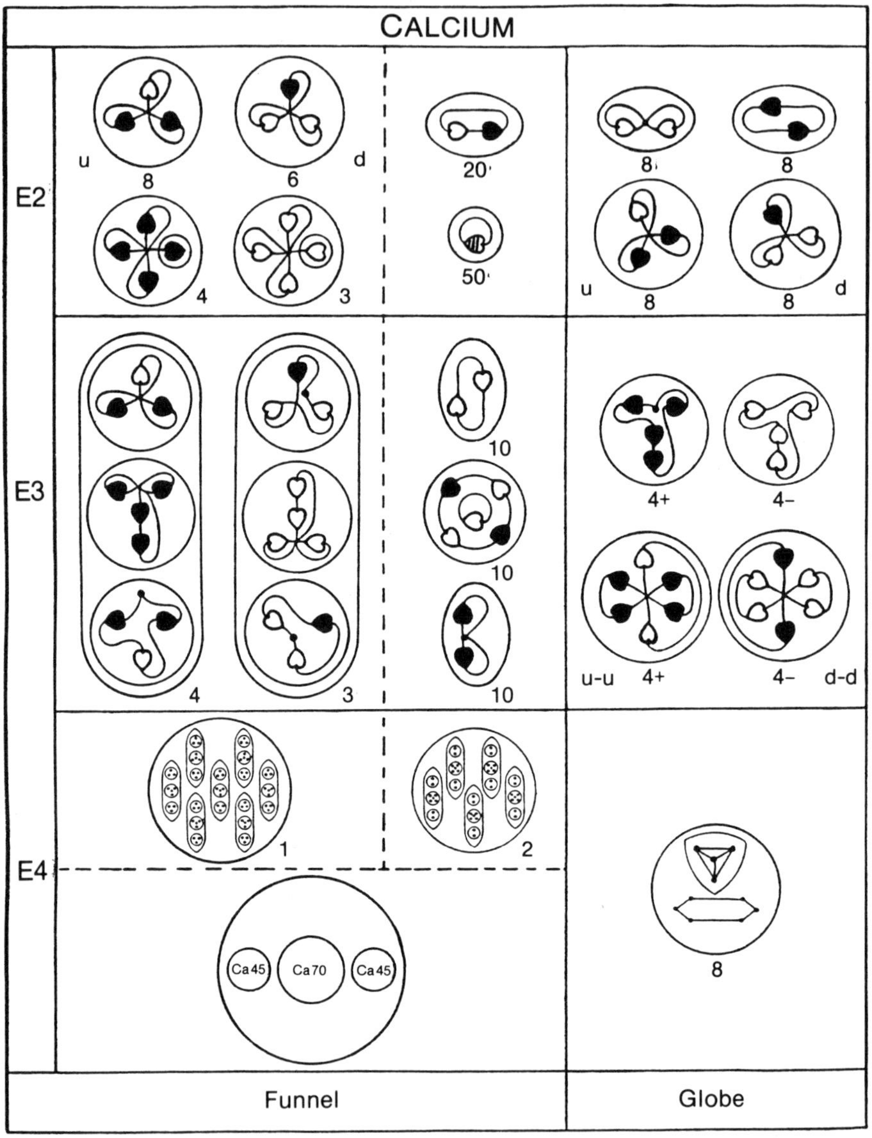
CALCIUM
E2
u
8
6
d
4
3
20
50
8
8
u
8
8
d
E3
4
3
10
10
10
4+
4–
u-u
4+
4–
d-d
E4
1
2
Ca 45
Ca 70
Ca 45
8
Funnel
Globe

Fig. 7.120

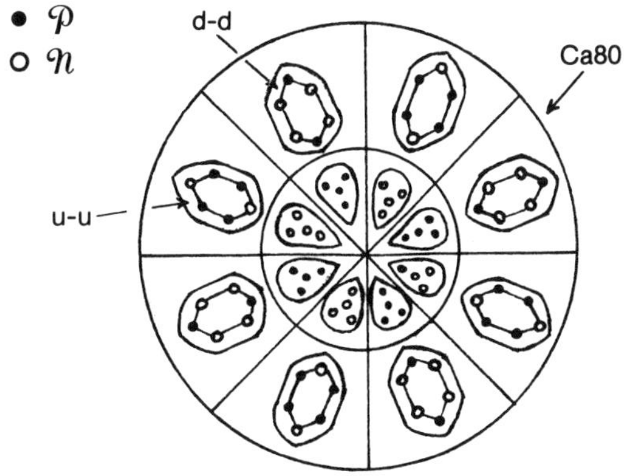

Fig. 7.121

Figure 7.120 confirms this, for it shows that there are four Be10 groups with two (+) triplets (u quarks) and three Be10 groups with two (−) triplets (d quarks). The two types of Be10 contain (+) and (−) groups of four U.P.A.'s. These are the bound states of four 𝒫 and four 𝒩 omegons. The composition of the two types is identical with that found for the (+) and (−) Be10 groups in the analysis of the beryllium M.P.A. Every detail in the M.P.A. of calcium is accounted for in a way consistent with the results of analysis of other M.P.A.'s.

REFERENCES

1. Occult Chemistry, 3d ed., p. 1.
2. Ibid., p. 372.
3. Ibid., p. 293.
4. Ibid., p. 377.
5. Ibid., p. 357.
6. The variation with distance of the coupling between colour-shade charges of omegons has not yet been determined. Quantum chromodynamical results for quark-antiquark pairs are irrelevant, since SU(9) (and not SU(3)) is the fundamental gauge group for omegon strong interactions.
7. Occult Chemistry, 3d ed., p. 15.
8. The Theosophist, vol. **30,** July 1909.
9. Ibid., vol. **54,** November 1932 ("85"); October 1932 ("87").
10. Ibid., vol. **54,** October 1932.
11. Occult Chemistry, 3d ed., p. 28.
12. S. M. Phillips (unpublished).
13. I. Khan, Phys. Rev., **D14,** 1648 (1976).
14. S. M. Phillips (unpublished).
15. Occult Chemistry, 3d ed., p. 38.
16. The Theosophist, vol. **54,** December 1932.
17. Occult Chemistry, 3d ed., p. 45.

18. Ibid.
19. Ibid., p. 207.
20. G. W. C. Kaye and T. H. Laby, Tables of Physical and Chemical Constants, 14th ed. (New York: Longmans, 1973).
21. Occult Chemistry, 3d ed., p. 173.
22. Ibid., p. 30.
23. Ibid., p. 66.
24. Ibid., p. 351.

CHAPTER 8

Micro-Psi Molecules

I do not know whether these things exist in Nature or whether they are made, so to speak—whether they exist in Nature made by the Logos or whether they exist only when made by men.

C. W. Leadbeater, examining alpha and beta naphthol

When a person with micro-psi vision examines the interior of materials in the solid or liquid state, he reports generally that he observes many M.P.A.'s in close association, instead of single, isolated M.P.A.'s. They may either be bound together in regular lattice arrays extending beyond the range of his vision or be separate, belonging to distinct groups. The composition and pattern of these groups seem to be characteristic of the compound under examination, inasmuch as the same combination of M.P.A.'s always becomes visible to the observer whenever he examines a sample of the chemical and irrespective of whether its identity is known to him. The M.P.A.'s belonging to these groups are those of the elements present in the compound. Their number is equal to the number of atoms of the corresponding element in the molecule (assuming that the compound has molecules). Because of this correspondence, such a group of M.P.A.'s will be called a "micro-psi molecule" (M.P.M.).

A confusing multiplicity of other phenomena may also confront the observer. The intense dynamical activity visible to him forces him to be incomplete and highly selective in his description, a human limitation that should be borne in mind by the reader when examples of micro-psi observation of molecules are discussed later in this chapter. A striking feature of this description is the common reference to what appear to be freely moving "minute points of light" that rush about in disorderly fashion, often shooting across the observer's field of vision with explosive force. When one of them is highly magnified, it is found to be a single U.P.A. (either (+) or (−)). If micro-psi observation were non-perturbative, this description of free U.P.A.'s inside matter would be irreconcilable with Hypothesis 1 (chap. 4), which identifies these particles as the normally confined constituents of quarks that make up the nucleons inside atomic nuclei. But the implication of Hypothesis 2 is that nuclei *are* perturbed by the act of micro-psi observation. Indeed, a dynamical picture of the transformation of observed nuclei into M.P.A.'s was outlined in the previous chapter, where it was proposed that the process (as yet unknown) whereby the micro-psi faculty furnishes information (when properly interpreted) about the microcosmos destabilizes nuclear matter by freeing its quark and omegon constituents. That numerous free and highly mobile U.P.A.'s should be reported to exist in matter, despite the lack of experimental evidence for free magnetic monopoles in nature, is not a serious problem but rather a feature to be expected, in view of the implications of Hypothesis 2, namely, that micro-psi observation of nuclei disintegrates them into free quarks/omegons that then recombine through their strong gluon-exchange forces into exotic elementary particles that do not escape but form quasi-nuclear bound states observed as M.P.A.'s. It is this initial state, created by the very act of his observation, that the investigator reports, although it must also be pointed out that no explicit observation of the formation of an M.P.A. has ever been reported in any study carried out to date.

An important feature of some (but not all) M.P.M.'s is that one or more of their M.P.A.'s may appear to be broken up and their constituent bodies either scattered over the entire system or mixed with the groups of particles in other M.P.A.'s (for example, see the M.P.M.'s of sodium chloride and of nitric acid). If M.P.M.'s were molecules and M.P.A.'s were nuclei, such overt nuclear disintegration would be impossible to reconcile with the scale of energy changes accompanying chemical reactions. Indeed, it would violate the most basic notions of chemistry. But M.P.M.'s are *not* molecules, according to Hypothesis 2. The following corollary is implied:

The M.P.M. of a chemical compound is the system that results from the transformation into M.P.A.'s (and their subsequent mutual interaction) of similar nuclei present in two molecules of the compound.

Just as two nuclei participate in the formation of an M.P.A., so corresponding nuclei belonging to two molecules form the various M.P.A.'s in an M.P.M. This transformation of similar nuclei will be called "pairing." It is illustrated in figure 8.1. It takes place in the local Coulomb field due to nearby atoms that happen to be outside the domain of micro-psi observation. In chapter 7 the correlation between the position of an element in the periodic table and the form of its M.P.A. was explained in terms of the electromagnetic coupling between the freed quarks/omegons and either the non-valence electrons occupying the outermost closed shells of the atom or the valence electrons shared with other atoms—a coupling that creates directions of preferred aggregation of these freed particles and a lack of spherical symmetry in the resultant quasi-nuclear bound state. For a given

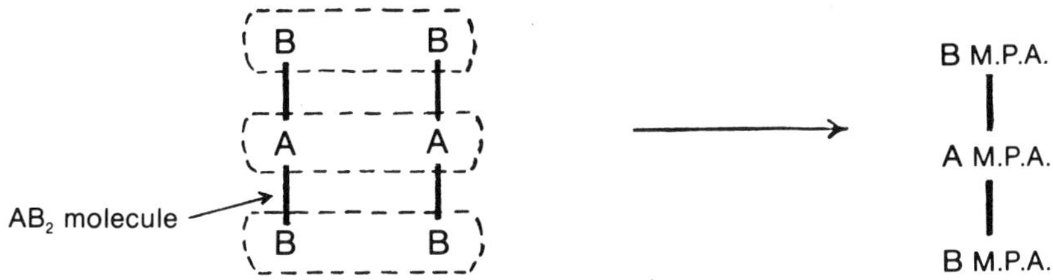

Fig. 8.1

nucleus of an atom bound to other atoms in a molecule, its pairing with a similar nucleus of a neighbouring molecule may be disturbed by the pairing of other nuclei and by the electromagnetic forces acting over atomic distances between the particles freed from one nucleus and those freed from other nuclei in the two molecules. It is proposed that this accounts for the disintegrated state of some M.P.A.'s in various M.P.M.'s and for the intermingling of M.P.A.'s in others. Rather than being thought of as broken up, these M.P.A.'s should be regarded as not having been completely assembled out of the matrix of free particles resulting from the disintegration of the nuclei in the two molecules. Other factors may be involved in what is a complex, dynamical process. The important point remains, however, that these otherwise puzzling features are compatible with the general picture of formation of M.P.A.'s, implied by Hypothesis 2.

In some cases the M.P.M. of a compound is apparently unstable, because it is not observed. Instead, only the products of its disintegration are detected. An example of this is the M.P.M. of ozone (discussed later). Instead of a triangular array of three M.P.A.'s of oxygen, the investigators Annie Besant and C. W. Leadbeater reported the existence of a triangular group of three *half*-M.P.A.'s, that is, three single spirals of the oxygen M.P.A. and not three double spirals. The oxygen M.P.A. must be an unstable system when formed as a result of micro-psi observation of ozone molecules. It is reported to be quite stable when oxygen itself is examined. The presence of the two other oxygen nuclei in each molecule forming the M.P.M. and their transformation into M.P.A.'s must destabilize each newly formed M.P.A., so that it splits in half and forms the two half M.P.M.'s that were recorded by the investigators.

With a few important exceptions, M.P.M.'s bear little resemblance in shape to their counterpart molecules. For example, the M.P.M. of benzene is octahedral, whereas its molecule is hexagonal. Remarkably, these structural differences can be traced back to the original arrangement of the atoms or to valence electron states in the molecule. The benzene M.P.M. has features that differentiate (albeit indirectly) between the localized sigma and delocalized pi electrons of the molecule, and the complicated micro-psi description of the diamond lattice can be related to the face-centred cubic array of its atoms. Graphite is reported to be composed of parallel sheets of carbon M.P.A.'s that are arranged at the corners of regular hexagons. This accords exactly with the known crystal structure.

Unlike their molecular counterparts, some M.P.M.'s are of nuclear size, according to the interpretation made for them in this book. An example is the M.P.M. of water (discussed later). Others are much larger. Lack of dynamical calculations makes even order-of-magnitude estimates of their sizes uncertain. Drawings provided by Besant and Leadbeater are unreliable as accurate representations (in terms of scale) of the spatial distribution of particles, whether of U.P.A.'s or of M.P.A.'s. This is because the investigators stated that these particles had rapid, complex motions, making only a time-averaged arrangement relative to one another at all meaningful (the diagrams probably depict this only very approximately), and also because investigators had to use different degrees of micro-psi magnification in order to discern groups of particles of widely varying size. Most of their work was completed before the era of modern nuclear physics, and they had no conception of either atomic or nuclear orders of size. Their subjective estimation of relative sizes is rendered untrustworthy by their use of varying magnification. In general, the laws of perspective pertaining to optical vision and the judgements of the human brain concerning intervals between events in space-time are not reliable guides to the world of micro-psi experiences. Only insight and the conceptual aids of modern physics and chemistry can make intelligible the fragmentary and highly personalized accounts of this world. That this actually proves to be possible is a strong argument for these experiences being just what they appear to be, namely, cognitive perceptions of physical, microscopic objects.

Examples of micro-psi descriptions of molecules are discussed in the following pages.

DIAMOND

Micro-Psi Description of the Diamond Lattice

Micro-psi observation of the diamond lattice reveals that the basic unit regularly repeated throughout lattice space consists of a group of five carbon M.P.A.'s. The funnels of one half of an octahedral M.P.A. face, in an 1:1 fashion, funnels that belong to four M.P.A.'s around it. Adjacent funnels overlap, and the groups of U.P.A.'s within any one funnel occupy gaps between corresponding groups in the funnel adjacent to it. This interlocking appears to create strong bonds between the M.P.A.'s. Figure 8.2 shows the basic unit, with funnels containing C27 and C26 groups indicated by adjacent dark and light faces of the octahedra. Figure 8.3 shows the reverse side of the group of five octahedra (hereafter referred to

Fig. 8.2

Fig. 8.3

as the "unit of 5"). There is, of course, no definite octahedral shape visible—only an octahedral array of funnels, each of which is directed towards one face of the octahedron. Twenty-five of these units of 5 are arranged in five rows, forming a square (fig. 8.4). Above this sheet is a smaller square of sixteen units of 5, and above this is a square of nine units of 5. Next, there is a square of four units of 5, and, last, a single unit of 5 is in the centre. A total of fifty-five units of 5 form a square pyramid with the square of twenty-five units as its base. Each of the sixteen dark squares shown in the base represents a gap between adjacent units of 5 that is formed by opposing

Fig. 8.4. Diamond lattice, according to micro-psi observation

octahedral faces of carbon M.P.A.'s. Into a gap fits the lower half of the central octahedron of a unit of 5 belonging to the sheet of sixteen units of 5 adjacent to the base. Into the gaps in the latter sheet fit the central octahedra of units of 5 making up the sheet next above this one, and so on. An identical set of square sheets lies below the base. The sheets form two square pyramids placed back to back, that is, an octahedron (fig. 8.5). It is not regular but appears to be slightly flattened. Lattice space is filled by the stacking of such octahedral units (fig. 8.6). A single carbon M.P.A. is observed to hover over the middle of each face of these octahedra, pointing away from the centre of the octahedron and with its bottom point almost touching the central point of a face (fig. 8.7). These M.P.A.'s distort the octahedral geometry, for the centre of each triangular face is raised slightly, dividing it into three equal triangles. The octahedral unit becomes a twenty-four-sided figure—the triakis octahedron, shown in figure 8.8.

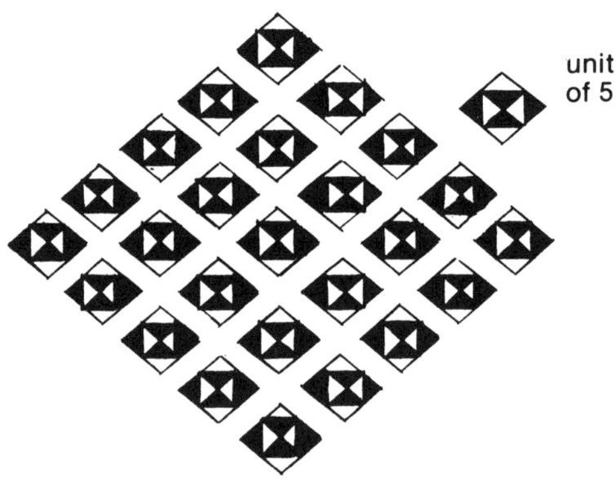

Fig. 8.5

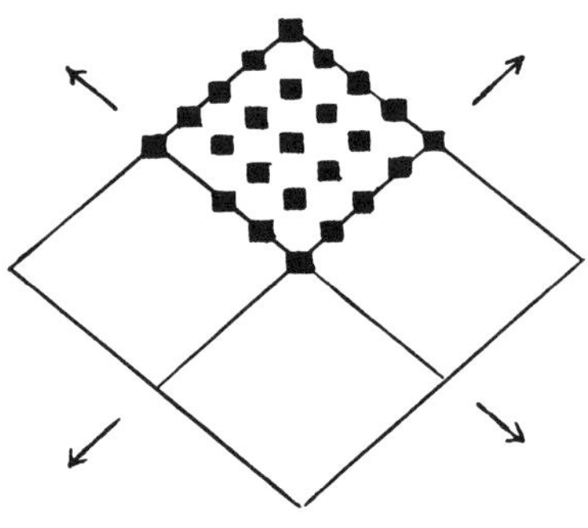
Fig. 8.6

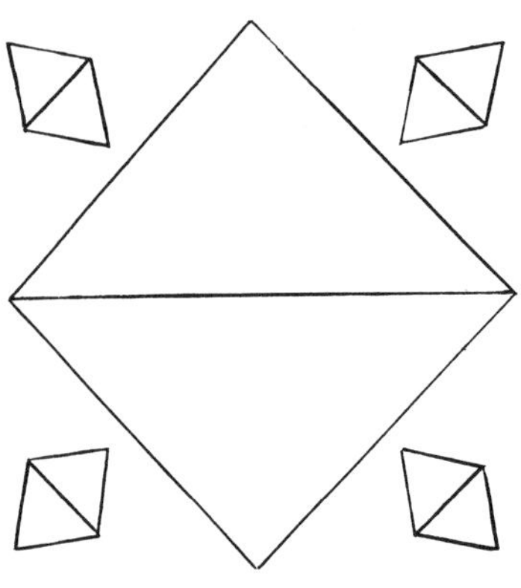
Fig. 8.7

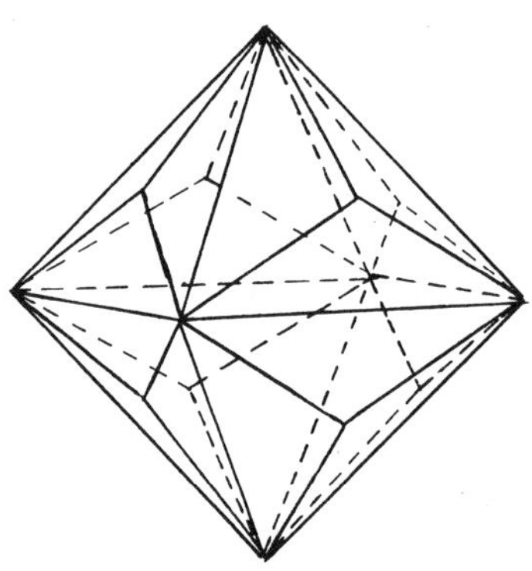
Fig. 8.8

Crystallographic Description of the Diamond Lattice

The diamond lattice has a face-centred cubic unit cell with sides of length 3.56 Å (fig. 8.9). Each carbon atom is covalently bonded to four others that are arranged tetrahedrally about it as the result of sp^3 hybridization of the four valence electrons of each atom. The C-C bond distance is 1.54 Å. There are eight atoms in the unit cell. A primitive basis of two identical atoms at $(X, Y, Z) = (0, 0, 0)$ and $(1/4, 1/4, 1/4)$ is associated with each lattice point. The unit cell contains four tetrahedra, the central atoms of which are at (1/4, 1/4, 1/4), (3/4, 1/4, 3/4), (1/4, 3/4, 3/4), and (3/4, 3/4, 1/4). They are marked x in figure 8.10. Atoms bonded to one

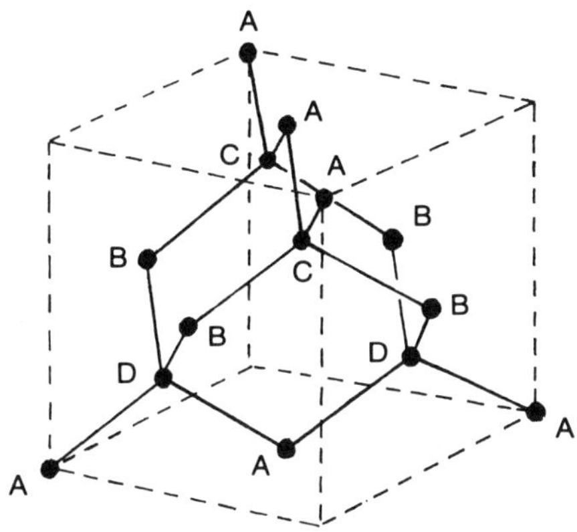

Fig. 8.9. Tetrahedral arrangement of carbon atoms in the unit cell of diamond

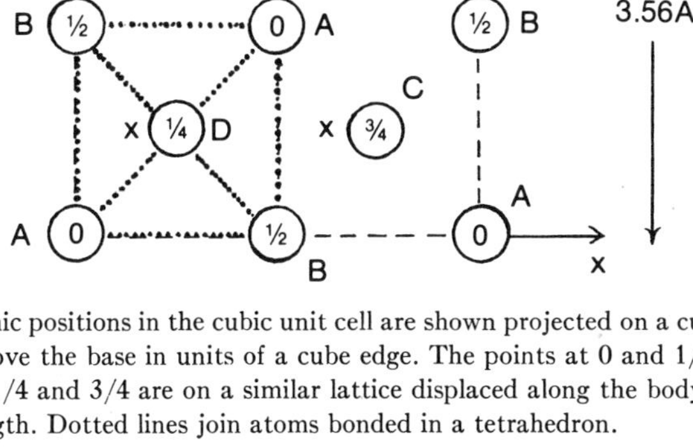

Fig. 8.10. Atomic positions in the cubic unit cell are shown projected on a cube face. Fractions denote height above the base in units of a cube edge. The points at 0 and 1/2 are on the f.c.c. lattice; those at 1/4 and 3/4 are on a similar lattice displaced along the body diagonal by one-quarter of its length. Dotted lines join atoms bonded in a tetrahedron.

another in one such tetrahedron are shown joined to an x atom by dotted lines. Since the crystal has a face-centred cubic unit cell, the spatial arrangement of atoms in the direction of the X-axis is the same as it is in the direction of the Y and Z-axes. The projection of the positions of atoms onto the XY plane is identical with the projection of atomic positions onto the YZ or XZ planes. Therefore, the choice of projection plane is arbitrary, provided that it is parallel to one of the faces of the cubic unit cell.

The connection between these two dissimilar descriptions of the diamond lattice is now determined. Micro-psi examination of a section of the diamond crystal leads to the formation of M.P.A.'s from pairs of nuclei, according to Hypothesis 2. This transformation extends over several dozen unit cells (or whatever happens to be the extent of the lattice coming under micro-psi observation). Consider a two-dimensional array of unit cells. Figure 8.11

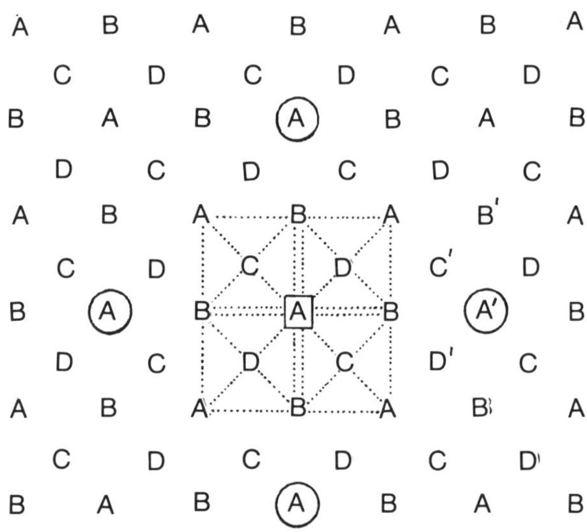

Fig. 8.11. Projection of atomic positions onto XY plane for an extended sheet of f.c.c. unit cells

shows the projection of the positions of carbon atoms onto the XY plane (which is arbitrarily chosen). The A atoms are lattice replications of the atoms at (0, 0, 0) and at (1/2, 1/2, 0); B atoms are the replications of atoms at (0, 1/2, 1/2) and at (1/2, 0, 1/2); C atoms are the lattice replications of the atoms at (3/4, 1/4, 3/4) and at (1/4, 3/4, 3/4); and D atoms are replications of atoms at (1/4, 1/4, 1/4) and at (3/4, 3/4, 1/4). The A atom in the box is bonded to atoms in four other tetrahedra, two being in one unit cell and two being in an adjacent unit cell. Since all carbon atoms occupy equivalent lattice sites, the process leading to formation of M.P.A.'s from pairs of carbon nuclei cannot differentiate between their positions in their unit cells. There is an odd number (5) of atoms in a tetrahedral array of bonded atoms, and so 1:1 pairing of nuclei bonded together in the same tetrahedron would leave one nucleus unpaired. Formation of M.P.A.'s in this way would not respect the geometrical equivalence of all carbon atoms. For reasons of symmetry imposed by the regular crystal lattice, one nucleus in a tetrahedral array cannot pair with a nucleus belonging to a different array, while at the same time the four other nuclei in the array form two M.P.A.'s by pairing among themselves. Formation of M.P.A.'s over many unit cells must take place in a manner that is geometrically equivalent for every nucleus. Since an A atom is bonded to atoms in four neighbouring tetrahedra (as shown in fig. 8.11), it cannot pair off with any atoms shared by these tetrahedra because, if it did, one atom would be left and forced to pair off with another in a way that would be different, geometrically speaking, from the other pairings. Hence, an A atom must form an M.P.A. with an atom that does not belong to the same tetrahedron. Atoms A′, B′, C′, and D′ are the nearest type of neighbours that are not bonded to it. But only A′ is connected with A by a lattice translation vector. Atom A is surrounded octahedrally by six such atoms. Four (shown in circles) are in the *XY* plane, and two (not shown) are directly above and below atom A. It has a choice of atoms with which to form M.P.A.'s, and these choices are all equivalent because of the cubic symmetry of the unit cell. Pairing of nuclei can take place in any one of three mutually perpendicular atomic planes.

In the plane $Z = 1/4$, B atoms have the same face-centred square arrangement and similar conclusions apply: M.P.A.'s are formed only from nearest, non-bonded B neighbours connected with a given B atom by a lattice translation vector in the plane $Z = 1/4$. Similarly for C and D atoms in the planes $Z = 1/2$ and $Z = 3/4$, respectively. Figure 8.12 shows successive stages in the conversion of the atoms A in the *XY* plane into M.P.A.'s. Pairs of neighbouring, non-bonded atoms at the corners of the face-centred squares form M.P.A.'s. Each face-centred atom pairs with another face-centred atom. The pairing proceeds in an identical way for all atoms because the face-centred atom of one square is also a corner atom of another square that is displaced relative to the first by one-half of a diagonal's length. The M.P.A.'s thus formed have initially an off-face-centred rectangular arrangement (fig. 8.12, *a*), but this is an unstable configuration because the M.P.A. in the off-face-centred position has a stronger electromagnetic coupling to its nearest neighbours. If an M.P.A. formed tetrahedral bonds with four others, a face-centred square array would result. This is actually favoured by their mutual Coulombic repulsion due to their charges of $+12$. But eight, not four, valence electrons are

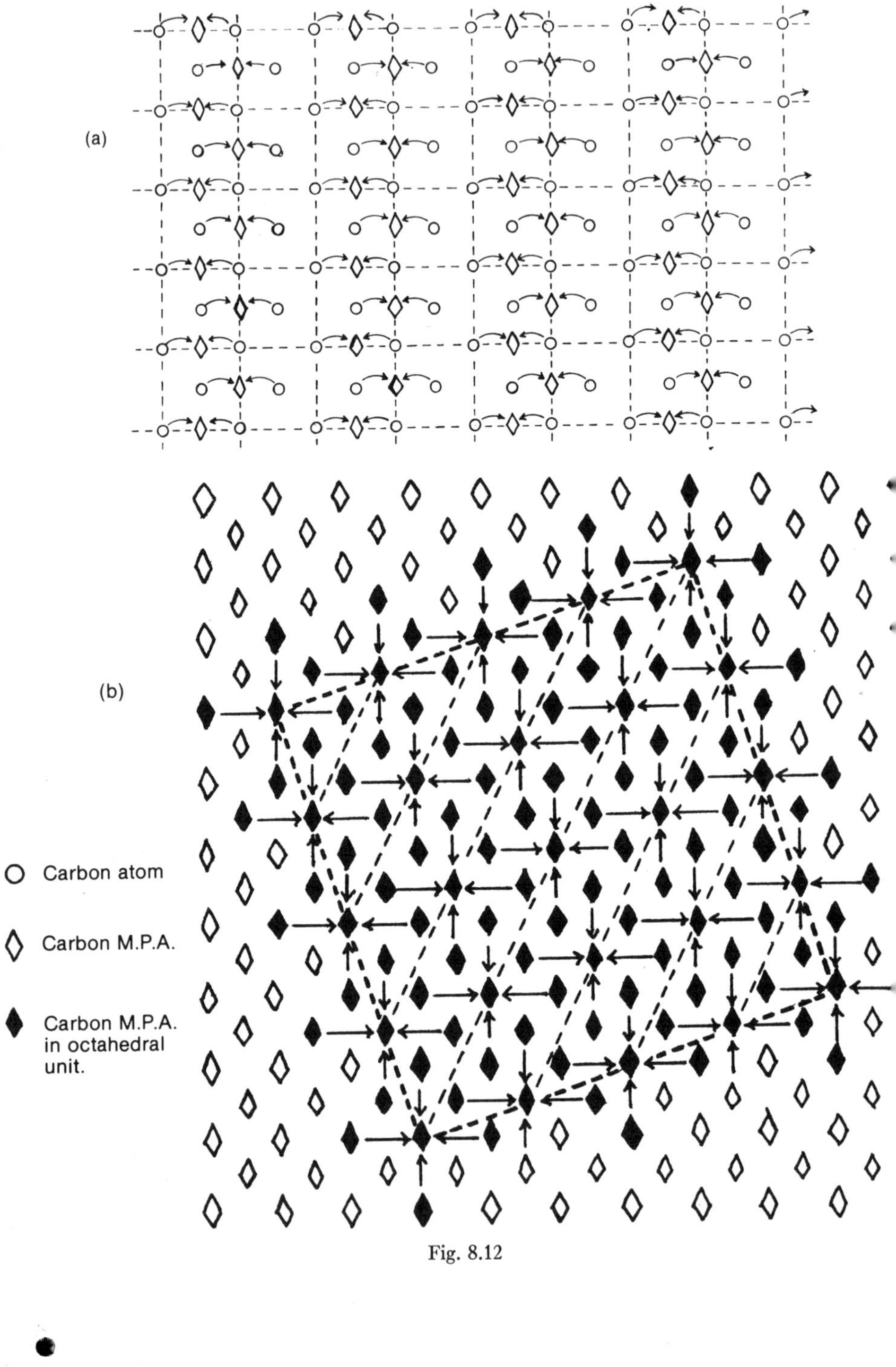

Fig. 8.12

involved in the formation of the M.P.A., and it bonds to eight other M.P.A.'s through its eight funnels. This bonding proceeds as shown in figure 8.12, *b*, four M.P.A.'s clustering around a given one in the same plane to form a unit of 5. Four others in an adjacent sheet complete its bonds. Note that this does not leave out any M.P.A. in the plane. The arrows indicate the directions of migration. The outer dotted lines enclose a face-centred square array of twenty-five units of 5. Comparison of this array (fig. 8.13) with

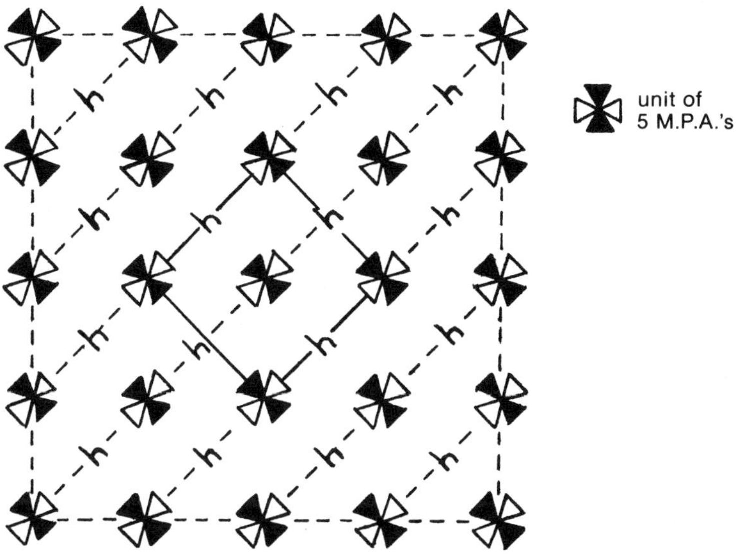

Fig. 8.13. Face-centred square array of twenty-five units of 5 M.P.A.'s. Gaps (*h*) between units are the dark square holes in fig. 8.4.

figure 8.4 indicates that the micro-psi depiction of the diamond lattice has been reproduced. The dark squares are gaps in the lattice of M.P.A.'s created by the migration of M.P.A.'s formed from face-centred carbon atoms to make the central unit of 5 in the face-centred square array of units of 5. Their presence directly reflects the original face-centred cubic arrangement of atoms in diamond.

Consider the stacking of the sheets of units of 5 in the planes $Z = 1/4$, $Z = 1/2$, and $Z = 3/4$. They, too, have a chess-board pattern, and units of 5 in the sheet $Z = 1/4$ are displaced relative to those in the XY plane by one-quarter of a unit-cell length in the direction of the X- and Y-axes. More generally, units of 5 in a sheet are displaced by one-quarter of a unit-cell length along the X- and Y-axes relative to those in the adjacent sheet below it. In the sheet $Z = 1$, units of 5 lie directly above units of 5 in the XY plane, since this sheet is the bottom face of the adjacent unit cell and atoms in it are directly above corresponding atoms in the XY plane. Figure

8.14 is an elevation view in the XZ plane of the pyramidal arrangement of units of 5 in sheets above the XY plane. A similar but inverted array lies below the XY plane, showing the same chess-board pattern. A similar pattern exists in the YZ plane because the lattice of M.P.A.'s has cubic symmetry. Therefore, the lattice is assembled from octahedral units made up of two square pyramids. This is just what micro-psi vision reveals. The base contains $5^2 = 25$ units of 5, the first sheet ($Z = 1/4$) contains $4^2 = 16$ units, the next sheet ($Z = 1/2$) has $3^2 = 9$ units, the sheet above this ($Z = 3/4$) has $2^2 = 4$ units, and, last, the apex ($Z = 1$) consists of one unit. There are four gaps between adjacent units in the base, making $4^2 = 16$ holes all together, in agreement with the number of holes depicted in figure 8.4. The octahedral unit contains eighty-five units of 5:

Plane $Z = 0$	$5^2 = 25$	
Plane $Z = \pm 1/4$	$4^2 + 4^2 = 32$	
Plane $Z = \pm 1/2$	$3^2 + 3^2 = 18$	Total = 85
Plane $Z = \pm 3/4$	$2^2 + 2^2 = 8$	
Plane $Z = \pm 1$	$1 + 1 = 2$	

C. W. Leadbeater, who originally investigated the lattice structure of diamond, regarded the octahedral unit as two pyramids with *separate* bases, for he calculated the number of units of 5 to be 110, twice that in one pyramid. This is incorrect, for the units in the XY plane ($Z = 0$) are shared by both pyramids, and the lattice of M.P.A.'s is assembled, not from two pyramids placed back to back, but from two pyramids with the *same* base. His calculation double-counts the M.P.A.'s in the base. Such an error was natural and forgivable in one who was not knowledgeable in crystallog-

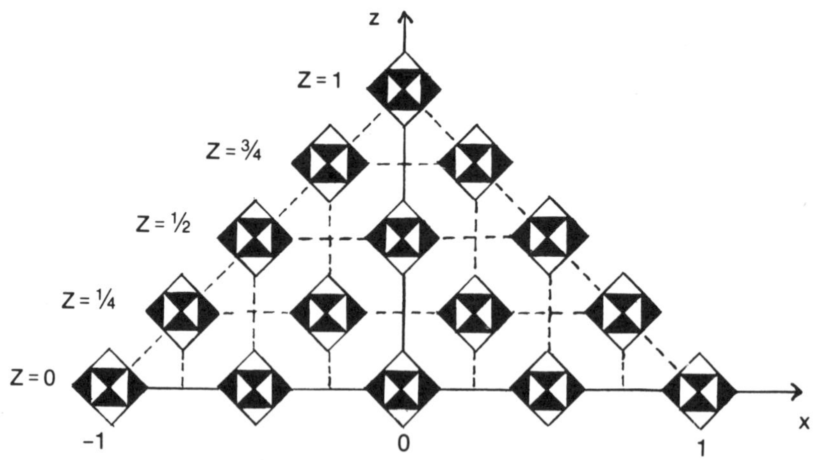

Fig. 8.14. Pyramidal array of units of 5 M.P.A.'s

raphy. In the region of the crystal affected by micro-psi observation, the lattice of M.P.A.'s is built up by the stacking of octahedral units containing eighty-five units of 5, as shown in figure 8.15.

The complex pattern in figure 8.4 that was recorded by Leadbeater was necessitated by his construction of the lattice from units of 5 M.P.A.'s instead of from single M.P.A.'s. This was unavoidable, since the face-centred cubic array of carbon atoms in diamond changes to a face-centred cubic array of units of 5 M.P.A.'s, when a region of the lattice (perhaps extending over several dozen unit cells) is subjected to the perturbative effect of micro-psi observation.

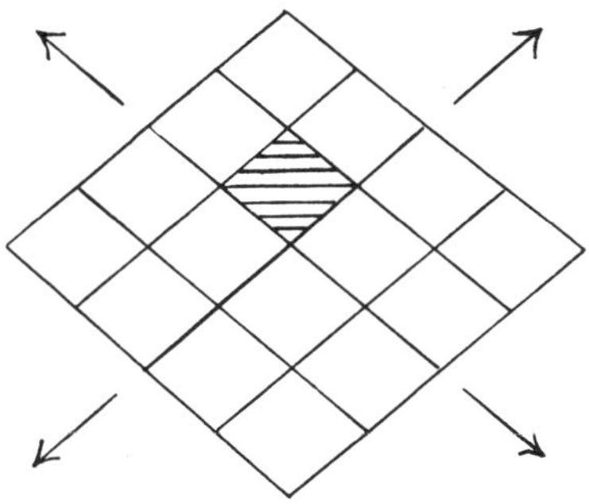

Fig. 8.15. Microscopic region of diamond lattice perturbed by micro-psi is filled with octahedral units.

The M.P.A.'s, reported to float over the faces of the octahedral unit, cannot be part of the unit cell of the lattice of M.P.A.'s. Why they are present, distorting the shape of the octahedral unit into a triakis octahedron, is not clear, although they may have arisen from interstitial atoms in the imperfect crystal specimen originally examined by Leadbeater.

It will now be shown that the micro-psi description of the diamond lattice cannot be the crystallographic description of the lattice of carbon atoms. Suppose that the unit of 5 were identified either with a tetrahedron of (4 + 1) bonded carbon atoms in the cubic unit cell (fig. 8.16, *a*) or with the four atoms at the corners of a face of the unit cell plus the face-centred atom (fig. 8.16, *b*). In order to obtain the reported face-centred square array of units of 5 in a given lattice plane, each of the four mutually unbonded atoms (marked *X* above) must shift towards the central atom and become bonded to it. There is no way that the lattice can be assembled by grouping in a chess-board pattern separate units of 5 atoms within the same unit cell. This is because each carbon atom is shared by four tetrahedra, and so, if the units of 5 were tetrahedra (and M.P.A.'s were chemical atoms), they would have to consist of five half-atoms. This is physically absurd as well as being in contradiction to the micro-psi report, which states that the units of 5 are made up of five M.P.A.'s, not five half-M.P.A.'s. The sharing

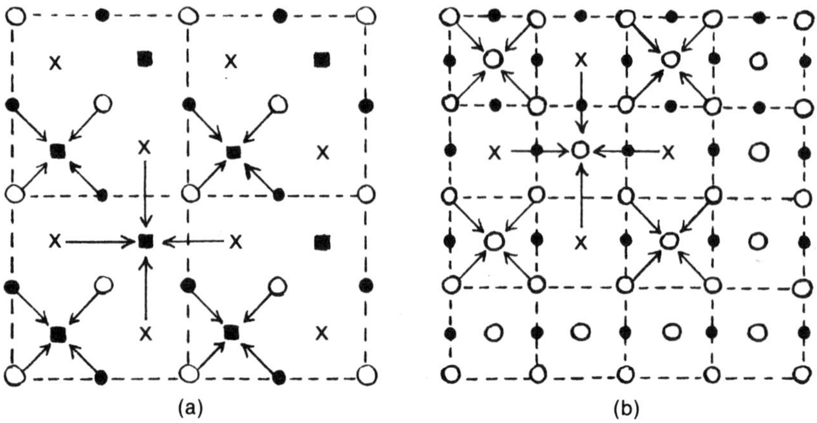

Fig. 8.16. If the M.P.A. is the atom, the observed face-centred square array of units of 5 M.P.A.'s requires a migration of carbon atoms.

problem would be solved by identifying adjacent units of 5 with corresponding (4 + 1) tetrahedra of atoms in adjacent unit cells (fig. 8.16, *a*). But then there are no atoms in a tetrahedron to constitute the central unit of 5 that creates the chess-board pattern. A shift of X atoms nearer to the centre is necessary, but then the micro-psi description no longer is one of an undisturbed lattice of carbon atoms. It is concluded that the identification of M.P.A.'s as atoms makes the crystallographic and micro-psi descriptions of the diamond lattice mutually incompatible.

GRAPHITE

The lattice consists of infinite plane sheets of carbon atoms arranged at the corners of regular hexagons with sides of length 1.42 Å. The sheets are parallel and 3.4 Å apart (fig. 8.17). They are stacked so that the centre of a hexagon in one sheet is directly above a carbon atom in the adjacent sheet below it.

Figure 8.18 is a model of the graphite lattice, observed with micro-psi vision. The carbon M.P.A.'s are arranged in hexagons, the C27 and C26

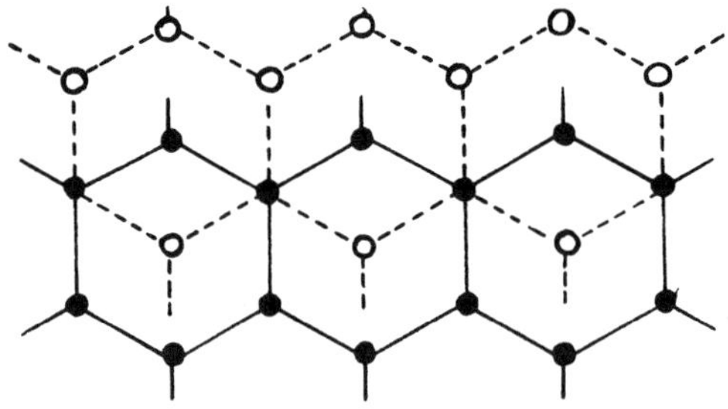

Fig. 8.17. Graphite lattice

funnels being denoted by light and dark faces of the octahedra that represent the M.P.A.'s. Pairs of corresponding carbon nuclei in adjacent sheets pair to form M.P.A.'s. This process, initiated by micro-psi observation, leaves unchanged the hexagonal arrangement of atoms.

Graphite was first examined in 1926,[1] two years after Bernal determined its crystallographic structure by X-ray diffraction methods.

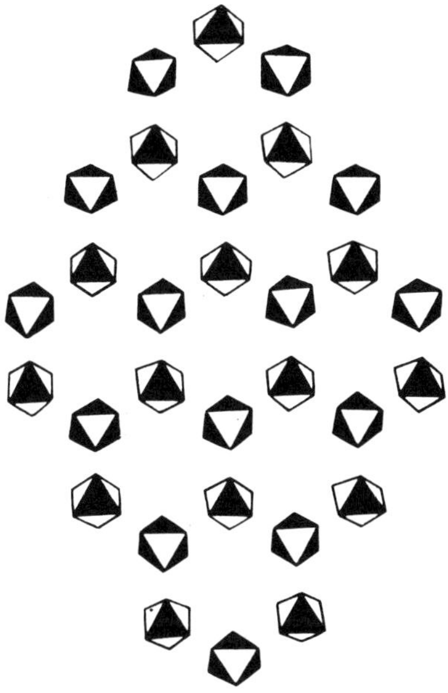

Fig. 8.18. Graphite lattice, according to micro-psi vision

WATER

The molecule is shown in figure 8.19. There is sp^3 hybridization of the *s* and *p* atomic orbitals of the oxygen atom, as evidenced by the bond angle of 104°28′, comparing with the theoretical value of 90° if the bonding electrons shared by the hydrogen and oxygen atoms occupied a *p* orbital. Consequently, two hydrogen atoms occupy two corners of a tetrahedron where their *s* orbitals overlap with singly occupied sp^3 orbitals of the oxygen atom, while two lone pairs of electrons not participating in bonding occupy two sp^3 orbitals whose axes are directed towards the other two corners. The OH bond length is 0.97 Å.

Figure 8.20 is a model of the M.P.M. of water (*a* and *b* are different perspectives). The central spiral column is an oxygen M.P.A., and a hydrogen M.P.A. is bound to each end. Figure 8.21 shows the proposed mode of formation of the M.P.M. from two molecules of water. The Hydrogen Triangles are protons that are bound to the oxygen M.P.A. by nuclear

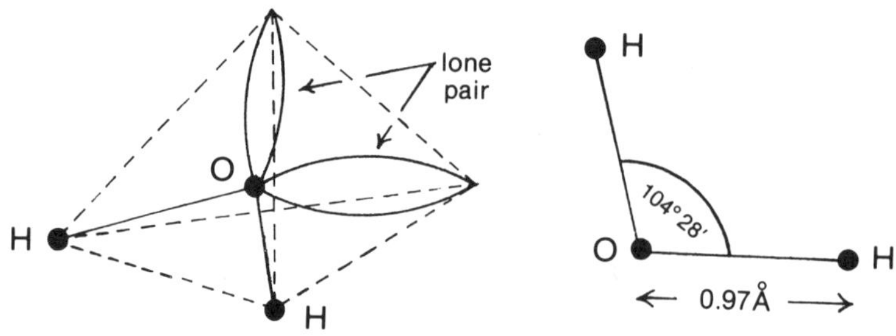

Fig. 8.19. Water molecule

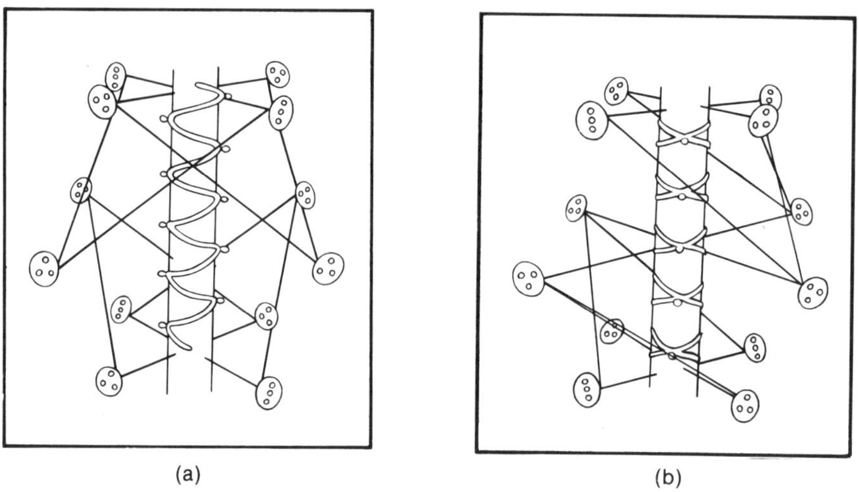

Fig. 8.20. M.P.M. of water

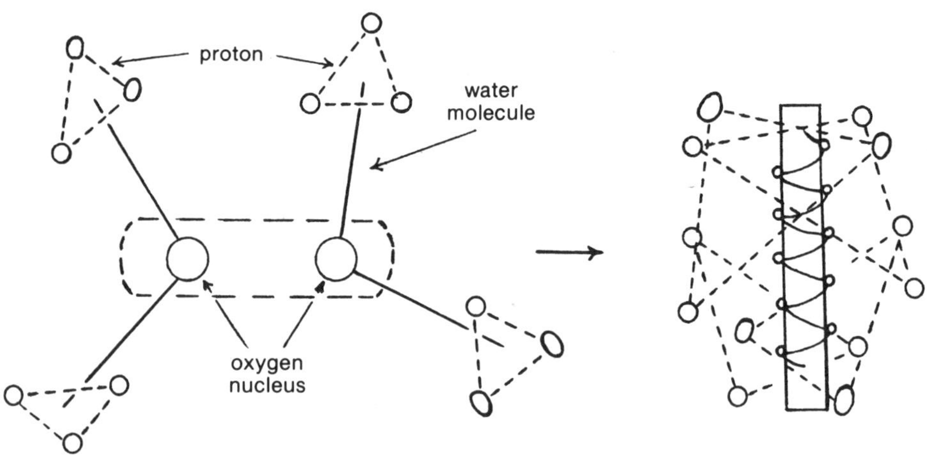

Fig. 8.21. Formation of water M.P.M.

forces. They cannot be bound by Coulomb forces because they would be repelled by the oxygen M.P.A., which has an electric charge of +16. Note that the M.P.M. is linear, not triangular, like the molecule. The overlapping of Hydrogen Triangles is a result of quarks from the protons in the hydrogen atoms becoming free at first but then strongly interacting with one another and recombining into protons, with the impenetrable SU(9) Nielsen-Olesen vortices between quarks interlocking so that the protons cannot separate. As in the hydrogen M.P.A., the pairs of protons are trapped, rather than bound together by nuclear forces. The object is of nuclear size, as may be inferred from the micro-psi description of the hydroxyl group (discussed next).

HYDROXYL GROUP

Micro-psi examination of chemical compounds containing the OH group does not reveal (as might have been expected) a hydrogen M.P.A. attached to an oxygen M.P.A. Instead, the hydrogen M.P.A. appears split up into its two Hydrogen Triangles, which float above and below the ends of the double spiral of the oxygen M.P.A. (fig. 8.22). The investigators remarked: "Though these two triangles of Hydrogen are separated, with Oxygen in between, they are still bound to each other, and a linking force goes through the middle of the Oxygen snake."[2] This is the attractive nuclear force acting between the protons and indicates that the oxygen M.P.A. has a

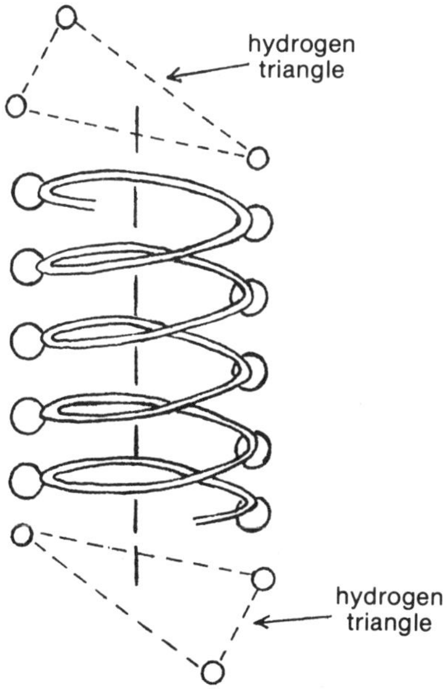

Fig. 8.22. OH group

typical nuclear size of one or two fermis.* It was also noted: "Each triangle rotates flat, and while rotating, sways a little up and down, as the lid of a pot rotates before it finally settles down."[3] This remarkable observation has the following explanation: according to Hypothesis 1, U.P.A.'s are $\mathcal{P}$ or $\mathcal{N}$ omegons, which carry electric charges of $+5/9$ and $-4/9$, respectively. According to the analysis of the oxygen M.P.A., the two rigid, spiral coils are made up of $\mathcal{P}$-$\mathcal{P}$ and $\mathcal{N}$-$\mathcal{N}$ diomegons and rotate rapidly in opposite directions about a common axis. The revolving electric charges constitute solenoidal electric currents providing a magnetic field along this axis (fig. 8.23). The Hydrogen Triangles are protons endowed with a magnetic dipole moment and spin. The magnetic field causes Larmor precession of the spin of each proton about its direction, that is, the Hydrogen Triangles perform a spinning-top motion about the axis of the oxygen M.P.A., the plane of each triangle periodically tilting. The significance of this observation was not realized at the time that it was made (1919). It is now clear: the two investigators Besant and Leadbeater described spin precession of a proton in a magnetic field, several years before Uhlenbeck and Goudsmit suggested that electrons might have an intrinsic spin and before the spin of the proton was itself measured.

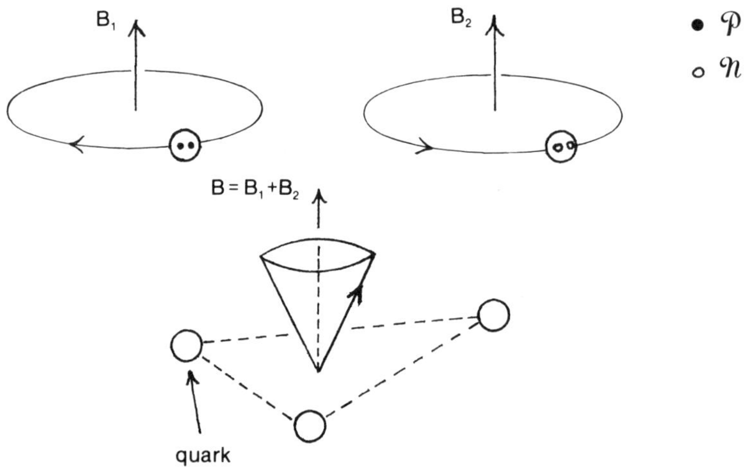

Fig. 8.23

COMMON SALT

The crystal lattice of sodium chloride is face-centred cubic, the basis consisting of one Na^+ ion and one Cl^- ion separated by one-half the body diagonal of a unit cube (fig. 8.24). Each ion has six ions of the opposite charge as nearest neighbours. The length of the unit cell is 5.63 Å.

When common salt is examined with micro-psi vision, no lattice of dis-

* One fermi is 10^{-13} cm.

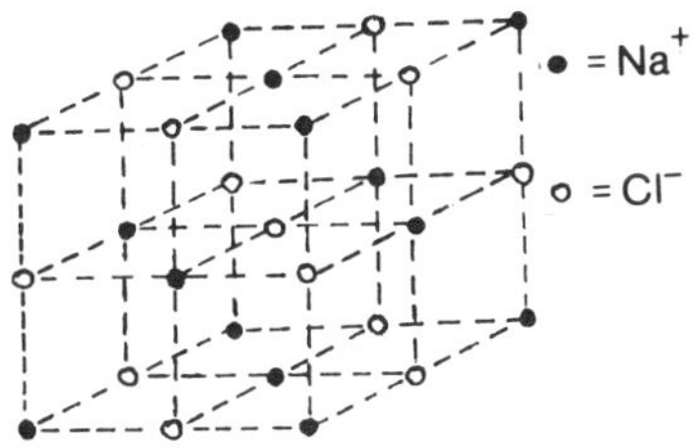

Fig. 8.24. Sodium chloride lattice

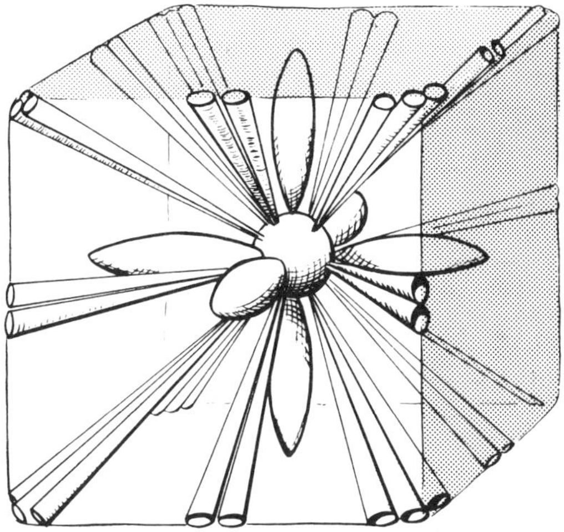

Fig. 8.25. M.P.M. of sodium chloride

tinct M.P.A.'s of sodium and chlorine is apparent to the observer. Instead, the M.P.A.'s of both elements appear broken up and their constituent bodies rearranged and intermixed (fig. 8.25). The twenty-four Na16 funnels radiate in pairs from the centre of a cube to the middle points of its twelve edges. The twenty-four Cl.25 funnels radiate from the same centre in groups of three to the eight corners of the cube. At the centre is a sphere containing a tetrahedral cluster of Na10 globes, each M.P.A. providing two of them (fig. 8.26). In the interior of this cluster is a single Hydrogen Triplet. Six "tulip-shaped" bodies made up of a triplet and a duad of U.P.A.'s radiate from the centre of the cube to its face centres (fig. 8.27).

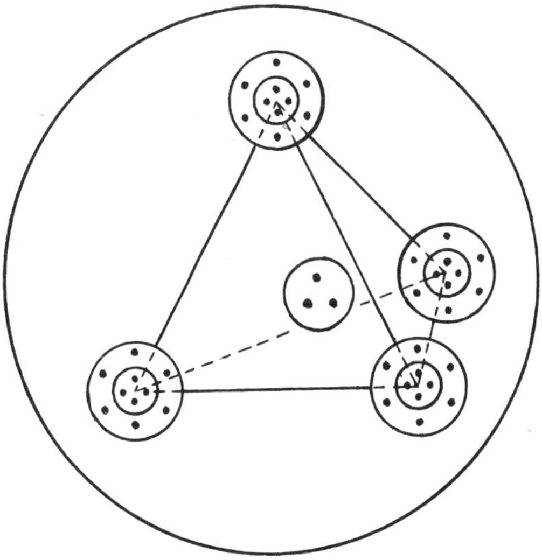

Fig. 8.26

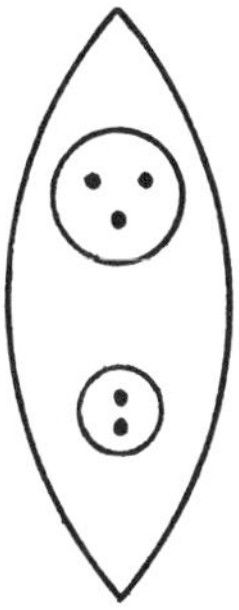

Fig. 8.27

The dumb-bell forms of the M.P.A.'s of sodium and chlorine are those of the free M.P.A.'s. In the crystal, however, formation of M.P.A.'s from pairs of nuclei takes place in an anisotropic electrostatic environment. It is proposed that the coupling of the charged bodies forming the funnels in each M.P.A. to the Coulomb fields of the surrounding ions in the crystal disturbs the normal process of their formation. The sodium M.P.A. is formed by the pairing of nearest Na^+ ion neighbours. Similarly, the chlorine M.P.A. is formed from nearest Cl^- ion neighbours. These ions are at the opposite corners of a cube (fig. 8.28). The mixing of the components of each M.P.A. arises because they are formed at the same point in space, namely, the centre of the cube. The electric fields at this point directed away from it towards the upper corners of the cube are due primarily to surrounding, undisturbed Cl^- ions (fig. 8.29). The fields directed towards this point from

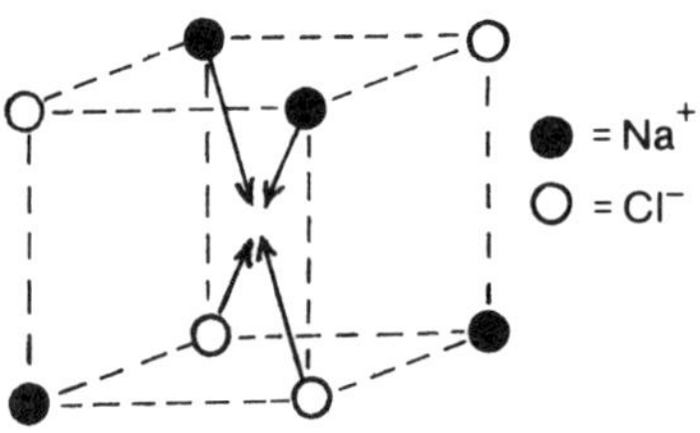

Fig. 8.28

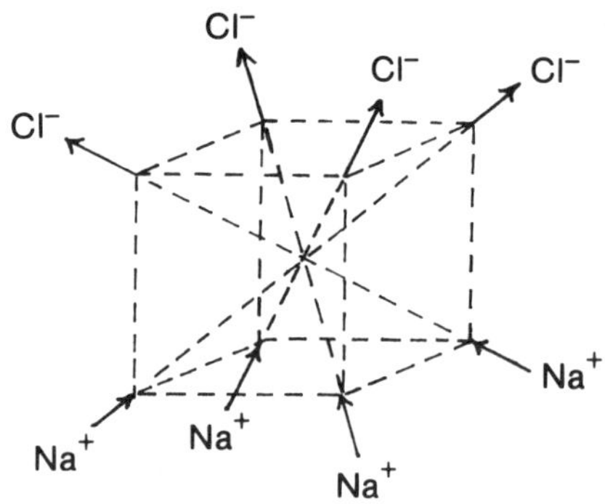

Fig. 8.29

the four lower corners of the cube are due primarily to nearest Na^+ ions. Two types of Na16 funnels were found in the analysis of the sodium M.P.A.: the Na16 (+) with twelve $\mathcal{P}$ and four $\mathcal{N}$ omegons has an electric charge of +44/9; the Na16 (−) with four $\mathcal{P}$ and twelve $\mathcal{N}$ omegons has a charge of −28/9. Two types of Cl.25 funnels were found in the analysis of the chlorine M.P.A.: the Cl.25 (+), made up of an Na16 (+) and a proton, has a charge of +53/9; the Cl.25 (−), made up of an Na16 (−) and a neutron, has a charge of −28/9. These charged bodies interact with the electric fields and take up the following rigid arrangement in space:

Cl.25 funnels: Four groups of three are attracted to Cl^- ions and are directed towards the upper corners of the cube of Na^+ and Cl^- ions. They have the composition (+, −, +) and a total charge of +26/3. Four groups of three funnels are attracted to Na^+ ions and are directed towards the lower corners of the cube. They each have the composition (−, +, −) and a total charge of −1/3.

Na16 funnels: Pairs of Na16 (+) funnels with a charge of +88/9 are oriented towards the middle of the edges of the top face of the cube. Pairs of Na16 (−) funnels with a charge of −56/9 are directed towards the middle of the edges of the bottom face. An Na16 (+) and an Na16 (−) funnel with a charge of +16/9 are directed towards the middle of the vertical edges of the cube (fig. 8.30).

During the formation of the M.P.M., the Na14 and Cl.19 groups break up and their constituents reassemble to form a system whose form is compatible with the cubic symmetry of the local Coulomb field (fig. 8.31):

$$\text{Cl.19} = 2\text{u} + 3(2\mathcal{P}\text{-}2\mathcal{N})\ ;$$

$$\text{Na14} = 2(2\mathcal{P}\text{-}2\mathcal{N}) + 2\mathcal{P}\text{-}4\mathcal{N}\ ;$$

$$\text{Cl.19} + \text{Na14} = 16\mathcal{P} + 16\mathcal{N}$$

$$\rightarrow 3(\text{u}, \mathcal{P}\text{-}\mathcal{N}) + 3(\text{d}, \mathcal{P}\text{-}\mathcal{N}) + \mathcal{P}\text{-}\mathcal{N}\ .$$

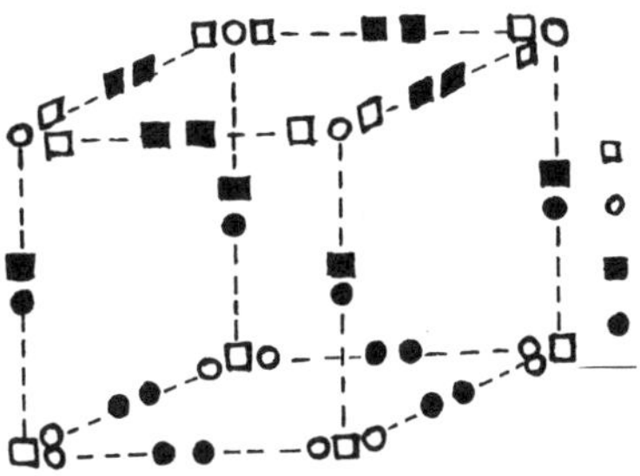

Fig. 8.30

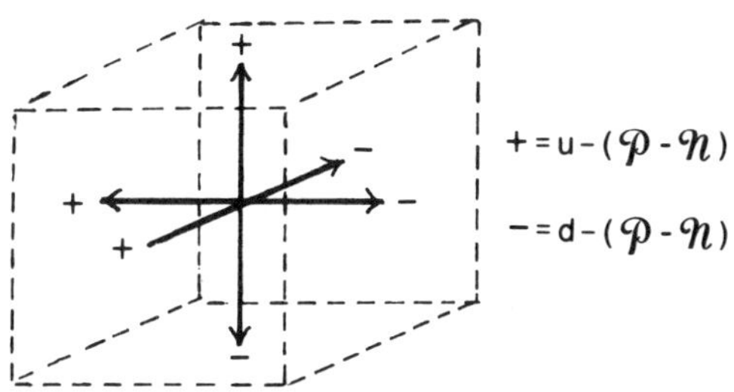

Fig. 8.31

A "tulip-shaped" body consists of a $\mathcal{P}$-$\mathcal{N}$ diomegon and either a u or a d quark. They are bound to the central sphere containing the four Na10 globes by strong forces. A duad of U.P.A.'s ($\mathcal{P}$-$\mathcal{N}$), not the reported triplet, is predicted to exist in the central sphere. This is because the Cl.19 rod should contain eighteen U.P.A.'s, not nineteen as recorded.

NITRIC ACID

The chemical molecule is planar, the nitrogen atom forming three equivalent, planar sp^2 bonds with the three oxygen atoms around it (fig. 8.32). The remaining $2p_z$ electron in the nitrogen atom enters into a delocalized π orbital, which spreads over the whole NO_3^-.

The form of the M.P.M. unequivocally demonstrates the fundamental difference between molecules and the systems formed from them as a result of the intervention of the micro-psi observer. The M.P.A.'s of hydrogen and nitrogen appear to be broken up and their constituents rearranged (fig. 8.33). At the centre of a triangular array of oxygen M.P.A.'s is the long ovoid belonging to the N110 group of the nitrogen M.P.A. It is surrounded, first, by a hexagonal array of N14 globes, then by six Hydrogen Triplets from the M.P.A. of hydrogen, and, finally, by seven N9 globes arranged at the corners of a heptagon. The two N20 and the two N24 groups move about outside these. In the chemical molecule, the hydrogen atom lies outside the triangle formed by the three oxygen atoms. In the M.P.M., however, the hydrogen M.P.A. is *inside* the triangular array of oxygen M.P.A.'s. Moreover, it is split up into its six triplets. The N14 globes have a charge of $+7/9$. Bound to the central long ovoid by short-range nuclear forces, they might be expected to form an octahedral array, owing to mutual Coulomb repulsion of identical charges. But they are also subject to the electromagnetic field due to the three oxygen M.P.A.'s, which have charges of $+16$. They are bound, instead, at the corners of a hexagon. It is not clear why the nitrogen M.P.A. is broken up into its constituent bodies. The Coulomb field due to the three oxygen M.P.A.'s may inhibit their aggregation into bound states during the formation of the M.P.M.

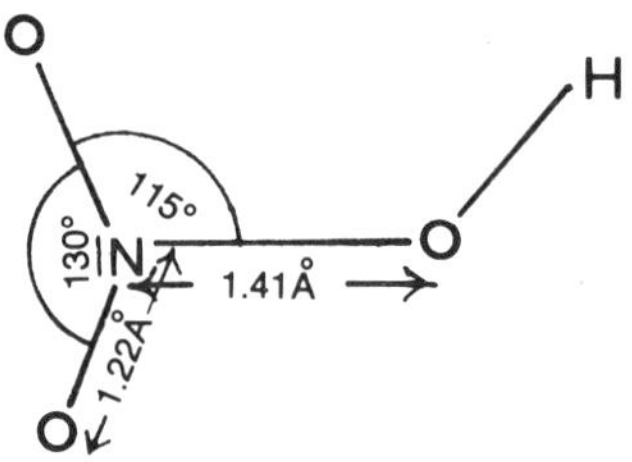

Fig. 8.32. Nitric acid molecule

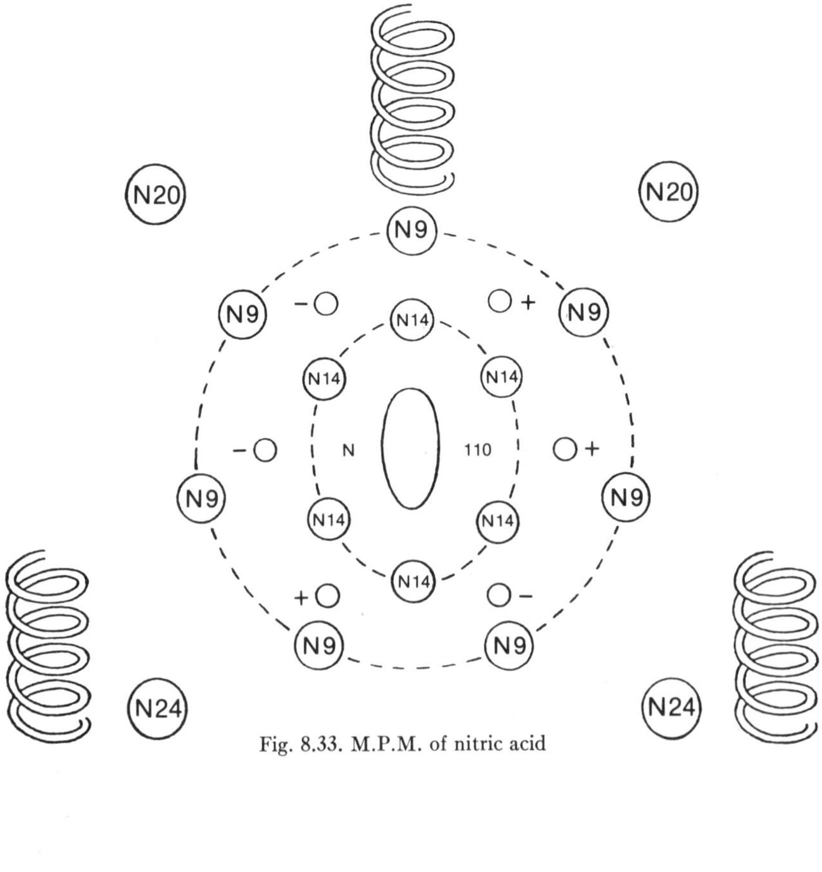

Fig. 8.33. M.P.M. of nitric acid

METHANE M.P.M.

The molecule of methane consists of four hydrogen atoms arranged tetrahedrally about a carbon atom. The four valence electrons of the carbon atom share sp^3 hybrid orbitals with the four 1*s* electrons belonging to the hydrogen atoms. The C-H bonds are arranged at the tetrahedral angle of 109°28′ (fig. 8.34).

Fig. 8.35 shows the M.P.M. of methane. Eight Hydrogen Triangles hover over the mouths of the funnels of the carbon M.P.A. The four U.P.A.'s at the centre of the octahedral array of funnels each act as nuclei binding together a C26 and a C27 group.

The M.P.M. is formed by the pairing of carbon nuclei belonging to two methane molecules (fig. 8.36). The eight carbon valence electrons are in the

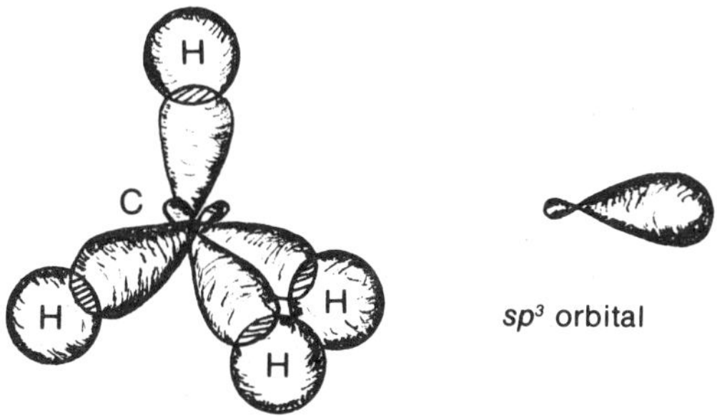

Fig. 8.34. Methane molecule

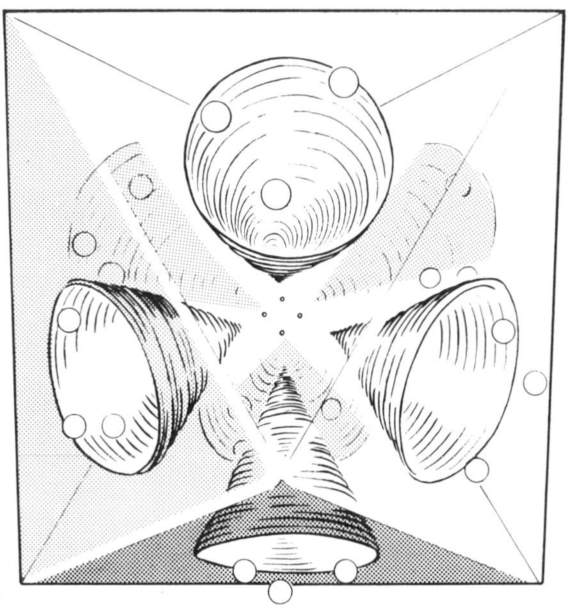

Fig. 8.35. M.P.M. of methane

same spin state and are bound to the carbon M.P.A. (which acts as a point electric charge of +12). In accordance with the Exclusion Principle, which forbids the occupation of the same orbital by two electrons in the same spin state, the eight electrons must occupy different regions of space. The geometrical configuration with least Coulomb interaction energy is an octahedral array of electrons. The four body diagonals of a cube are the directions along which the eight electrons are most likely to be (fig. 8.37). These are the directions of the axes of the funnels. The eight protons provided by the hydrogen atoms in two methane molecules are attracted by their electromagnetic interaction with these electrons to the funnel axes. For this reason, the Hydrogen Triangles are observed to hover over the mouths of the funnels. Figure 8.35, which is a reproduction of a drawing provided by the investigators,[4] clearly portrays the triangular group of three quarks that constitute a proton.

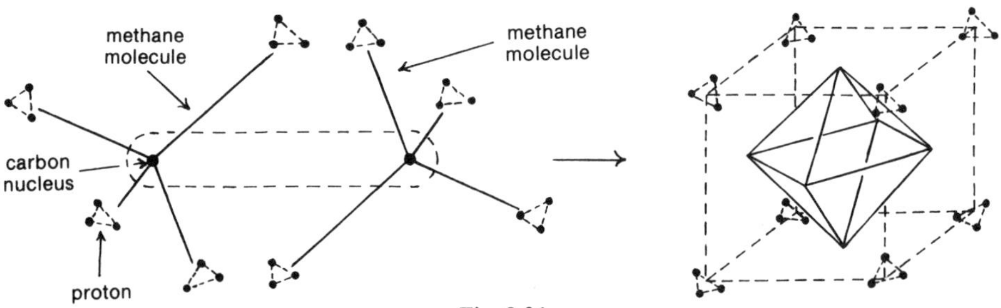

Fig. 8.36

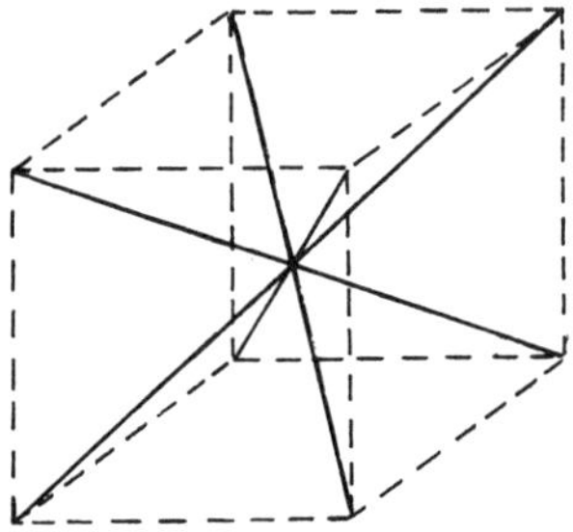

Fig. 8.37

ACETYLENE M.P.M.

In the linear acetylene molecule H-C-C-H, each carbon atom is in an *sp* hybridized state. Two electrons occupy this hybrid orbital, one being shared by a hydrogen atom, the other being shared by the other carbon atom in a σ bond. Overlap between the unpaired electrons in the $2p_Z$ orbitals of each carbon atom and between the electrons in $2p_Y$ orbitals gives two π bonds (fig. 8.38).

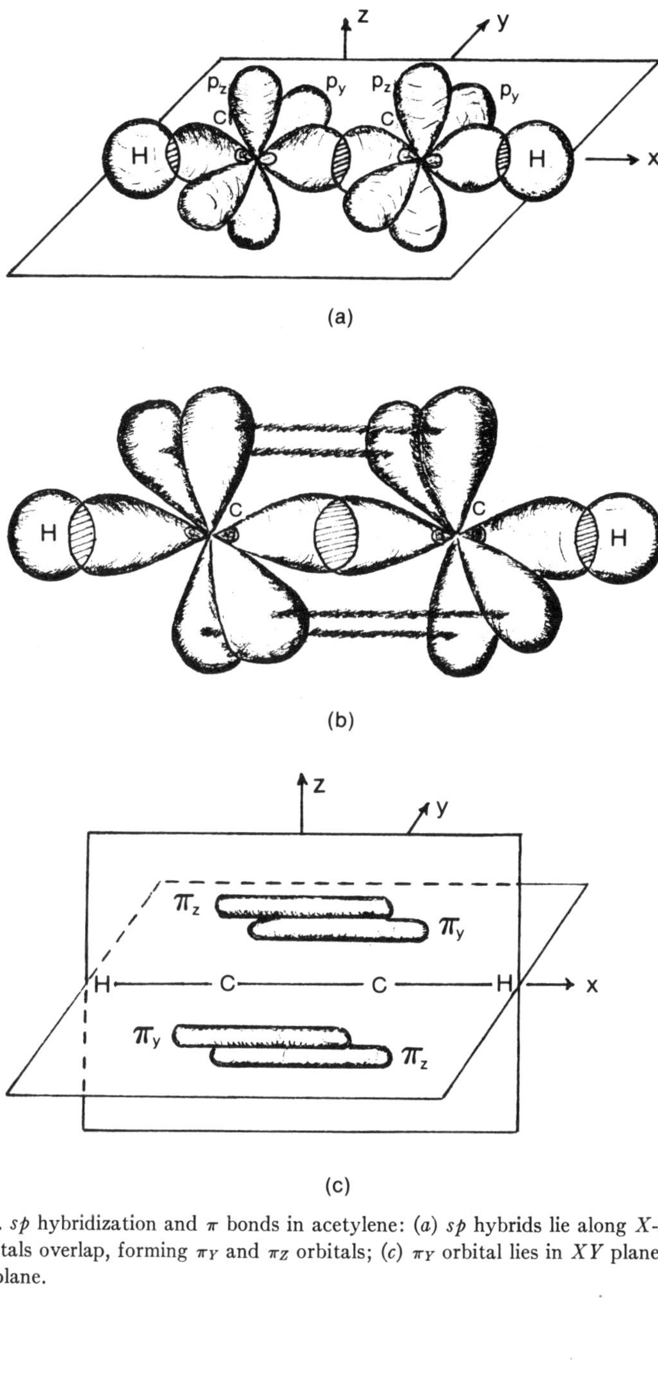

Fig. 8.38. *sp* hybridization and π bonds in acetylene: (*a*) *sp* hybrids lie along X-axis; (*b*) p_Y and p_Z orbitals overlap, forming π_Y and π_Z orbitals; (*c*) π_Y orbital lies in XY plane, π_Z orbital lies in XZ plane.

Micro-psi observation of aliphatic organic compounds reveals that, corresponding to the original single C-C bond, pairs of carbon M.P.A.'s appear to be linked in the manner shown in figure 8.39. Curved lines of force appear to flow out from two adjacent funnels of one M.P.A. and down into adjacent funnels of the other, producing a rigid link. In the M.P.M. of acetylene, two carbon M.P.A.'s are bonded in this way. Twelve Hydrogen Triplets provided by two hydrogen M.P.A.'s hover over the remaining twelve funnels that do not participate in this bonding (fig. 8.40).

The M.P.M. is formed by the pairing of carbon nuclei in two molecules (fig. 8.41). As three-quark bound states, protons are stable in a superconducting Higgs vacuum with spontaneously broken SU(9) symmetry. In a vacuum with broken SU(3) symmetry, protons are unstable and dissociate

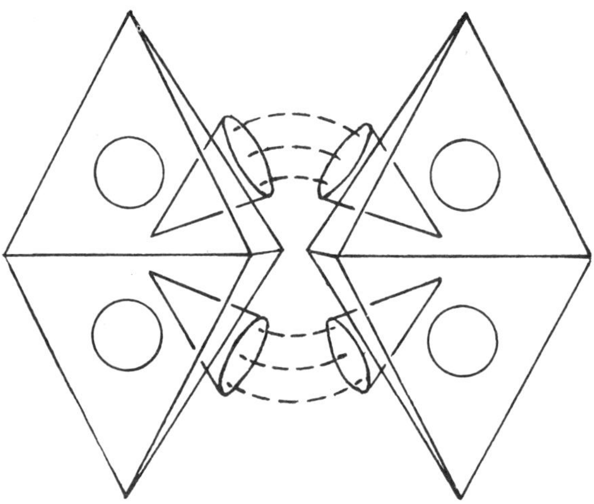

Fig. 8.39. "Lines of force" between funnels bind carbon M.P.A.'s

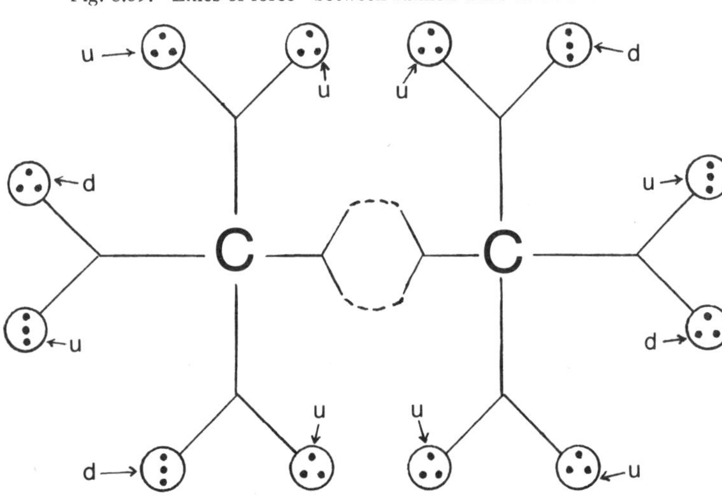

Fig. 8.40. M.P.M. of acetylene

into quarks, which behave as free, unconfined hadrons. Such a vacuum state is apparently created locally by the act of micro-psi observation of an acetylene molecule, for the Hydrogen Triangles (protons) are broken up into free triplets (quarks), which are then attracted to the groups of particles inside the funnels, to which they become bound by nuclear forces.

Despite the fact that a triple bond exists between the carbon atoms in the molecule of acetylene, the two carbon M.P.A.'s in the M.P.M. of acetylene are linked by pairs of funnels belonging to each one in the same way that carbon M.P.A.'s corresponding to a single C-C bond are linked. This is an example of the crucial differences between chemical molecules and their M.P.M.'s.

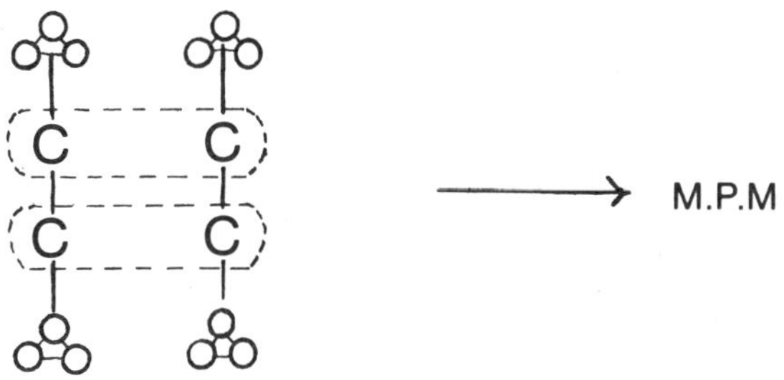

Fig. 8.41

DIETHYL ETHER M.P.M.

Figure 8.42 shows the form of the chemical molecule and the structure of the M.P.M. The double helix M.P.A. of oxygen links two bonded pairs of carbon M.P.A.'s. The ends of each spiral chain of duads of U.P.A.'s open out and point to adjacent pairs of funnels in each octahedron. Twenty Hydrogen Triangles hover over the mouths of funnels not involved in binding carbon M.P.A.'s together. Curved lines of force flow between adjacent pairs of funnels in neighbouring M.P.A.'s and bind them together.

The M.P.M. is formed by the pairing of corresponding carbon nuclei and of oxygen nuclei in two chemical molecules (fig. 8.43). The oxygen M.P.A. has an electric charge of +16. In the analysis of the oxygen M.P.A. in chapter 7, the (+) and (−) spirals were shown to have positive and negative electric charges, respectively. The carbon M.P.A. has a charge of +12. There should be strong Coulomb repulsion between the two M.P.A.'s, the carbon M.P.A. tending to compress the double helix of the oxygen M.P.A. (Prediction 1, fig. 8.44). The two spirals have charges of opposite sign and so tend to be pulled in opposite directions by the positively charged carbon M.P.A. (Prediction 2, fig. 8.45). The positively charged oxygen M.P.A. repels the positively charged carbon M.P.A., tending to destabilize it (Prediction 3, fig. 8.46). These predictions are confirmed by the following observations:

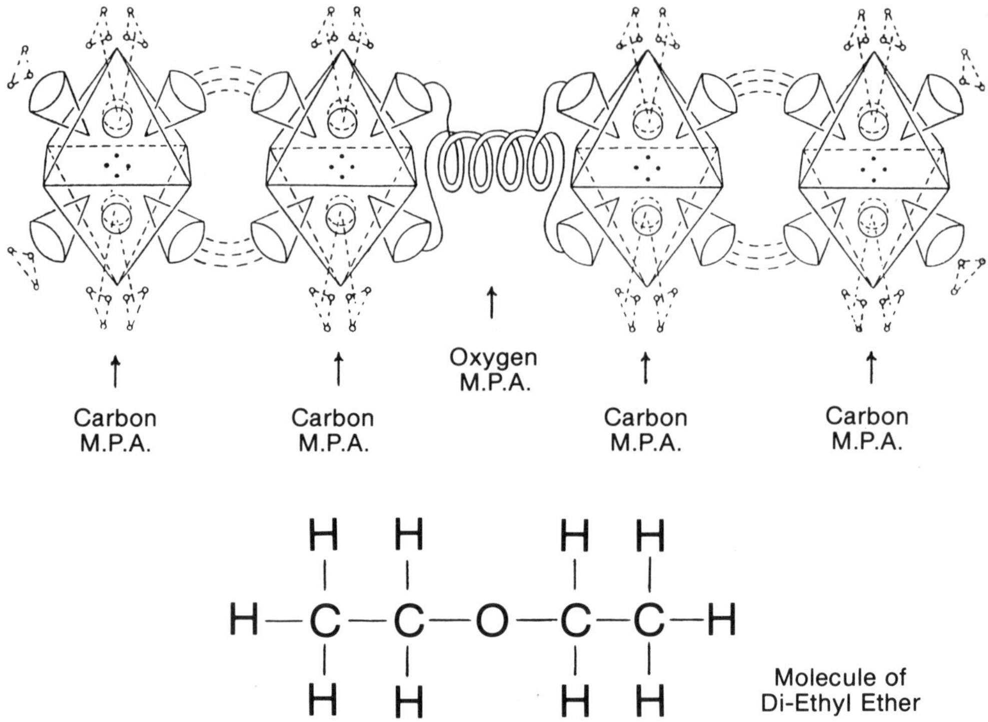

Fig. 8.42. M.P.M. of diethyl ether $(C_2H_5)_2O$

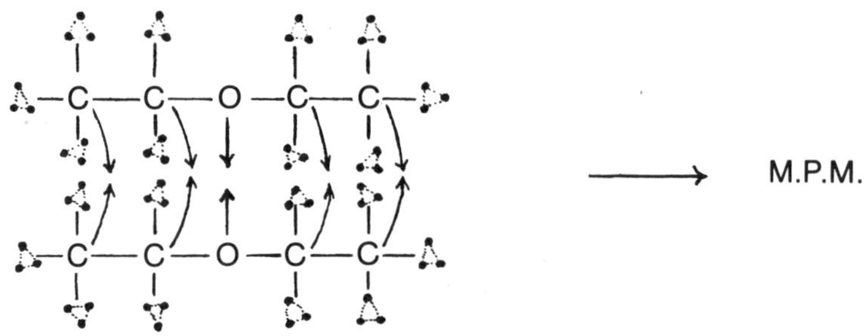

Fig. 8.43

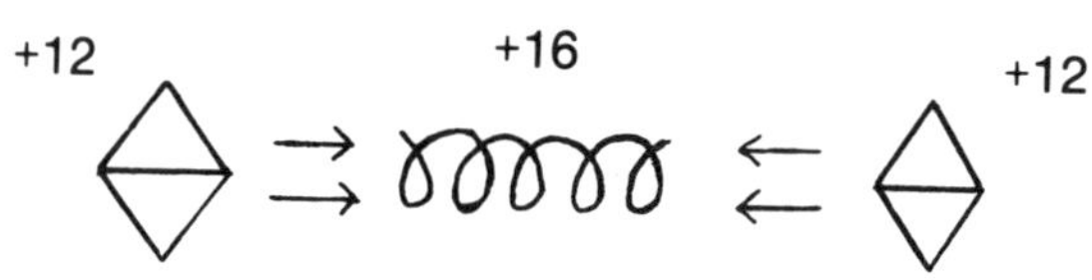

Fig. 8.44. Positively charged carbon M.P.A.'s compress and shorten positively charged oxygen M.P.A.

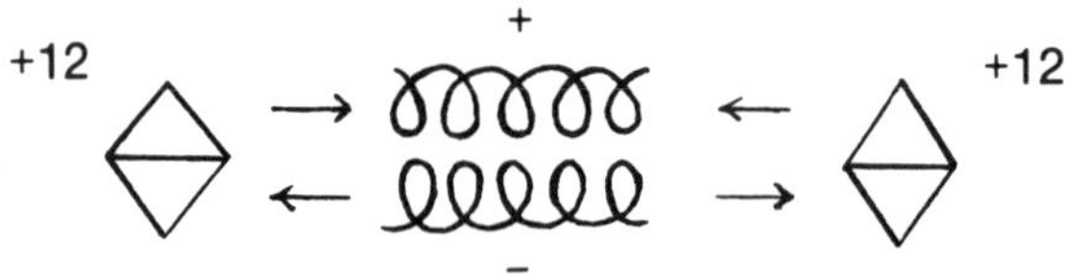

Fig. 8.45. Positively and negatively charged spirals of oxygen M.P.A. are pulled in opposite directions by carbon M.P.A.'s.

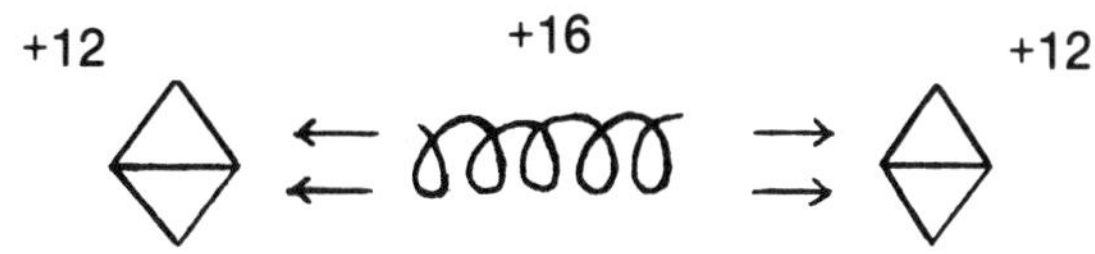

Fig. 8.46. Positively charged oxygen M.P.A. tends to push apart the carbon M.P.A.'s, so destabilizing the M.P.M.

PREDICTION 1: "The Oxygen atom is thicker and shorter than usual. . . ."

PREDICTION 2: ". . . the two parts of the molecule hold together because the snakes* are pulled in opposite ways because one is negative and the other positive."

PREDICTION 3: "If it were not held strongly the molecule would fall apart."[5]

The chemical molecule is stable, and this last remark concerning the potential instability of the M.P.M. illustrates how molecules and M.P.M.'s are quite different objects, the stability of the former not necessarily guaranteeing the stability of the latter.

BENZENE M.P.M.

Figure 8.47 shows the form of the molecule. Its plane hexagonal shape arises from sp^2 hybridization of the atomic orbitals of valence electrons in each of the six carbon atoms. Of the four valence electrons of each atom, three occupy sp^2 orbitals. One of these is shared by the $1s$ electron of a hydrogen atom, and two are shared by valence electrons from neighbouring carbon atoms. The three sp^2 electrons form three σ bonds in one plane, their directions making an angle of 120° with one another. The fourth valence electron occupies a $2p_z$ orbital perpendicular to the plane of the molecule. These overlap in pairs to give localized π orbitals. Overlapping can occur in two alternative ways, with the result that six electrons are completely delocalized, being shared equally by all six carbon atoms in the benzene molecular ring. The delocalized π orbital extends above and below the plane of the molecule, and the bond between the carbon atoms is neither a single nor a double bond but a bond of intermediate strength.

Figure 8.48 is a drawing of a model of the M.P.M. of benzene. It is not hexagonal but octahedral—another illustration of the fundamental difference between molecules and M.P.M.'s. Four semicircular "fans" of six funnels form the sides of a square. Above and below this square are similar fans. A dodecahedral array of twelve funnels is at the centre of the square. Hydrogen Triplets float over the mouths of every funnel in the fans. In

* This refers to the spiral coils of the oxygen M.P.A.

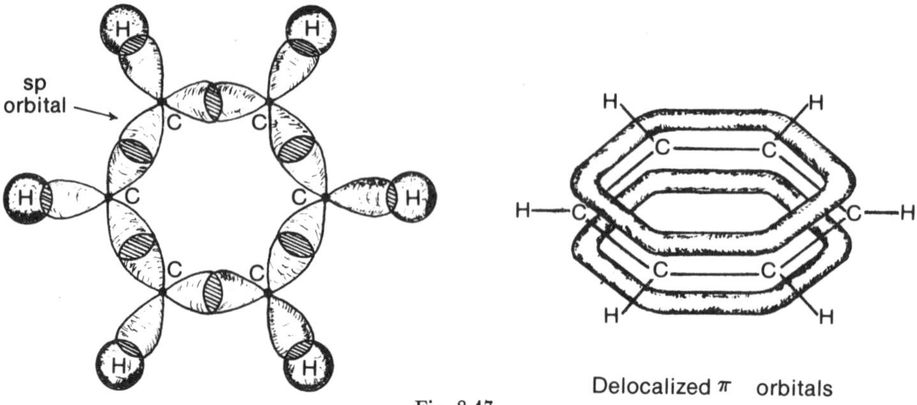

Fig. 8.47

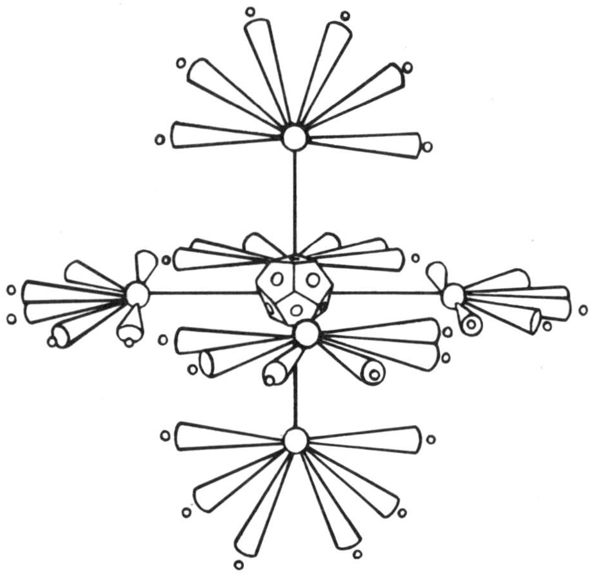

BENZENE C_6H_6

Fig. 8.48. M.P.M. of benzene

order to counter naive acceptance by the reader of this model as a realistic representation of the M.P.M., the following remark made by the investigators is quoted: "We must remember that no model can ever adequately represent the reality, since first the distances between Anu and between groups of them, and their relative sizes, cannot be correctly represented in any model, and secondly each funnel which looks solid is not solid at all but is only a whirlpool of force created by the Anu as they revolve."[6]

The M.P.M. is formed by the pairing of carbon nuclei in two molecules of benzene and by the dissociation into free quarks of the twelve protons belonging to the hydrogen atoms in these molecules (fig. 8.49). The three sp^2 valence electrons shared by a carbon atom in each molecule are responsible for a fan of six funnels. The twelve funnels in the dodecahedral array are due to the (6 + 6) delocalized π valence electrons present in the two benzene molecules forming the M.P.M. The geometrical symmetry of their

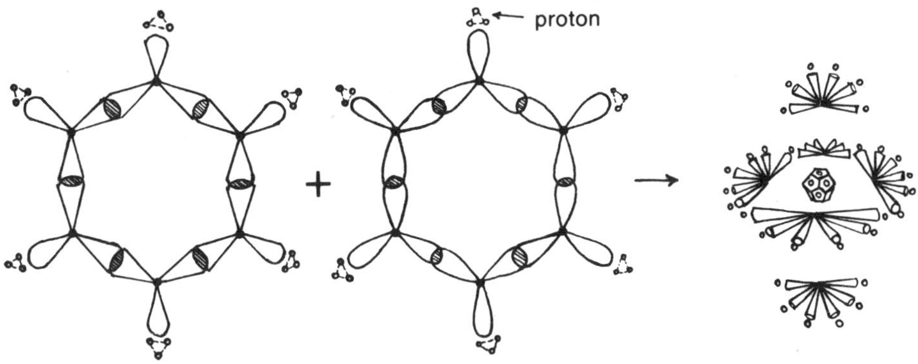

Fig. 8.49

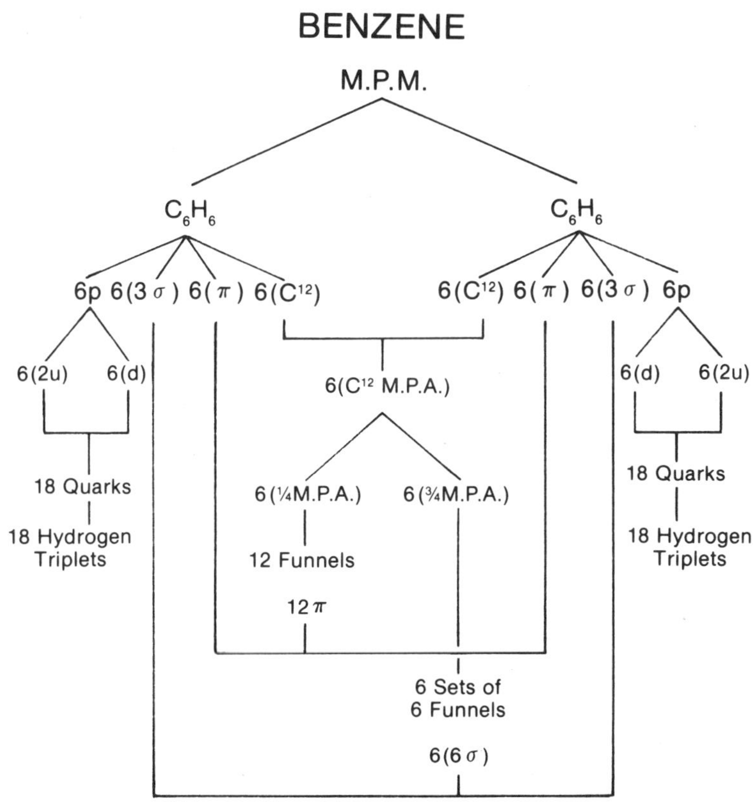

Fig. 8.50

bound-state configuration is determined by the Coulomb repulsion of the charged bodies inside the funnels. The hexagonal symmetry of the molecule is determined by the trigonal symmetry of the sp^2 bonds between neighbouring carbon atoms. In the M.P.M. these bonds no longer exist and the six positively charged fans are arranged at the corners of an octahedron. The space-time picture of its formation from two hexagonal benzene molecules will not be discussed here. What is important to note is that micro-psi observation of the molecule (orginally made in 1924)[7] appears to distinguish (albeit indirectly) between the three σ bonds and the π bond of each carbon atom, as figure 8.50 illustrates. The observed difference in the structural function of two funnels with respect to the other six of each carbon M.P.A. reflects the association of one π electron and three σ electrons with each carbon atom in the molecule and the different formative effect that localized and delocalized electrons have. The Hydrogen Triplets are the thirty-six quarks released during the formation of the M.P.M. from the protons in the twelve hydrogen atoms belonging to the two molecules of benzene (fig. 8.51). They are bound to the groups of particles inside the funnels by nuclear forces.

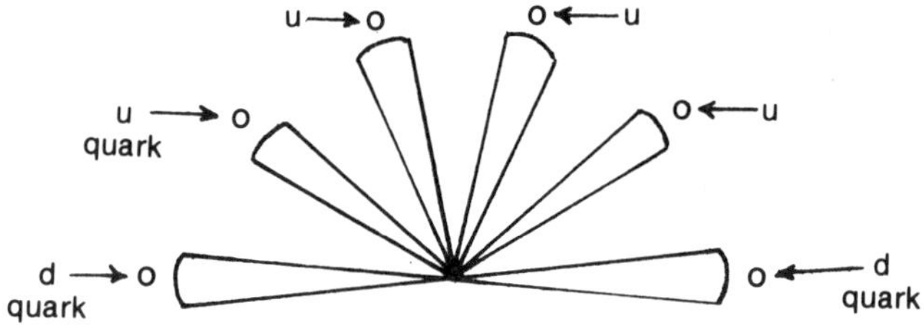

Fig. 8.51. Set of 6 funnels

OZONE M.P.M.

A molecule of ozone consists of three oxygen atoms bound to one another at the corners of an isosceles triangle (fig. 8.52).

When ozone is examined with micro-psi vision, an object is observed that consists of three single "oxygen spirals" (see discussion of the oxygen M.P.A. in chap. 7 for their detailed description). They are at the corners of an equilateral triangle and revolve in the same plane about its centre. Figure 8.53 shows a model of the object. Two varieties are reported: one consists of two (+) and one (−) oxygen spiral; the other is made up of one (+) and two (−) spirals. The investigators noted a tendency for the object to break up into an oxygen M.P.A. and a free spiral.

The fact that three *half*-M.P.A.'s of oxygen (and not three complete M.P.A.'s) comprise the unit of ozone that is discernible to the observer provides strong support for Hypothesis 2, as now explained. If a double

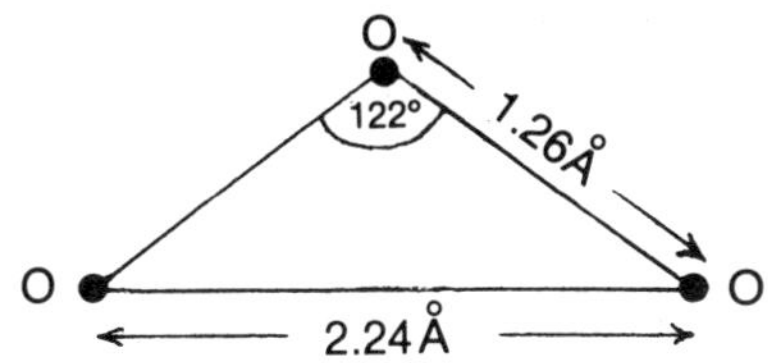

Fig. 8.52. Ozone molecule

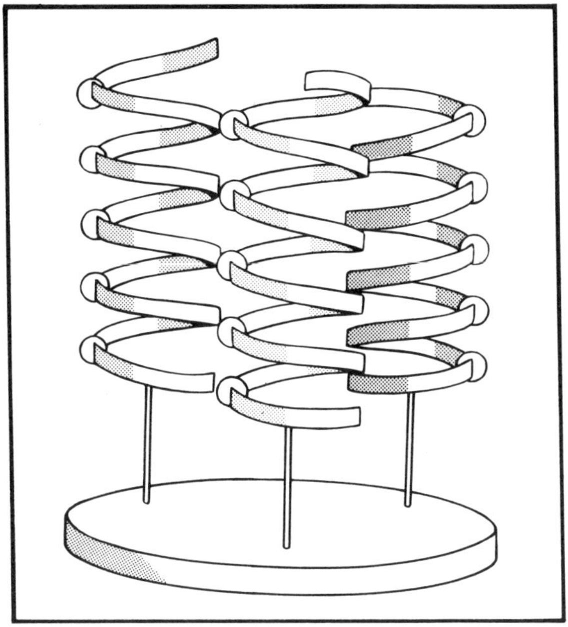

Fig. 8.53. Ozone M.P.M.

spiral M.P.A. of oxygen were an oxygen nucleus (or some perturbed form thereof), the micro-psi observer would report seeing a triangular array of three double spirals instead of three single spirals. The investigators worked according to the assumption that M.P.A.'s and M.P.M.'s were chemical atoms and molecules, respectively, and they wrote the formula of the unit of ozone as $\frac{1}{2}(O_3)$. This designation was consistent, of course, with this assumption—the expression $(\frac{1}{2}O)_3$ would have been more apt—but they appear not to have commented on the serious problem that it created, namely, that it implied that the ozone molecule consisted of three *half*-atoms!

The M.P.M. of ozone is formed by the pairing of nuclei in two molecules of ozone (fig. 8.54). It consists of three oxygen M.P.A.'s, each with a net positive electric charge of +16. Coulomb repulsion between them makes the M.P.M. unstable. But, instead of splitting up into three separate oxygen M.P.A.'s, it dissociates into two halves (fig. 8.55). One (A) is made up of two (+) and one (−) oxygen spiral and has a total charge of (62 +

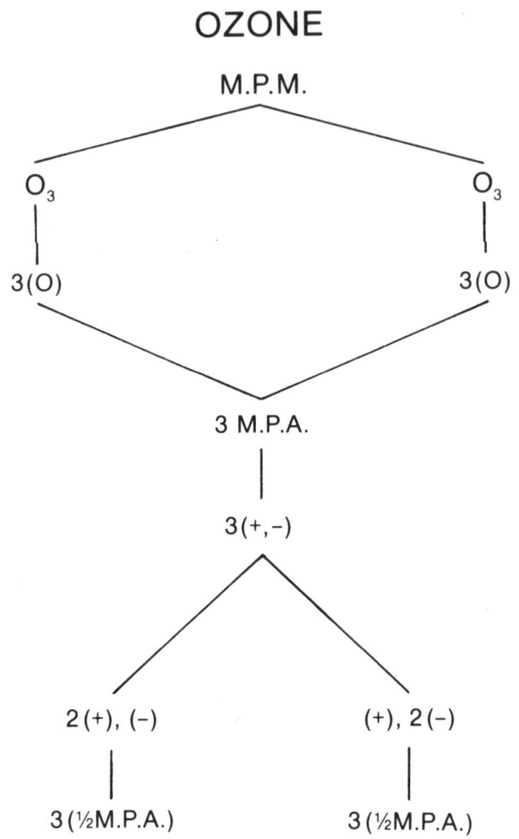

Fig. 8.54

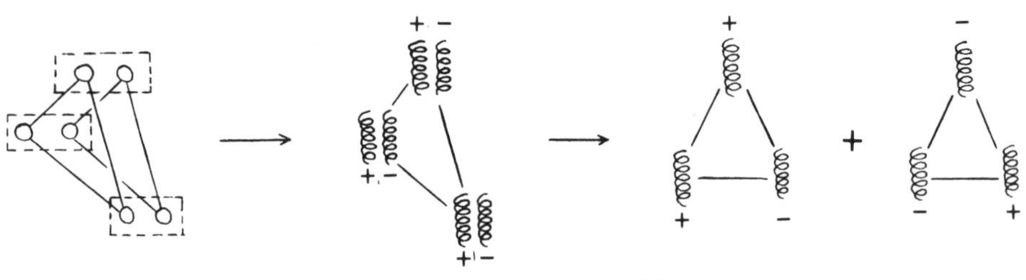

Fig. 8.55. Formation of ozone M.P.M.

X) (see the analysis of the oxygen M.P.A. for the definition of X); the other (B) consists of two ($-$) spirals and one ($+$) spiral and has a total charge of $-(14 + X)$. The two halves have a total charge of $+48$, that is, six times the charge of an oxygen nucleus, as required by Hypothesis 2. These are the two varieties described by the investigators, who did not observe the M.P.M. itself but noticed only the two more stable objects into which it dissociated. In both objects, there are two spirals with charge of the same sign and a spiral whose charge is of the opposite sign. In principle, therefore, they can form a stable system of charged particles that are bound by Coulomb forces. But in A, Coulomb repulsion between the two (+) spirals, each with charge $(46 + X)$, makes either one less strongly bound than the (−) spiral with charge $-(30 + X)$, which is attracted to both of them. Similarly for B, repulsion between the two (−) spirals makes them less strongly bound than the (+) spiral. On general dynamical grounds, it is not unreasonable that either bound system of spirals should be observed to split up into an oxygen M.P.A., at the same time releasing a free (+) or (−) spiral. The investigators remarked on this dissociation as follows: "It was noted that Ozone $\frac{1}{2}(O_3)$ has a tendency to revert to Oxygen, leaving one snake to go and find a mate for itself."[8]

REFERENCES

1. The Theosophist, vol. **47,** June 1926.
2. Occult Chemistry, 3d ed., p. 266.
3. Ibid.
4. The Theosophist, vol. **45,** March 1924.
5. Occult Chemistry, 3d ed., p. 321.
6. Ibid., p. 322.
7. The Theosophist, vol. **45,** April 1924.
8. Occult Chemistry, 3d ed., p. 96.

Conclusion

The most incomprehensible thing about the universe is that it is comprehensible.
Albert Einstein

A new theory of matter has been proposed to account for a large body of observations made over a period of thirty-eight years by Annie Besant and C. W. Leadbeater, who claimed to possess a siddhi of Indian yoga that enabled them to experience and to describe highly magnified images of atoms and molecules present in any material that they chose to examine. Two versions of their claim may be distinguished: first, that they did describe atoms and molecules in their normal state (this is what they believed), and second, that they had some form of cognitive perception of real, microscopic objects, which, however, were not atoms or molecules. The first version must be rejected because (1) it means that micro-psi accounts of the composition of nuclei contradict fundamental concepts of nuclear physics (e.g., the existence of nucleons as basic constituents of nuclear matter); (2) it disagrees with the highly successful Quark Model by requiring the hydrogen nucleus (or proton) to consist of six (not three) distinct bodies; (3) it has physically absurd implications for molecular structure, examples of which are that some M.P.A.'s appear to be in a state of disintegration when observed, supposedly, as part of the molecule of a chemical compound, and that the supposed ozone molecule is described as made up of three half-M.P.A.'s of oxygen, instead of three whole M.P.A.'s, while the "molecule" of benzene is described as octahedral in shape—not hexagonal, as every student of chemistry knows.

But, if these microscopic objects are *not* atoms, what are they? This question was answered in chapter 4 by hypothesizing, first, that the U.P.A. —the most elementary unit of physical matter that is visible to micro-psi vision—is a $\mathcal{P}$ or $\mathcal{N}$ omegon (the constituents of u and d quarks, according to the Omegon Model) and, second, that the M.P.A. of an element is a quasi-nuclear system of quarks and omegons released from pairs of atomic nuclei by the act of micro-psi observation, which must be regarded as a strongly perturbative interaction between the observer and those atoms with which he establishes some kind of causal link. Close agreement was found between the predicted populations of M.P.A.'s and the numbers counted by the investigators. Such agreement could not have been fabricated intentionally by them, because nearly all the numbers were published many years before mass spectrographs began to be used to measure isotopic weights and mass numbers of the commonly occurring isotopes. It may be argued that such levels of agreement lack the significance that is claimed for them because they are obtained by selecting just those nuclides that give the best match of numbers. But this criticism misses the crucial point. The fact remains that a linear relationship exists between the populations of M.P.A.'s and the mass numbers of appropriate nuclides. Remarkably, commonly occurring nuclides *do* exist with mass numbers providing (in significantly many cases) exact agreement between predicted and counted populations—numbers that may amount to several thousands. The only satisfactory explanation for such matching (at a time when tables of nuclides and their mass numbers were unavailable to the investigators) is that the nuclides must have participated in the formation of the M.P.A.'s.

In chapter 5, accounts of the "lines of force" linking U.P.A.'s were compared with the String Model picture of hadrons. Their striking similarity suggested that U.P.A.'s are non-Abelian magnetic monopoles—the end-points of Nielsen-Olesen vortex excitations of the superconducting Higgs vacuum—and that the lines of force are quantized bundles or tubes of colour flux. These colour-field lines confine U.P.A.'s in groups by being expelled from the surrounding vacuum—the non-Abelian analogue of the Meissner Effect.

In chapter 6, various firsthand accounts made by a micro-psi observer twenty years ago were related to modern elementary particle physics, in particular to research ideas that are only a few years old. The striking correspondences between them provided strong evidence for the objective character of micro-psi images.

A general picture of the process of formation of M.P.A.'s was formulated in chapter 7 and then tested by a detailed analysis of twenty-three M.P.A.'s. There emerged a high degree of consistency between theory and observation that cannot be plausibly explained as due to chance. Once Hypotheses 1 and 2 were assumed and a few simple groups of U.P.A.'s were given a fixed physical interpretation, the Quark and Omegon Models imposed severe restrictions on the type ("positive" or "negative") and number of duads and triplets of U.P.A.'s in the M.P.A. of a given element. Despite this, it was shown that the disintegration diagrams complied very considerably with such theoretical demands. Those discrepancies that occurred occasionally were usually attributable to miscounting by the investigators rather than to mistakes in their assignment of "positivity" or "negativity" to groups. It was found that these properties (whenever explicitly noted) nearly always matched the predicted electrical polarity of the group, that is, the (+) and (−) signs assigned by the investigators to the groups always matched theoretical expectation. Such remarkable consistency is explainable only if an M.P.A. is formed from two nuclei and if quarks are composite, with u and d quarks having the flavoured subquark composition predicted by the Omegon Model.

In chapter 8, the complex description of the diamond lattice given by C. W. Leadbeater was related (with the aid of Hypothesis 2) to the arrangement of its carbon atoms. Structural features of some M.P.M.'s were related to their corresponding molecules.

Pictorial evidence presented in this book is neither ambiguous nor reasonably explainable in terms of fraud or hallucinations that accompany some types of mental disorders. The parapsychologist may feel tempted to seek explanation in terms of varieties of extra-sensory perception that are familiar to him (e.g., clairvoyance or precognition) rather than accept the claims of micro-psi observers at their face value. But it must be pointed out that real understanding is not advanced by substituting for one form of E.S.P. another that is no less incomprehensible. Furthermore, the suggestion that micro-psi may be a precognitive experience of future concepts of physics fails to explain why the observer does not describe nuclei, atoms, and molecules but, instead, refers to quite different systems, which have been shown in this book to be derived from them.

The reader will, undoubtedly, experience psychological difficulty in accepting the objective character of micro-psi vision, however much he may be persuaded by the evidence presented here. Indeed, there are many problems left undiscussed. He will ask: How can the human mind induce at a distance the physical transformation of nuclei that is hypothesized? How can two individuals describe the microcosmos in a way that is found, fifty or more years later, to harmonize with both experimental and theoretical elementary particle physics? The author is unable to answer such questions. But the incompleteness of the present work gives the reader no justification for rejecting the evidence on the grounds that science cannot account for the way in which micro-psi yields valid information about the microcosmos. It is philosophically unsound (as well as intellectually dishonest) to argue that certain E.S.P. phenomena do not exist because, if they did, science would have no plausible explanation for them.

No attempt has been made here to formulate a complete, dynamical theory of the formation of M.P.A.'s. Many problems and questions concerning these systems of quark/omegon matter remain as yet unsolved. For example, why do the M.P.A.'s of the elements display such a wide and exotic variety of constituents? Are M.P.A.'s absolutely stable, so that those which are formed and observed during an examination of a substance exist indefinitely after the observer ceases his examination and returns to normal awareness? It may be argued that, since an M.P.A. represents a state of matter which (like the interior of a neutron star) cannot be reproduced in the laboratory by purely technological means, all discussion about it must remain speculative and unverifiable by any feasible experiment. But such criticism can be refuted when it is pointed out that micro-psi experiments in the past have demonstrated conclusively that the experimenter, while working with the subject, can alter what the latter experiences at the very time when the experimental conditions are changed (by, for example, the experimenter's varying the strengths of applied magnetic fields or of electric currents passing through the material under study). Moreover, the way in which micro-psi images are disturbed by the experimenter is found to agree with theoretical expectation (of which the subject has no knowledge). This strongly suggests that the images depict real objects, discussion about which can be tested experimentally. Past investigations, such as those quoted in chapter 6, were preliminary and rudimentary in design, although they did achieve a number of valuable results. More elaborate experiments in the future should be seriously encouraged. They should test whether the micro-psi observer does, indeed, act at a distance on matter and initiate physical changes on a microscopic scale. Chapter 7 presented many examples in which slight deviations from the reported structure of some M.P.A.'s were predicted. These can be checked by fresh observations by suitable subjects who have no knowledge of the predictions made in this book.

It should be emphasized that, although U.P.A.'s have been identified here as omegons, the strength of the case for extra-sensory perception of quarks by Besant and Leadbeater is independent of the particular subquark model serving as a basis for explaining their observations. Even if future experiments in high-energy physics invalidate the Omegon Model, the main thesis of this book is not weakened thereby, for it rests on the aptness of micro-psi descriptions in relation to the Quark and String Models of particle physics and not on the correctness of a newly proposed theory. U.P.A.'s remain subquarks even if the latter (when detected) are found not to be omegons.

The aim of the present work has been to relate past micro-psi research to contemporary particle physics. The work carried out by Annie Besant and C. W. Leadbeater over fifty years ago has been shown to be consistent with many ideas that are currently being discussed by physicists. In view of this, the reader is invited to ask himself: How could it have proved possible eventually to demonstrate such consistency if they had not been able to do just what they claimed?

Appendix

A CHRONOLOGY OF MICRO-PSI INVESTIGATIONS*

1895 Lucifer, London, November: H, N, O.

1907 *The Theosophist*, vol. **29**, part 1: H, Na, Cl, Cu, Br, Ag, I, Au.

1908 *The Theosophist*, vol. **29**, part 2: "Oc", Li, Be, B, C, N, O, F, Na, Mg, Al, Si, P, S, Cl, K, Ca, Sc, Ti, V, Cr, Mn, Fe, Co, Ni, Cu, Zn, Ga, Ge, As, Se, Br, Rb, Sr, Y, Zr, Nb, Mo, Ru, Rh, Pd, Ag, Cd, In, Sn, Sb, Te, I, Os, Ir, Pt, "Pt B," Au.

The Theosophist, vol. **30**, part 1: Li, C, F, Si, K, Ti, Mn, Fe, Co, Ni, Ge, Rb, Zr, Ru, Rh, Pd, Sn, Os, Ir, Pt, Ra, He, Ne, "meta Ne," Ar, "meta Ar," Kr, "meta Kr," Xe, "meta Xe," "kalon," "meta kalon."

Occult Chemistry, by Annie Besant and C. W. Leadbeater, Theosophical Publishing House (Adyar, Madras, India).

1909 *The Theosophist*, vol. **30**, part 2: Cs, Ba, La, Ce, Pr, Nd, Pm (called "A," then "Illinium"), "meta Illinium," "X," "Y," "Z," Sm, Gd, Tb, Dy, Er, Tm, Ta, W, Os, Ir, Pt, Hg, Tl, Pb, Bi, Ac (called "C"), Th, U.

1919 *Occult Chemistry*, 2d ed., by Annie Besant and C. W. Leadbeater, edited by A. P. Sinnett, Theosophical Publishing House (London, England).

1924 *The Theosophist*, vol. **45**: U, NaCl, CH_4, H_2O, OH, H_2O_2, CH_3OH, CH_3COOH, C_6H_6, NaOH, HCl, CO, CO_2, NA_2CO_3, Cl isotope.

1925 *The Theosophist*, vol. **46**: $Ca(OH)_2$, CaC_2, C_2H_2, CH_3Cl (and isomer), $CHCl_3$, CCl_4, $C_{10}H_8$, $C_{14}H_{10}$, O_3, diamond.

1926 *The Theosophist*, vol. **47**: graphite, Te isotope.

1932 *The Theosophist*, vol. **54**: "Adyarium," "Occultum," Li, F, Na, Cl, K, Mn, Fe, Co, Ni, Cu, Br, Rb, Tc (called "Masurium"), Ru, Rh, Pd, "X," "Y," "Z," Ag, I, Cs, Pm (called "Illinium"), Gd, Er, Yb, Re, Os, Ir, Pt, "Pt B," Au, At (called "85"), Fr (called "87").

1933 *The Theosophist*, vol. **54**: He, Ne, "meta Ne," A, "proto A," "meta A," Kr, "meta Kr," Xe, "meta Xe," "kalon," "meta kalon," Rn, "meta Rn," two varieties of O_3, three varieties of O, two varieties of H.

1951 *Occult Chemistry*, 3d ed., by Annie Besant and C. W. Leadbeater, edited by C. Jinarajadasa and E. W. Preston, Theosophical Publishing House (Adyar, Madras, India).

1957–59 Investigations by Geoffrey Hodson and Dr. D. D. Lyness (unpublished).

* I thank Dr. M. G. Hocking and Dr. V. Vasantasree for their assistance in collating references.

Glossary

ATOMIC NUMBER. Number of protons in the nucleus of an atom.

BARYON. Hadron with half-integral spin and non-zero baryon number (an additive quantum number B that is unchanged by all known interactions; $B = 1$ for protons and neutrons).

BOSON. Particle possessing integral spin.

CHIRAL PAIRS. Pairs of objects whose shapes are mirror images of each other (e.g., a pair of gloves).

COLOURS. The three different states of a quark resulting from its exchange of gluons with other quarks.

COULOMB FIELD. The electric field of a particle due to its electric charge.

DIQUARK. A strongly bound pair of quarks.

DYON. A particle carrying both electric and magnetic charge.

EXCLUSION PRINCIPLE. A principle stating that two fermions of the same type cannot occupy the same quantum state at one time.

FERMION. A particle that obeys the Exclusion Principle; all particles, such as electrons and protons, whose spin is half-integral are fermions.

FLAVOURS. Different species of quarks (and subquarks).

FLUX QUANTUM. Unit of magnetic flux, in terms of which the flux in vortices is quantized as integer multiples.

GAUGE FIELDS. A set of particles whose exchange between subatomic particles gives rise to the interaction between the latter.

GLUONS. A set of particles whose exchange between quarks is responsible for their strong interaction.

HADRON. A particle exhibiting strong interactions with other subatomic particles.

HIGGS FIELD/PARTICLE. An as yet undetected particle whose coupling to gauge fields gives the latter mass.

HYDROGEN TRIANGLE. Triangular cluster of three hydrogen triplets; half of the hydrogen M.P.A.

HYDROGEN TRIPLET. Linear or triangular cluster of three U.P.A.'s.

ISOTOPES. Atoms whose nuclei contain the same number of protons but different numbers of neutrons.

LEPTON. A class of particles that exhibit no strong interactions and are subject only to weak and electromagnetic interactions.

MAGNETIC MONOPOLE. A particle carrying magnetic charge.

MASS NUMBER. The number of nucleons in a nucleus.

MESON. Hadron with integral spin and zero baryon number.

MICRO-PSI. A siddhi of Indian yoga, allowing quasi-visual perception of microscopic objects.

MICRO-PSI ATOM (M.P.A.). The "atomic" unit of matter, according to micro-psi vision; its form and composition are uniquely characteristic of an element.

MICRO-PSI MOLECULE (M.P.M.). The micro-psi analogue of a chemical molecule; it consists of the M.P.A.'s of those elements whose atoms make up a molecule.

NEUTRINO. Electrically neutral lepton that interacts only weakly with other particles.

NUCLEONS. The charged (proton) and neutral (neutron) constituents of nuclei.

NUCLIDE. A species of nucleus of an element, characterized by its atomic number and mass number.

OMEGON. Hypothetical constituent of quarks; it is predicted to have ten flavours.

QUARK. The fermion constituent of hadrons.

SPIN. The intrinsic angular momentum of a particle; it is quantized in units of $\frac{1}{2}\hbar$ ($\hbar = h/2\pi$, where h is Planck's constant); if half-integral, the spin is an odd-integer multiple of $\frac{1}{2}\hbar$; if integral, it is an even-integer multiple of $\frac{1}{2}\hbar$.

SPIN PRECESSION. Wobbling, spinning-top motion of the direction of spin of a particle subject to a magnetic field.

ULTIMATE PHYSICAL ATOM (U.P.A.). The most elementary discrete unit of matter visible to micro-psi vision.

VORTEX. A vortical excitation of superfluids and superconductors to which a magnetic field of strength higher than the critical value is applied. In type 2 superconductors, the magnetic flux in a vortex is quantized.

ZITTERBEWEGUNG. Oscillatory motion of a free particle with spin $\frac{1}{2}\hbar$ (e.g., an electron) that is superimposed on its uniform, rectilinear motion.

Bibliography

A simple (though now slightly dated) introduction to particle physics is *Elementary Particles*, by I. S. Hughes (Penguin Library of Physical Sciences, 1972). The following articles in *Scientific American* provide useful background material:

August 1969	The Size and Shape of Atomic Nuclei. *Baranger and Sorenson.*
June 1971	The Structure of the Proton and the Neutron. *Kendall and Panofsky.*
July 1974	Unified Theories of Elementary Particle Interaction. *Weinberg.*
February 1975	Dual-Resonance Models of Elementary Particles. *Schwarz.*
June 1975	Electron-Positron Annihilation and the New Particles. *Drell.*
October 1975	Quarks with Color and Flavor. *Glashow.*
January 1976	The Search for New Families of Elementary Particles. *Cline, Mann, and Rubia.*
November 1976	The Confinement of Quarks. *Nambu.*
October 1977	Fundamental Particles with Charm. *Schwitters.*
March 1978	Heavy Leptons. *Perl and Kirk.*
October 1978	The Upsilon Particle. *Ledermann.*
December 1978	Nuclear Molecules. *Bromley.*
February 1979	Solitons. *Rebbi.*
July 1979	The Bag Model of Quark Confinement. *Johnson.*

Index